Generative AI for Cybersecurity

This book lays a systematic foundation for professionals, researchers, and industry readers who are interested in the applications and implications of generative AI for cybersecurity. It covers the latest advances in generative AI and its applications, risks, and opportunities in cybersecurity.

The authors first introduce the fundamental background of generative AI, the latest cybersecurity issues, and related potential applications in cybersecurity systems. Following this, they comprehensively review the state-of-the-art research and development, covering various aspects of generative AI applications in this area and related challenges and issues, such as training data availability, computational complexity, generalization to different scenarios, AI governance, quantum-empowered AI and many more. These discussions provide a strong understanding of recent advances in the two fields of generative AI and cybersecurity and the convergence of these domains, which will help readers to shape the field as it matures. Hands-on experiments presented throughout will also give them the practical skills for success. By leveraging its capabilities, readers can overcome challenges, understand the risks, enhance performance, and unlock new opportunities for handling the challenges of cybersecurity with generative AI. Consequently, they will be able to apply their knowledge to utilize generative AI in cybersecurity applications to prevent economic and other losses due to cyber risks such as phishing, fake news, deepfake-based fraud, and other cyberattacks.

The contents of this book are appropriate for a wide range of readers from general readers to industry experts and scientists. Because it bridges the gap between generative AI and cybersecurity, experts from both fields will benefit from the information presented within. Students with a background in either area will also benefit from the approach that leads from general to specific applications.

Generative AI for Cybersecurity

Fundamentals, Applications, Risks, and Opportunities

Diep N. Nguyen, Ly Vu, Quang Uy Nguyen, and Dinh Thai Hoang

CRC Press
Taylor & Francis Group
Boca Raton London New York

CRC Press is an imprint of the
Taylor & Francis Group, an informa business

Designed cover image: Shutterstock

First edition published 2026
by CRC Press
2385 NW Executive Center Drive, Suite 320, Boca Raton FL 33431

and by CRC Press
4 Park Square, Milton Park, Abingdon, Oxon, OX14 4RN

CRC Press is an imprint of Taylor & Francis Group, LLC

ISBN: 978-1-032-82009-5 (hbk)
ISBN: 978-1-032-82010-1 (pbk)
ISBN: 978-1-003-50253-1 (ebk)

DOI: 10.1201/9781003502531

Typeset in Nimbus Roman font
by KnowledgeWorks Global Ltd.

Publisher's note: This book has been prepared from camera-ready copy provided by the authors.

Dedication

This book is supported in part by Australia-Vietnam Strategic Technologies Centre, which is supported by the Australian Government and Vietnamese Government

To my family, PhD students, and you who may find this book useful—Diep N. Nguyen

To my family and students—Ly Vu

To my family, to my parents, Nguyen Van Hien and Nguyen Thi May, for working so hard to bring up their son; to my wife, Nguyen Thi Minh Hang, for sharing a lot of happiness and difficulty in life with me; and to my daughter, Nguyen Bao Ngoc, and my son, Nguyen Huu An, for their effort in studying to become good students and giving me a lot of happy and meaningful moments in my life—Quang Uy Nguyen

To my family—Dinh Thai Hoang

Contents

Foreword

In our increasingly interconnected world, cybersecurity stands as a critical pillar supporting global stability, economic resilience, and national security. As technologies rapidly evolve, so too do the threats that challenge our digital infrastructures. In this context, the publication of *Generative AI for Cybersecurity: Fundamentals, Applications, Risks, and Opportunities* is both timely and significant.

Generative AI, particularly since the remarkable emergence of tools such as ChatGPT, has captured global attention, thanks to its profound capability to generate sophisticated content/data. Such a capability not only promises revolutionary advancements in cybersecurity but also highlights critical vulnerabilities that malicious actors might exploit. Navigating this dual potential requires a deep and nuanced understanding of Generative AI as well as its applications/implications and even risks to cybersecurity.

Authored by renowned experts in the fields of artificial intelligence and cybersecurity, this book systematically offers essential insights into how Generative AI can proactively address and mitigate complex cybersecurity threats, from malware detection and phishing prevention to safeguarding cyber-physical systems, including IoT and cloud computing platforms. At the same time, the authors responsibly acknowledge and explore the significant risks posed by Generative AI, emphasizing the need for vigilant and innovative defensive strategies as well as governance.

By providing rich case studies, practical implementations, and discussions on ethical and practical challenges, this book serves as an invaluable resource not only for researchers and industry professionals but also policymakers and global leaders. It equips readers with the necessary knowledge to harness Generative AI's potential securely, fostering international cooperation to enhance cybersecurity at a global scale.

Australia remains deeply committed to promoting a secure and resilient cyberspace through international collaboration and innovation in critical technology. I commend the authors for their significant contribution, offering clarity, vision, and practical guidance in harnessing the transformative power of Generative AI safely and effectively. It is my privilege to endorse this timely publication. May it inspire continued global cooperation and proactive advancement in securing our shared digital future.

Foreword by Ambassador Dowling
Australia's Ambassador for Cyber Affairs and Critical Technology
https://www.dfat.gov.au/about-us/our-people/homs/ambassador-for-cyber-affairs

Foreword

Preface

The release of ChatGPT by OpenAI in 2022 drew paramount attention to Generative AI due to its enormous potential in different practical applications. For cyber-physical systems and digital infrastructures, Generative AI has been seen as a powerful solution to cybersecurity. Cybersecurity threats have reached alarming proportions on a global scale, with devastating social and economic impacts. In fact, the global annual cost of cybercrime exceeded $1 trillion, as reported by Cybersecurity Ventures. The fallout from cyber threats extends far beyond financial burdens, affecting critical infrastructure, national security, and personal privacy. The World Economic Forum's Global Risks Report highlighted cyberattacks as one of the top five global risks in terms of likelihood, ahead of pandemics and environmental disasters, underscoring an urgent need for robust cybersecurity measures on a global scale.

GenAI, thanks to its exceptional "creative/generative" ability, has been emerging as a promising solution to effectively address various cybersecurity issues, e.g., detecting unknown attacks, malware, phishing, adversarial attacks, deepfakes, and identity theft, as well as visualizing network traffic, security logs, and testing the effectiveness of security policies and response strategies. These potentials of GenAI, including vulnerability assessment/scanning, allow organizations to proactively identify and address security weaknesses. Moreover, GenAI also empowers different types of AI, such as deep learning and deep reinforcement learning, by generating high-quality training data. This capability becomes particularly valuable when data acquisition in cybersecurity is costly or limited. However, the idea of implementation of Generative AI in cybersecurity is still at an early stage. Remaining key challenges, such as training data availability, computational complexity, generalization to different scenarios, robustness to noisy and incomplete data, new/unknown cyberattacks, ethical and privacy concerns, and real-time adaptation and learning, must be addressed. Overcoming these challenges is crucial to enabling practical applications of Generative AI in this domain.

While holding the above immense promise, GenAI also introduces significant cybersecurity risks. For instance, attackers can harness Generative AI to craft convincing deepfake content for identity theft, misinformation campaigns, or impersonation. It also enables the automated generation of sophisticated phishing emails that bypass traditional email filters, making users more vulnerable to social engineering attacks. Moreover, GenAI can create novel and evasive malware variants that challenge traditional antivirus detection methods. Adversarial attacks, exploiting GenAI-created adversarial examples can undermine the accuracy of machine learning-based security systems. The misuse of GenAI in cyberattacks underscores the imperative to bolster defenses against these emerging threats, necessitating ongoing innovation in cybersecurity to counteract the capabilities it bestows upon malicious actors.

Given the above, this book, titled *Generative AI for Cybersecurity: Fundamentals, Applications, Risks, and Opportunities*, aims to lay a systematic foundation for the research community and industry that are interested in Generative AI and its applications, as well as implications/risks to cybersecurity. To this end, we first provide a fundamental background of Generative AI, the tutorial on the latest issues in cybersecurity, and their state-of-the-art solutions. We then provide an overview of potential applications of Generative AI to address problems in cybersecurity, as well as the risks to cyberspace brought up by Generative AI. After that, the book delves into specific challenges of cybersecurity, e.g., imbalanced training dataset, lack of attacks/training data, and deepfakes, and discusses how Generative AI can potentially tackle them. To provide readers with hands-on experience, the book also presents typical study cases where Generative AI can be used to secure cyber-physical systems, e.g., Internet of Things, Blockchains, and Cloud Computing platforms. Finally, we highlight the latest research and emerging interests in applying Generative AI to safeguard future cyberspace. The book concludes by detailing challenges and open issues related to Generative AI in the global cybersecurity landscape. To summarize, the key features of this book are:

- Provides fundamental background, tutorial of Generative AI and its applications, implications to cybersecurity
- Provides the latest advances in applying Generative AI to tackle practical challenges in cybersecurity
- Provides case studies together with detailed programming implementation to help the readers learn and practice Generative AI to safeguard specific scenarios (e.g., for intrusion detection in IoT networks, threats/attack detection in cloud computing, blockchains)
- Discusses potential risks caused by Generative AI to cyber systems and potential mitigations, especially for emerging applications/issues
- Highlights challenges, open issues, and future research directions of Generative AI in the global cybersecurity landscape

Author

Diep N. Nguyen (Senior Member, IEEE) received the M.E. degree in electrical and computer engineering from the University of California at San Diego (UCSD), La Jolla, CA, USA, in 2008, and the Ph.D. degree in electrical and computer engineering from the University of Arizona (UA), Tucson, AZ, USA, in 2013. He is currently the Head of 5G/6G Wireless Communications and Networking Lab, Director of Agile Communications and Computing group, Faculty of Engineering and Information Technology, University of Technology Sydney (UTS), Sydney, Australia. Before joining UTS, he was a DECRA Research Fellow with Macquarie University, Macquarie Park, NSW, Australia, and a Member of the Technical Staff with Broadcom Corporation, CA, USA, and ARCON Corporation, Boston, MA, USA, and consulting the Federal Administration of Aviation, Washington, DC, USA, on turning detection of UAVs and aircraft, and the U.S. Air Force Research Laboratory, USA, on anti-jamming. His research interests include computer networking, wireless communications, and machine learning applications, with emphasis on systems' performance and security/privacy. Dr. Nguyen received several awards from the U.S. Congress (Vietnam Education Foundation), the U.S. National Science Foundation, and the Australian Research Council. He has served on the Editorial Boards of the IEEE Transactions on Mobile Computing, IEEE Communications Surveys & Tutorials (COMST), and the IEEE Open Journal of the Communications Society.

Ly T Vu received her B.E. degree in Information Technology Engineering from Le Quy Don Technical University, Vietnam, in 2011, her master's degree in information and communication engineering from Inha University, Korea, in 2014, and her Ph.D degree from the University of Technology Sydney, Sydney, NSW, Australia in 2024. She is a Faculty Member of the Department of Computer Science, Institute of Information and Communication Technology, Le Quy Don Technical University, Vietnam. Her research interests include anomaly detection, machine learning, and deep neural networks.

Quang Uy Nguyen received his Bachelor of Engineering degree in informatics and Master of Science in computer science at Le Quy Don Technical University, Vietnam. He received the Ph.D. degree from University College Dublin, Dublin, Ireland, in 2011. He is currently an Associate Professor and a Senior Lecturer with Le Quy Don Technical University, where he is also the Deputy Head of the Computer Science Department and the Director of the Intelligent Computing Research Group. Prof. Nguyen is the principal investigator of two research projects funded by the Vietnam National Foundation for Science and Technology Development (NAFOSTED) and one research project funded by the Vingroup Innovation Foundation (VinIF), and his research interests include machine learning, computer vision, information security, and evolutionary algorithms.

Dinh Thai Hoang received his Ph.D. degree from the School of Computer Science and Engineering, Nanyang Technological University, Singapore, in 2016. He is currently a faculty member at the University of Technology Sydney (UTS), Australia. Over the last ten years, he has significantly contributed to advanced wireless communications and networking systems. His excellent record evidences this with one patent filed by Apple Inc., three books, eight book chapters, more than 120 IEEE Q1 journals, and more than 80 flagship IEEE conference papers in communications and networking. Most of his journal papers have been published in top IEEE journals, including IEEE JSAC, IEEE TWC, IEEE COMST, and IEEE TMC. Furthermore, his research papers have had a high impact, evidenced by nearly 20,000 citations with an h-index of 53 (according to Google Scholar) over the last ten years. Since joining UTS in 2018, he has received more than AUD 6 million in external funding and several precious awards, including the Australian Research Council Discovery Early Career Researcher Award for his project "Intelligent Backscatter Communications for Green and Secure IoT Networks," IEEE TCSC Award for Excellence in Scalable Computing for Contributions on "Intelligent Mobile Edge Computing Systems" (Early Career Researcher), and the IEEE TCI Rising Star Award for "Technical Contributions on the Internet." Alternatively, he is the lead author of two authored books, *Ambient Backscatter Communication Networks*, published by Cambridge Publisher in 2020, and *Deep Reinforcement Learning for Wireless Communications and Networking*, published by IEEE-Wiley Publisher in 2022. He is currently an Editor of IEEE TMC, IEEE TWC, IEEE TCOM, and IEEE TNSE.

Part I

Fundamental Background

Part I first provides a fundamental background of Generative AI and a tutorial on cybersecurity, different types of attacks/vulnerabilities, key challenges, and research trends in this domain. We then provide an overview of potential applications of Generative AI to address challenges in cybersecurity, as well as the risks to cyberspace brought up by Generative AI. Part I has three chapters as follows.

1 Generative AI and Its Related Components

Generative AI is defined as a class of AI models used to create data that is similar to the data in our real world. Specifically, Generative AI models can create data in forms like text, images, audio, videos, 3D models, etc. Since the booming of ChatGPT in 2023, generative AI has received a paramount interest from the community. It is believed that generative AI will be the most influential AI technology in the next decade.

The reinvention of Generative AI is mainly thanks to the development and advancement of deep learning, a class of multiple-layer neural networks that are specifically designed and trained to solve many complex problems. In this chapter, we will discuss different components of deep learning and generative AI. This chapter is the foundation for the next chapters in the book.

1.1 DEEP NEURAL NETWORK

Deep learning is a branch of machine learning that utilizes neural networks, typically characterized by numerous layers, to identify patterns and make informed decisions based on data [1]. This approach draws inspiration from the human brain's architecture, consisting of various interconnected nodes that handle and interpret information. Deep learning algorithms can be trained for diverse applications, such as recognizing images and speech, translating languages, detecting anomalies, and playing games. They are especially effective in tasks that demand high precision, like object recognition in images or language translation.

A major advantage of deep learning is its capacity for ongoing learning and adaptation. As it ingests increasing volumes of data, a deep learning model can progressively enhance its accuracy and effectiveness. This makes it particularly suitable for tasks that involve large datasets, especially when it's challenging to establish explicit decision-making rules. Deep learning has already made a substantial impact across various sectors, including healthcare, finance, transportation, retail, and entertainment [2]. Its influence is anticipated to expand even further in the future as data availability continues to rise and more organizations seek to harness deep learning's capabilities for improved decision-making and operational efficiency.

In a more specific definition, deep learning can be understood as the extension of traditional neural networks with some new advancements. The three most important improvements of deep learning compared to the traditional neural network include the network structure, the activation function, and the optimization algorithms for training models.

DOI: 10.1201/9781003502531-1

First, there have been many new network structures developed in the last decade. Some of the most popular deep network structures include Convolutional Neural Networks, Recurrent Neural Networks, Autoencoders, Encoder-Decoder, Attention, and Transformer.

Second, due to the limitations of sigmoid and tanh functions relating to the gradient vanishing problems, there have been a large number of new active functions proposed for deep learning. Some of the most popular activation functions consist of Relu, LeakyRelu, Exponential Relu etc.

Last but not least, there have been many improvements from the traditional stochastic gradient descent (SGD) algorithm. The most effective algorithms include SGD with momentum, Nesterov Accelerated Gradient (NAG), Adaptive Gradient (AdaGrad), AdaDelta, RMSprop, and Adam.

In the following subsections, we will further study the popular deep network architectures (convolutional neural network, recurrent neural network, and autoencoder) and optimization algorithms. Other deep network architectures will be discussed and presented in the coming chapters.

1.1.1 CONVOLUTIONAL NEURAL NETWORKS

Perhaps, the most popular deep network structure is Convolutional Neural Network (CNN) [3]. CNN is very popular and widely applied to a wide range of problems, particularly in image and video processing. Compared to the traditional neural networks, i.e., Multi-layer Perceptron (MLP), CNN has two main differences. The first difference is that the input to MLP is a vector of features, while the input to CNN is often the whole/raw image in two or three dimensions. In other words, MLP only works with features extracted from the original data, while CNN can perform directly on the raw data. Thus, CNNs are powerful tools for feature or representation learning [4]. The second difference between CNN and MLP is the objective of transforming the feature in each layer. While MLP aims to extract the global features, CNN, on the other hand, focuses on extracting the local features in images.

Generally, a CNN consists of three main components: Convolution layers, Pooling layers, and Dense networks. The convolution layers are used to extract the local features from the input data. In other words, the convolution layer maps the input data into a new feature using a local convolution operator. For a single channel input image, $I_{[i,j]}$, a convolution operator of size $[W,H]$ can be defined by $K_{[x,y]}$ where $x = 1,...W$ and $y = 1,..,H$. The output feature map f when applying K on I is calculated as follows:

$$F[i,j] = (I * K)_{[i,j]} = \sum_{x=1}^{w} \sum_{y=1}^{h} K_{[x,y]}.I_{[i-x,j-y]} \tag{1.1}$$

The pooling layers are used to extract the abstract/general features from the input. They are often implemented by creating/selecting a smaller number of important features from a greater number of input features. There are two popular pooling operators in CNN. The first pooling operator is the Max-Pooling, where the max operator is applied to calculate the output from the input. For a feature map $F[i,j]$, the output of applying a Max-Pooling $P_{x,y}$ with kernel size of (W,H) on this feature is calculated as follows:

$$Max - Pooling_{[i,j]} = max_{x=1,y=1}^{W,H}(F_{[i-x,j-y]}) \tag{1.2}$$

The second pooling operator is Average-Pooling, where the average operator is used to calculate the output from the input. Average-Pooling is often used at the last layer of the Convolution block in CNN, in which a large number of feature maps are flattened into a single vector before inputting into the dense layers.

For a feature map $F_{[i,j]}$, the output of applying a Max-Pooling $P_{x,y}$ with kernel size of (W,H) on this feature is calculated as follows:

$$Average - Pooling_{[i,j]} = \frac{\sum_{x=1,y=1}^{W,H}(F_{[i-x,j-y]})}{W+H} \tag{1.3}$$

The dense layers are often used at the end of a CNN to map the output to the target problem. The dense layers often include a number of layers in an MLP. The intuition is that after a number of convolution and pooling layers, the raw data has been projected to more general/abstract features. Thus, the target problem can be easily solved using a few dense layers. A typical CNN often consists of a dozen to a hundred convolutional and pooling layers with only one or a few dense layers.

Since the development of Alexnet [5], there have been many CNN networks developed for image, video processing, and other problems. Following, we will discuss some of the most popular CNN models.

AlexNet: AlexNet [5] was proposed by Alex and his co-workers in 2012. Comparing to the traditional MLP, Alexnet has three important improvements. The first improvement is that it has a deeper structure (more layers) than the traditional MLP. Traditionally, MLP has only one or a few layers. Alexnet improves this by successfully trains a network of 8 layers. This is the first time researchers have successfully train such a deep network. The second improvement is the replacement of the Sigmoid activation function by the Relu function. Using the ReLU function helps to mitigate the vanishing gradient problem, thus allowing for the training of networks with more layers. The last improvement of Alexnet is the use of the Dropout layer to regularize the network. The Dropout layer is implemented by randomly removing some connections in one or some layers of the network during the training process. This technique, thus, helps to prevent the over-fitting problem.

The performance of Alexnet is very impressive. This model achieves top five error of 15.3% vs 26.2% of second place in the 2012 Imagenet challenge. Moreover, since Alexnet has much deeper structure than the conventional MLP, this model is called a deep network. Until now, Alexnet is still considered the most influential research paper in the field of computer vision. The success of Alexnet has inspired the birth and development of the field of deep learning.

VGG: VGG was developed by the Visual Geometry Group at the University of Oxford in 2014 [6]. It won second place in the 2014 Imagenet challenge. Compared to Alexnet, VGG has two major improvements. The first improvement is that VGG uses a smaller kernel size compared to Alexnet. Specifically, VGG uses only a 3x3 kernel while Alexnet uses 11x11, 5x5, 3x3 kernels. Using the smaller size helps VGG reduce the model size and the computational time compared to Alexnet. Moreover, the authors also demonstrate that we can achieve the same result of the larger kernel size by using multiple smaller-sized kernels. The second improvement is that VGG uses a deeper architecture (double or more) than Alexnet. In fact, there are two versions of VGG: VGG16 and VGG19, which have 16 and 19 convolutional layers, respectively. Another interesting property of VGG is that it has a very standard structure: VGG is simply the stacking of convolutional layers. Thus, VGG has been the foundation for many following convolutional neural networks. VGG achieves a top 5 error of 10.2% on the Imagenet dataset.

GoogLeNet: Googlenet [7], also known as Inception-v1, was proposed by Google researchers at the Asia Center in 2015. It won first place in the 2015 Imagenet challenge. The novelties of Googlenet compared to the previous models include four folds. First, this network is built by combining many modules/blocks instead of stacking convolutional layers. In other words, the authors proposed the idea of using building blocks for developing deep convolutional neural networks. This idea is very important and it is used by many researchers for developing the following deep network models. Second, 1x1 kernels are used for dimensionality reduction to remove computational bottlenecks. This is the first research that used such a small kernel size in CNN. Third, the model has a parallel architecture that includes a convolution layer with filters of different sizes, 1×1, 3×3, and 5×5. After that, the paralleled branches are concatenated. Using multiple kernel sizes in parallel allows Googlenet to extract features of different-grained levels. Last, the model is designed to have one main classifier and two auxiliary classifiers. The main classifier is used as the final one during testing/inferencing. The two auxiliary classifiers are used to encourage the discrimination between different classes in the lower stages. Moreover, the auxiliary classifiers also help to directly transform the gradients of the error samples to the lower layers in the model. Thus, they mitigate the gradient vanishing problem.

Googlenet achieved top-1 accuracy on the Imagenet dataset of 78.2% and top-5 accuracy of 94.1%. This is the first deep network model that achieves the accuracy as competitive as human performance on Imagenet. In fact, the top-5 accuracy of human beings on the Imagenet dataset is about 95% [8].

ResNet: Restnet [9] is perhaps the most influential CNN model until date. Resnet was presented at the 2016 IEEE Conference on Computer Vision and Pattern Recognition (CVPR), and it won first place in the Imagenet competition in 2016. The novelties of Restnet include three folds. First, this model uses skip connect (residual connect) to address the gradient vanishing problem. Although this is not the first time the skip connection is used in neural networks. However, the success of Restnet has inspired the popularization of skip connections in designing many subsequent CNN architectures. Second, Restnet has a much deeper architecture (up to 152 layers) compared to the previous CNN models. The interesting thing is that even though Restnet has a much deeper architecture, it does not suffer from gradient vanishing problems. Thus, its generalisation power is significantly improved.

This improvement is generally thanks to the skip connection used in the model. Last, Restnet is one of the first models that use batch normalisation to combat the distribution shifting problem. Batch normalization is a technique used to improve the training process of deep network models through normalization of the input of some layers by re-centering and re-scaling [10].

Resnet achieved the top-1 accuracy of 87.0% and the top-5 accuracy of 96.3%. This is the first deep learning model that achieved higher accuracy on the Imagenet dataset compared to the human performance. Due to the highly effective performance, Resnet has been widely used in many recently developed deep network models. In fact, the Resnet paper is the most cited paper among the deep network publications to date.

MobileNet: After the success of Resnet in achieving higher accuracy than human beings, the focus of the research community has moved to applications of CNN models instead of designing more sophisticated models to achieve higher accuracy on the Imagenet dataset. One of the most desired objectives is to develop lightweight deep network models that can be deployed on mobile devices. This leads to the introduction of Mobilenet [11] in CVPR 2017. MobileNet brings two key innovations. First, it introduces a novel convolutional operator known as depthwise separable convolution. This type of convolution is specifically engineered to extract features from individual channels, leading to a significant reduction in both the model's parameter count and computational expense. Second, MobileNet incorporates two hyperparameters that allow for a balance between latency and accuracy. The first is the width multiplier, which uniformly reduces the number of channels across each layer, while the second is the resolution multiplier, which decreases the size of the model's input.

Mobilenet achieved the top-1 accuracy of 70.6% on the Imagenet dataset. This is even better than Googlenet and VGG although the number of operators and the number of parameters in Mobilenet are much less than Googlenet and, VGG. Mobilenet and its variants are perhaps the most popular models applied to edge and embedded systems today [12].

EfficientNet: EfficientNet [13] was introduced at the ICML in 2019 and remains one of the top convolutional neural network (CNN) models for image classification today. A notable feature of EfficientNet is that it was created using a search algorithm rather than through manual design by researchers. Additionally, EfficientNet employs an innovative scaling approach that uniformly adjusts the network's depth, width, and resolution, rather than scaling each dimension separately. This enables the model to be adapted to various sizes based on available computational resources and specific tasks. Consequently, there exists a series of eight models, ranging from B0 to B7, each offering increasing levels of complexity.

EfficientNet models deliver cutting-edge performance across various computer vision tasks, such as image classification, object detection, and semantic segmentation, all while keeping model size and computational costs relatively low. This efficiency makes them ideal for deployment on diverse devices, including mobile and embedded systems, where computational resources and memory are constrained. Overall, EfficientNet models signify a major breakthrough in efficient deep learning, providing an impressive balance between performance and resource efficiency, making them an invaluable asset for numerous computer vision applications.

After the introduction of EfficientNet models, the field of Convolutional Neural Networks (CNNs) has continued to evolve, with researchers and engineers exploring new ways to enhance the performance, efficiency, and versatility of these powerful deep learning architectures. One of the objectives is to design a new network architecture that leverages the performance of CNNs to compete with the transformer models [20]. Another approach is to combine the strengths of CNNs and transformers, leveraging the spatial locality of convolutions and the global context modeling capabilities of transformers [15]. Furthermore, researchers also pay attention to designing specialized CNN architectures like structured sparsity, dynamic kernel invocation, and hardware-aware neural architecture search to create more compact and efficient CNN models that are suitable for edge and mobile applications [16].

1.1.2 AUTOENCODERS

Autoencoders are unsupervised neural network architectures designed to uncover the fundamental features of input data while compressing it into a lower-dimensional representation. Essentially, an autoencoder is a neural network that learns to reconstruct its input at the output [17]. This architecture consists of two primary components: an encoder and a decoder, as illustrated in Figure 1.1.

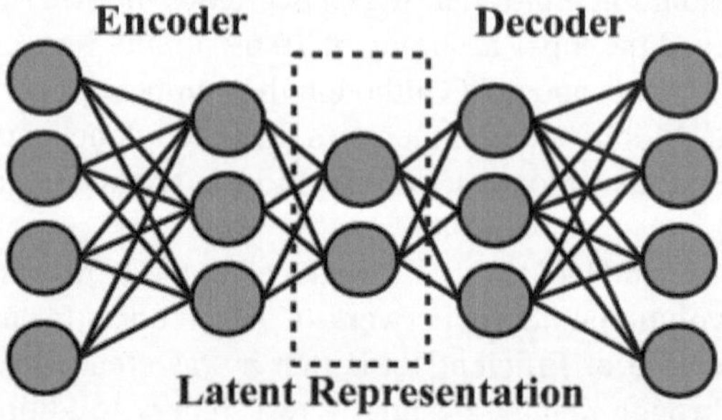

Figure 1.1 Autoencoder Architecture

The encoder converts the input into a latent representation, known as the bottleneck layer, while the decoder reconstructs the original input from this representation. The parameters for both the encoder and decoder are optimized using a reconstruction loss function, which measures the difference between the input and the output, as shown in Equation 1.4. Here, x_i and x_i' represent the data samples at the input and output of the autoencoder, respectively. The bottleneck layer typically has a lower dimensionality than the original data, which helps prevent trivial mappings and compels the model to focus on retaining essential information necessary for accurately reconstructing the data at the output.

$$L_{AE} = \sum_{i=1}^{N} (x_i - x_i')^2 \tag{1.4}$$

To date, there have been a large number of AE models developed and applied to various application domains [18]. In general, these AEs can be divided into two categories: regularized AEs and generative AEs. The regularized AEs aim to drive the model to learn a richer and more expressive representation while still reconstructing the input well (i.e., reducing the reconstruction error). This is achieved by imposing a constraint on the structure of the AEs to trade off between the bias and the variance. Several types of regularized AEs have been introduced [19, 20].

The generative AEs aim to train an AE model for generating new data samples. A typical generative AE model is Variational AE (VAE) [21]. A VAE attempts to project the input data into a Gaussian distribution in the latent space and then samples from this distribution to reconstruct the input. The KL-divergence loss is used to force the latent representation to follow the standard normal distribution. After training, new data samples are generated by sampling from the latent vector and then inputting this vector to the decoder. The more detailed discussion about VAE will be presented in the next section.

1.1.3 RECURRENT NEURAL NETWORKS

A Recurrent Neural Network (RNN)[22] is a type of neural network that utilizes the output from the previous time step as input for the current step. RNNs are particularly effective for addressing problems involving temporal dependencies. For instance, when predicting the next word in a sentence, the context provided by the preceding words is crucial. The key element of an RNN is the hidden state, which retains information from earlier steps in the sequence, often referred to as the memory state. A typical unfolded structure of an RNN is illustrated in Figure 1.2.

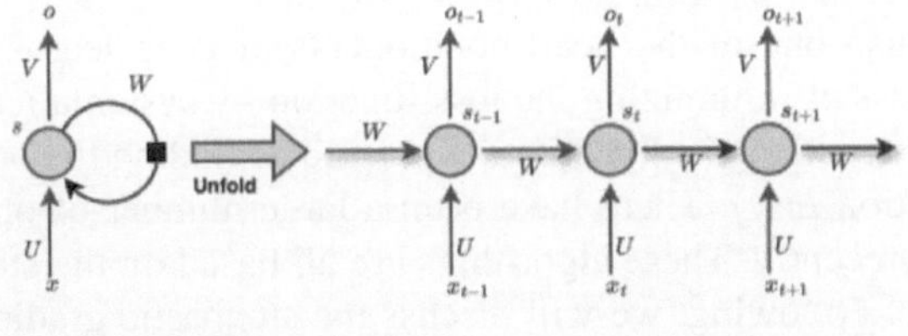

Figure 1.2 An unfold RNN.

Let us consider the sentence classification problem, and x_t be the input at time step t (x_t could be a one-hot vector corresponding to the word at step t); let s_t be the hidden state at time step t, then s_t is calculated based on the previous hidden state and the input at the current step as follows:

$$s_t = f(U.x_t + W.s_{t-1}) \tag{1.5}$$

where U is a matrix containing the connection weights for the inputs of the current timestep. W is a matrix containing the connection weights for the outputs of the previous time step. f is the activation function (tanh or Relu, etc) and s_0 is typically initialized to zeros.

Let o_t be the output at step t. For example, o_t is a vector of probabilities across our vocabulary in case we want to predict the next word in a sentence. ot can be calculated as following:

$$o_t = softmax(V.s_t) \tag{1.6}$$

A RNN can be thought of as multiple copies of the same network, each passing a message to a successor. An unrolled recurrent neural network is presented in the Figure 1.2.

An RNN is trained using the Backpropagation Through Time (BPTT) algorithm. BPTT is the adaptation of the backpropagation algorithm for RNNs to train recurrent models. This approach trains an RNN by passing the gradient through an unfold RNN. Using the chain rule of derivatives, it is proven that the gradient of one step is calculated by power of the weight matrix of hidden neurons [23].

Although RNNs are specifically designed for temporal dependent problems, they often suffer from two issues: vanishing or exploding gradients. This is because the gradient of the output of RNN with respect to its hyper-parameters is calculated by multiplying the hidden matrix over time steps. Thus, if there are small values (< 1), then the gradient will be diminished after some steps. Conversely, if we have large values (> 1), then the gradient will be explored over time. Although these issues can be softened by using truncate methods, they can not be solved completely. Thus, RNNs are not able to learn the long relationship in the sequence data. This problem motivated the introduction of the long short term memory units (LSTMs) to particularly handle the vanishing gradient problem. The detailed discussion about LSTMs and other RNN variants will be presented in Chapter 6.

1.1.4 OPTIMIZATION ALGORITHMS

Optimization is also one of the heart components of deep learning. Optimization refers to the process of minimizing the loss function by systematically updating the network weights. Mathematically, this is expressed as finding the optimal weight w^* given a loss function $L(w)$. There have been a large number of optimization algorithms developed recently. These algorithms are all based on the stochastic gradient descent technique. Following, we will discuss the stochastic gradient descent algorithms and its improvements for training deep network models.

Gradient Descent: The first and most popular optimization algorithm for neural networks is gradient descent. This algorithm is developed based on a strong analytic foundation: For a given differential function, starting at any initial point P_0 and traveling in the opposite direction to the derivation, we will reach the local optimum. For a neural network with w weights and L loss function, the formula for updating its weights using gradient is follows:

$$w = w - \alpha.\frac{\partial L}{\partial w} \tag{1.7}$$

where α is the learning rate that determines the step size of updating the weights in each step.

It should be noted that there are three versions of the gradient descent algorithm. The first version is batch gradient descent (BGD). In batch gradient descent, the gradient is calculated for the entire dataset on each training step. After that, the weight matrices are updated one per epoch. The limitation of batch gradient descent is that it is very expensive for large datasets. The second version is stochastic gradient descent (SGD) that aims to address the issue of batch gradient descent. Instead of calculating the gradient over all training examples, SGD updates the weights for each training example. Thus, SGD is much faster and more computationally efficient than BGD. However, since SGD updates the weights frequently, it can lead to the oscillation of the loss function, which makes the training process highly unstable. The third version is mini-batch stochastic gradient descent (MSGD). MSGD attempts to combine the best of both BGD and SGD. It randomly selects N training examples, thus called a mini batch, from the whole dataset and computes the gradients only from the selected samples and then updates the weights.

It should be noted that MSGD is the general version of both BGD and SGD. If the mini-batch size N is equal to one, then MSGD becomes SGD. Conversely, if this value is set to the size of the whole dataset, then MSGD becomes BGD. In many deep learning frameworks, such as Tensorflow, Pytorch, MSGD is implemented, and this version is named as SGD in these frameworks.

In practice, the mini-batch version of SGD is the most frequently used version because it is both computationally effective and results in more robust convergence. However, this algorithm still contains the following three main limitations.

- Oscillation problem: If the loss function changes quickly in one direction and slowly in another, the optimization process using SGD is often highly oscillatory resulting in a very slow training process.
- Local optimum: SGD is often trapped into the local optimum if the loss function has a local minimum or a saddle point. In these scenarios, SGD is often not able to "jump out" and proceed in finding a better minimum.
- Fixed learning rate: In SGD, the same learning rate is used for all parameters during the training process. This is not suitable in many problems, like the sparse datasets, where the features have different frequencies or significance.

Due to the limitations of SGD, there have been many improvements proposed to leverage its effectiveness. Following, we will discuss some of the most popular improvements of SGD.

SGD with Momentum: The first improvement of SGD comes from adding momentum to the updating equation. The motivation is to increase the optimization updating size if the gradient and the momentum sign are the same. Conversely, if the momentum and the gradient have opposite signs, then the momentum will cancel the gradient, and the loss function will move back to the optimal. The equation of updating model weights as follows:

$$\begin{cases} g_t = \frac{\partial J(\theta)}{\theta} \\ m_t = \gamma m_{t-1} + \alpha g_t \\ \theta_{t+1} = \theta_t - m_t \end{cases} \tag{1.8}$$

Momentum also helps to diminish the noise of the gradients and follows a more direct walk down the optimal value. Moreover, it also helps to reduce the oscillation of the gradients because the velocity vectors can smooth out the highly changed gradient.

Nesterov momentum: Another version of momentum is Nesterov momentum which calculates the gradient in the future position instead of the current position. In other words, we attempt to predict the gradient vector if we move according to the previously built-up velocity. The equation of updating model weights is as follows:

$$\begin{cases} g_t = \frac{\partial J(\theta - \gamma m_{t-1})}{\theta} \\ m_t = \gamma m_{t-1} + \alpha g_t \\ \theta_{t+1} = \theta_t - m_t \end{cases} \tag{1.9}$$

The Nesterov momentum helps to avoid going too fast and results in increasing convergence speed.

Adagrad: Adagrad, introduced in 2011 [24], is designed to adjust the learning rate for each parameter individually. This algorithm applies smaller updates to parameters corresponding to frequently occurring features while providing larger updates to those associated with less common features. As a result, Adagrad is particularly effective for sparse datasets. The formal equation for updating parameters in Adagrad is as follows:

$$\theta_{t+1,i} = \theta_{t,i} - \frac{\alpha}{\sqrt{G_{t,ii} + \varepsilon}} \cdot g_{t,i} \tag{1.10}$$

where $\theta_{t,i}$ is the parameter i^{th} at time step t, and $G_{t,ii}$ is a diagonal matrix in which each diagonal element t, ii is the sum of the squares of the gradients with respect to θ_i up to time step t.

The advantage of AdaGrad is that it eliminates the necessity for manual adjustment of the learning rate, allowing us to keep it at the default value of 0.01. However, a significant drawback of Adagrad is its accumulation of squared gradients in the denominator. Because each new term is positive, this accumulated sum continues to increase throughout training, resulting in a very small learning rate in the later epochs.

Adadelta: Adadelta [25] was proposed in 2012 by Matthew D. Zeiler. Its objective is to address the shrinking learning rate in Adagrad. In other words, Adadelta attempts to reduce the aggressive, monotonically decreasing learning rate. Adadelta is implemented using an exponentially decaying average of the squared gradients as follows:

$$E[g^2]_{t,i} = \gamma.E[g^2]_{t-1,i} + (1-\gamma).g_{t,i}^2 \tag{1.11}$$

The update rule in Adadelta now becomes:

$$\theta_{t+1,i} = \theta_{t,i} - \frac{\alpha}{\sqrt{E[g^2]_{t,i} + \varepsilon}}.g_{t,i} \tag{1.12}$$

Since Adadelta is able to adjust the learning rate on a per-parameter basis, this algorithm often leads to faster convergence and better overall performance.

RMSprop: RMSprop [26] was independently proposed by Geoff Hinton in 2012. The objective of RMSprop is similar to that of Adadelta, that is to modify Adagrad to address the decay of the learning rate. In fact, the updating equation of RMSprop is mostly identical to that of Adadelta. The only difference is that the value of the hyperparameters γ and α is pre-determined. Specifically, the updating equation of RMSprop is as follows:

$$\begin{cases} E[g^2]_{t,i} = 0.9.E[g^2]_{t-1,i} + 0.1.g_{t,i}^2 \\ \theta_{t+1,i} = \theta_{t,i} - \frac{0.001}{\sqrt{E[g^2]_{t,i}+\varepsilon}}.g_{t,i} \end{cases} \tag{1.13}$$

Adam: Adam (Adaptive moment estimation) [27] is perhaps the most popular optimizer nowadays. It has been used extensively in both research and business applications. The motivation of Adam is to combine the two best previous ideas: Momentum and adaptive learning rate. Specifically, Adam combines the idea of SGD with Momentum, Adagrad and Adadelta. Adam first estimates the first moment (the mean) and the second moment (the uncentered variance) of the gradients, respectively:

$$\begin{cases} m_t = \beta_1.m_{t-1} + (1-\beta_1).g_t \\ v_t = \beta_2.v_{t-1} + (1-\beta_2).g_t^2 \end{cases} \tag{1.14}$$

where m_0 and v_0 are initialized as vectors of 0.

However, the above equation often causes the value of m_t and v_t to be biased towards zero. Thus, the author proposes the bias-corrected first and second moment to avoid that problem as follows:

$$\begin{cases} m_t' = \frac{m_t}{1-\beta_1^t} \\ v_t' = \frac{v_t}{1-\beta_2^t} \end{cases} \tag{1.15}$$

After that, the update rule in Adam becomes:

$$\theta_{t+1,i} = \theta_{t,i} - \frac{\alpha}{\sqrt{v_t'} + \varepsilon}.m_t' \tag{1.16}$$

Since there have been a large number of optimization algorithms developed for deep network models, this leads to an important question of selecting an appropriate optimization algorithm for a specific problem. In fact, there is no obvious answer to this question. In other words, there is no technique/approach to determine the best optimizer for training a specific neural network. In a recent publication [28], the author compares a large number of optimizers for three well-known bench-marking problems in computer vision. The experiment results in evidence for the effectiveness of adaptive learning methods like Adam and its variants.

1.2 GENERATIVE MODELS

Generative AI is regarded as one of the most groundbreaking innovations of the 2020s, poised to change how we live, work, and engage with technology. This field has sparked extensive discussions, debates, and forecasts in recent times. As shown in Figure 1.3, which outlines the evolution of AI along with key technological advancements over the decades, generative AI is anticipated to emerge as the most significant development in the AI landscape following the rise of deep learning.

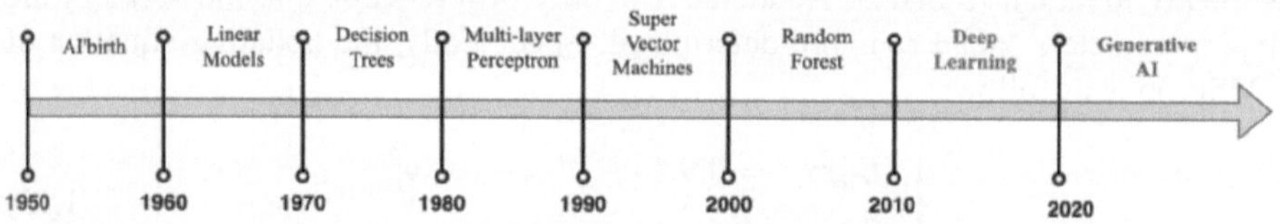

Figure 1.3 History of AI and its breakthrough technologies

Generative AI encompasses a branch of artificial intelligence dedicated to training machine learning models that can produce new data similar to an existing dataset. For example, by using a collection of cat images, we can develop a generative model that learns the underlying characteristics of these images. This model can then generate entirely new, realistic cat pictures that were not included in the original dataset. To build an effective generative model, it is crucial to compile a comprehensive dataset with a large number of samples, known as training data, where each individual sample is termed an observation. Figure 1.4 depicts the workflow involved in generative models.

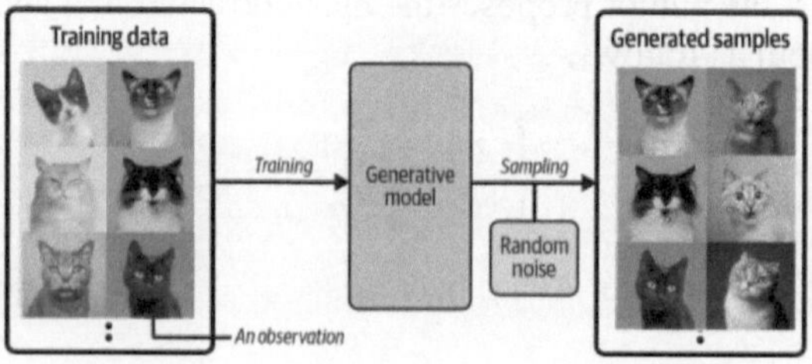

Figure 1.4 A generative model generates photos of cats

A generative model is a probabilistic model rather than a deterministic one. This type of model enables us to produce various outputs instead of yielding the same result every time. Consequently, a generative model must incorporate a random element that affects the individual samples it generates. Formally, generative models can be defined mathematically as follows: *Generative modeling aims to represent the probability of observing an observation X and then sampling from this distribution to create new observations.*

This is different from the discriminative models that attempt to model the probability of a label y given some observations X. In other words, generative models attempt to estimate the data distribution while discriminative models aim to estimate the conditional distribution.

We can imagine that there is some unknown probabilistic distribution that generates samples in the training dataset. A generative model attempts to present that unknown probabilistic distribution by learning from the samples in the datasets. After learning/training, the new, distinct observation can be created by sampling from the model. The new observations look similar to the observations in the dataset, as if they could have been included in the original training set. Formally, the framework of building a generative model includes the following four steps:

1. We have a dataset of samples $\mathbf{X}$.
2. We assume that the samples in the dataset have been generated from some unknown distribution P_{data}.
3. We want to construct a generative model P_{model} to estimate P_{data}.
4. We sample from P_{model} to create new data samples that are similar to the samples in the original dataset.

To effectively generate data samples, the generative model P_{model} should have the following properties:

- *Accuracy*: This is, the P_{model} should have high value (close to 1) for the generated samples that are similar to the original samples and low value (close to 0) for those that are dissimilar.
- *Generation*: This should be easy for us to sample from P_{model} to generate new data samples.
- *Representation*: This generated data should be sampled from a new representation space that is easier to understand/analyze than the original space.

Generally, discriminative models have been the main technology used in many application domains. The reason is that it is often easier to build a deep network model to solve a discriminative problem than a generative problem. For example, it is often easier to classify between a poem and a story than generating a poem or a story. However, as the deep learning technologies have matured, generative problems are becoming more comfortable to address. Subsequently, there have been many interesting applications of generative models such as ChatGPT, AlphaCode, and Dall-E2 emerged recently. In the following section, we will present a taxonomy of generative models and the principles of various generative models.

1.3 TAXONOMY OF GENERATIVE MODEL

There are several approaches to classify generative models such as based on their applications (LLMs, VLMs, etc). In this book, we categorize generative models, based on the probability density function assumed to generate the training data. The probability density function (PDF), or simply density function, is a function $p(x)$ that indicates the relative likelihood that x would be equal to that output. More precisely, the PDF is used to define the probability of a random variable falling within a certain range of values, rather than assuming a single value.

For a specific dataset that is generated by an unknown true density function $P_{data}(x)$, there are infinitely many density functions $P_{model}(x)$ available to estimate $P_{data}(x)$ to estimate $P_{data}(x)$. The goal of a generative model is to identify a suitable $P_{model}(x)$ that closely matches the true density function $P_{data}(x)$. Generally, there are three possible approaches to approximate $P_{data}(x)$.

1. Explicitly model the predefined form of the density function: This approach assumes the true density function has a certain form that can be presented by a set of parameters θ and then attempt to find the optimal parameters $\theta*$ using the training dataset. Variational Autoencoders, Energy-based models, and Diffusion models are the generative model in this class.
2. Constrain the model structure to estimate the density function: This method involves imposing restrictions on the model architecture to facilitate the calculation of the density function. For instance, autoregressive models arrange the input features are in a specific order, allowing the output to be generated sequentially—such as word by word or pixel by pixel. Another example in this category is the Normalizing flow model, which uses a sequence of manageable, invertible functions applied to a simple distribution to create more complex distributions.
3. Implicitly model the density function: This method involves a different approach that doesn't aim to estimate the probability density directly. Instead, it focuses on developing a stochastic process that generates data samples. A prominent example of this type of generative model is the generative adversarial network (GAN).

The taxonomy of generative models can be presented as Figure 1.5. It's important to note that these categories are not mutually exclusive, as there are models that combine elements from two different above approaches.

1.3.1 VARIATIONAL AUTOENCODERS

Variational Autoencoder (VAE) was introduced by Diederik P. Kingma and Max Welling in 2013 [21]. This network is among the most essential and recognized deep learning architectures used for generative modeling. A VAE is an extension of an Autoencoder (AE) to leverage the generative capability of an AE. Although an AE can be used to generate data samples using its latent vector, this vector is deterministic. Thus, AE can not generate diverse data samples, and it is rarely used as a generative model.

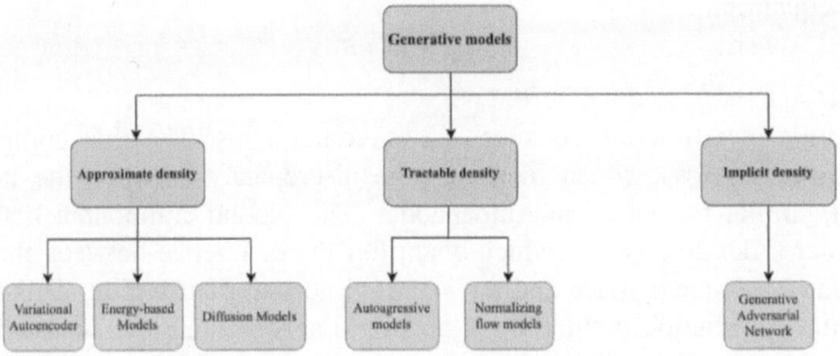

Figure 1.5 Taxonomy of generative models

VAE is proposed to improve AE and create a more sophisticated generative model. VAE also consists of two subnetworks, i.e., an encoder and a decoder, as in Figure 1.6. However, it is different from AE in two folds. First, the encoder of VAE maps each input to a multivariate normal distribution in the latent space instead of a latent vector. Second, the decoder of VAE generates a new data sample by stochastically sampling from the normal distribution in the latent space instead of a deterministic latent vector.

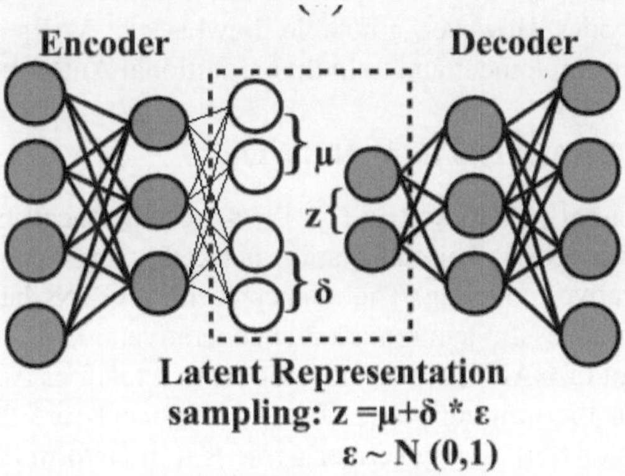

Figure 1.6 Variational Autoencoder architecture

In probability theory, a multivariate normal distribution is characterized by a mean vector and a covariance matrix. However, in the context of Variational Autoencoders (VAEs), it is assumed that the dimensions in the latent space are uncorrelated. As a result, the covariance matrix simplifies to a vector. Additionally, since variance values must always be positive, the VAE encoder aims to map the input to the logarithm of the variance, which can represent any real number. Specifically, if we denote the mean as μ and the logarithm of the variance as z_{log} in the latent space of the VAE, we can sample a point z from the distribution defined by these parameters using the following equation:

$$z = \mu + \sigma * \varepsilon \tag{1.17}$$

where $\sigma = e^{z_{log} * 0.5}$ and $\varepsilon \sim N(0,1)$.

The VAE's loss function consists of two components. The first component is the reconstruction loss, which measures the discrepancy between the input and the output, similar to that in an Autoencoder. The second component is the Kullback–Leibler (KL) divergence, which quantifies the difference between the normal distribution in the latent space and the standard normal distribution. This term encourages the distribution defined by μ and σ to approximate the standard normal distribution closely. The KL divergence between these two distributions can be expressed in closed form as follows:

$$KL_{loss} = D_{KL}[N(\mu,\sigma)||N(0,1)] = -\frac{1}{2}\sum(1+\log(\sigma^2)-\mu^2-\sigma^2) \tag{1.18}$$

The sum is calculated across all dimensions in the latent space. The KL_{loss} reaches its minimum value of 0 when both μ and z_{log} are equal to 0 for all dimensions. If either of these values deviates from 0, the KL_{loss} increases. After the training process, new data samples are generated by sampling z according to Equation 1.17, which is then passed to the decoder. Because these samples originate from a Gaussian distribution, the output from the VAE tends to exhibit greater diversity compared to that of an Autoencoder. However, a notable drawback of VAEs is that they can be more challenging to implement and train than traditional Autoencoders.

1.3.2 GENERATIVE ADVERSARIAL NETWORKS

Generative Adversarial Networks (GANs) [5] were introduced by Ian Goodfellow and his colleagues in 2014. This innovation marked a significant milestone in the evolution of generative modeling. The concept behind GANs has sparked the creation of numerous successful and remarkable generative models.

The core concept of GANs is to create a model that features two competing components throughout the training phase. These components are the Generator (Ge) and the Discriminator (Di). The generator's role is to transform random noise z into new data samples $\hat{x}$ that resemble those in the original dataset x. Meanwhile, the discriminator's task is to determine whether a given sample is from the original dataset (real) or produced by the generator (fake). The structure of a GAN is illustrated in Figure 1.7.

At the beginning of the training, the generator produces noisy outputs while the discriminator makes random predictions. Throughout training, we alternate between training the two subnetworks, allowing the generator to adjust in order to deceive the discriminator, while the discriminator improves its ability to differentiate between real and fake samples. This ongoing process encourages the generator to discover new methods to trick the discriminator, and the cycle repeats. Once training is complete, the generator is utilized to create new data samples by sampling noise (typically from a Gaussian distribution) and feeding it into the generator's input.

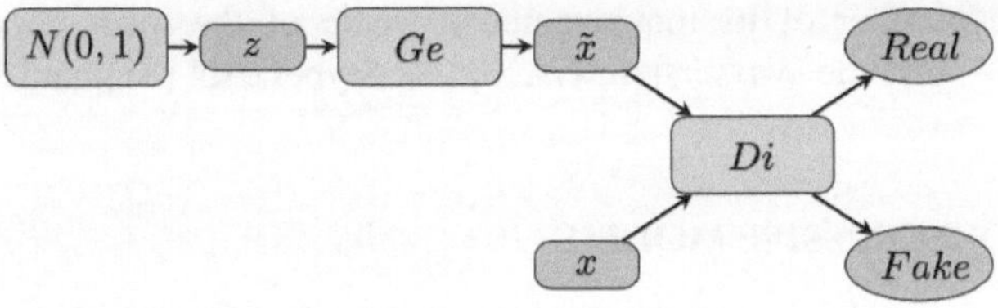

Figure 1.7 Generative Adversarial Network architecture

The loss function of GAN also includes two terms as Equation 1.19.

$$L_{GAN} = -[E_x[logD(x)] + E_z[log(1 - D(G(z)))]] \tag{1.19}$$

where $D(x)$ represents the likelihood that the discriminator Di identifies a real data instance x as authentic, $G(z)$ denotes the output produced by the generator Ge when given random noise z, $D(G(z))$ indicates the probability that the discriminator predicts the generated instance $G(z)$ to be real. E_x and E_z are the expected values across all real and generated instances, respectively. The discriminator aims to minimize this equation, while the generator focuses on minimizing the second term, i.e., $E_z[log(1 - D(G(z)))]$.

While GANs represent a significant advancement in generative modeling, they face the challenge of having a training process that can be hard to stabilize. Typically, there are four key obstacles encountered during the training of GANs:

1. Discriminator dominates the generator: During training, the discriminator might become overly strong, causing the signal from the loss function to weaken and fail to enhance the generator. In the worst-case scenario, the discriminator perfectly distinguishes between real and fake images, resulting in vanishing gradients and no further training progress for the generator.
2. Generator dominates the discriminator: If the discriminator lacks sufficient strength, the generator will easily deceive it with a limited set of samples. This phenomenon is referred to as mode collapse. In this situation, the gradients of the generator's loss function will drop to 0, preventing any further improvement. This often occurs when the generator is trained for multiple epochs without simultaneously updating the discriminator.
3. Uninformative loss: Another challenge in training GANs is the weak correlation between the generator's loss and the quality of its outputs. There are instances when the generator's loss increases even as the quality of its outputs improves. This complicates the monitoring and analysis of the GAN training process.
4. Hyperparameter sensitivity: GANs have numerous hyperparameters and are highly sensitive to even minor adjustments in these values. As a result, identifying a suitable set of parameters often demands significant trial and error.

Due to the above limitations of the canonical GAN model, researchers have proposed a large number of its improvements. Some of the most popular improvements of GANs include Wasserstein GAN [30], StypeGAN [31], and Self-Attention GAN [50].

1.3.3 AUTOREGRESSIVE MODELS

Autoregressive models represent a category of models that streamline the generative modeling process by treating it as a sequential task. They generate new data samples based on the conditional dependencies of prior values in the sequence. This method contrasts with earlier models like variational autoencoders (VAEs) and generative adversarial networks (GANs), which create new data from a latent random variable. Essentially, autoregressive models focus on directly estimating the data distribution through structured constraints.

The foundational autoregressive model is the recurrent neural network (RNN), as mentioned in Subsection 1.1.3. Among these, the Long Short-Term Memory network (LSTM) is perhaps the most well-known autoregressive model. In recent years, many new autoregressive models have been introduced across different application areas. A selection of the most commonly utilized autoregressive models will be explored further in Chapter 6.

1.3.4 NORMALIZING FLOW MODELS

In this subsection, we explore into a new category of generative models known as normalizing flow models [34]. These models exhibit characteristics of both variational autoencoders (VAEs) and autoregressive models. Normalizing flows can effectively and explicitly capture the data distribution similar to autoregressive models. They also convert data into a simpler distribution, akin to VAEs. However, a key distinction lies in the fact that normalizing flows enforce an invertible constraint on the transformation function, meaning that the decoder functions as the inverse of the encoder. This structure facilitates straightforward decoder computation and helps maintain the tractability of normalizing flows.

Developing normalizing flows is challenging because standard neural networks are not inherently invertible. To overcome this limitation, coupling layers have been introduced as a specialized type of neural network. For each input element, a coupling layer outputs a scale factor and a translation factor. For any given input x, it produces two tensors of identical size: one representing the scale factors and the other representing the translation factors. These tensors facilitate invertibility and ensure efficient transformations within normalizing flow architectures. The process of passing an input $x_{1:D}$ through a coupling layer is presented in Figure 1.8.

The first d dimension of x is split into two branches. In the first branch, they are kept the same. In the second branch, they go through two neural networks, i.e., s and t. The results of two branches are then applied element-wise to the rest, i.e.,

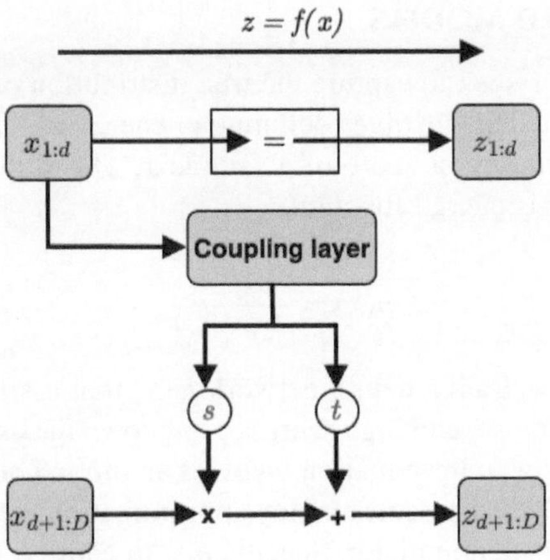

Figure 1.8 Process of passing the input x through a coupling layer

the $D-d$ dimensions, of the input. In summary, the output of the coupling layer is updated using the following equation.

$$z_{1:d} = x_{1:d} z_{d+1:D} = x_{d+1:D} \otimes exp(s(x_{1:d})) + t(x_{1:d}) \tag{1.20}$$

where $s(.)$ and $t(.)$ are two subnetworks that are named as the scale and translation function, and the $\otimes$ operator is the element-wise product.

The log likelihood loss function is used to train normalization flows. Let x be the random variable at the input and z be the random variable at the output, and $x = g(z)$; the log likelihood loss function of the density function of x is calculated as following:

$$log(p_x(x)) = log[P_z(g^{-1}(x))|det(\frac{\partial g^{-1}(x)}{\partial x})|] \tag{1.21}$$

In other words, we can train normalization flows by minimizing the negative function of the log likelihood loss as follows:

$$-log(p_x(x)) = -log[P_z(g^{-1}(x))] - log[|det(\frac{\partial g^{-1}(x)}{\partial x})|] \tag{1.22}$$

where $J = \frac{\partial g^{-1}(x)}{\partial x}$ is the Jacobian matrix of nxn volume. During the training, the target output distribution $P_z(z)$ is selected to be a standard Gaussian. Therefore, sampling from this distribution becomes easy. Moreover, since the function g is invertible, the determinant of the Jacobian matrix can be calculated as following:

$$det(J) = exp(\sum_{j=1}^{D-d} (s(x_{1:d})_j)) \tag{1.23}$$

1.3.5 ENERGY-BASED MODELS

These classes of models seek to capture the true distribution of data by employing a Boltzmann distribution. This involves defining an energy function, denoted as $E(x)$, which represents the energy or score of a sample x. The Boltzmann distribution is then computed using this energy function.

$$p(x) = \frac{e^{-E(x)}}{\int_x e^{-E(x)}} \tag{1.24}$$

In essence, we aim to train a neural network $E(x)$ that assigns low scores to data samples from the training set and high scores to unknown data samples. However, the normalizing denominator in the equation involves an intractable integral. To circumvent this issue, energy-based models employ approximation methods to eliminate the need for calculating this denominator. Specifically, to address the problem of the intractable denominator, energy-based models use the contrastive divergence function for training, and they use the Langevin dynamics technique for sampling.

The gradient of the energy function with respect to its input is determined using the Langevin dynamics method. This process begins at a random location within the sample space and progresses in the direction opposite to the calculated gradient, leading to a gradual reduction of the energy function. As a result, if the neural network is trained properly, initiating with random noise and following this procedure will yield a data sample that closely mirrors those in the training set.

It is important to emphasize that we maintain the neural network weights fixed while calculating the gradient of the output in relation to the input. The input is subsequently modified slightly in the opposite direction of the gradient, which helps in reducing the output (the energy score). This procedure is iterated several times. The formal update rule for the input is defined as follows:

$$x^k = x^{k+1} - \alpha \frac{\partial E(x)}{\partial x} + w \tag{1.25}$$

where $w \sim N(0,1)$ and w is the hyperparameter that needs to be tuned. This is similar to the learning rate in stochastic gradient descent. Moreover, at the beginning, x^0 is drawn from $U[-1,1]$.

To train Energy-Based Models, we will apply a technique by Geoffrey Hinton called contrastive divergence. Specifically, we attempt to minimize the negative function of the log-likelihood loss.

$$L_{EBM} = -E_{x \sim data}[log p_\theta(x)] \tag{1.26}$$

Where $p_\theta(x)$ is the Boltzmann distribution with energy function $E_\theta(x)$ and this loss is calculated as follows [35].

$$\frac{\partial L_{EBM}}{\partial \theta} = \mathbb{E}_{x \sim data}[\frac{\partial E_0(x)}{\partial \theta}] - \mathbb{E}_{x \sim model}[\frac{\partial E_0(x)}{\partial \theta}] \tag{1.27}$$

Thus, the parameter θ of the energy function is updated as following:

$$\theta_{t+1} = \theta_k - \alpha * [\mathbb{E}_{x \sim data}[\frac{\partial E_0(x)}{\partial \theta}] - \mathbb{E}_{x \sim model}[\frac{\partial E_0(x)}{\partial \theta}]] \tag{1.28}$$

Once the training is complete, new data samples are produced from the energy-based model (EBM) by running the Langevin sampler for numerous iterations, starting from a point of random noise.

1.3.6 DIFFUSION MODELS

Diffusion models have emerged as one of the leading and most effective generative models for creating images. They frequently outperform GANs in producing visual samples. The term "diffusion" is derived from the concept of thermodynamic diffusion in physics. A prominent example is the Denoising Diffusion Probabilistic Model (DDPM), introduced by Ho et al. in 2020, which integrates diffusion models with score-based generative techniques.

The idea of these models is to train a neural network to progressively denoise an image through a sequence of very small steps. The new data samples are then generated by starting with random noise and iteratively applying the model over multiple steps, as illustrated in Figure 1.9.

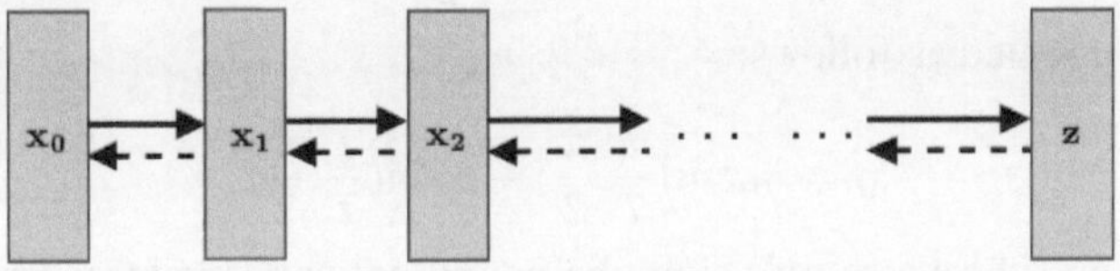

Figure 1.9 Process of passing the input x through Diffusion Model

Assume we have a data sample x_0 that we wish to gradually corrupt into Gaussian noise over many steps. This is achieved by adding a small amount of noise with variance b_t to produce a new data sample x_t at each step. The update process for this transformation can be expressed as follows:

$$x_t = x_{t-1} \cdot \sqrt{1 - b_t} + \varepsilon_t \cdot \sqrt{b_t} \tag{1.29}$$

where ε_t represents a standard Gaussian distribution. Moreover, we will scale the input, i.e., x_{t-1} so that the variance of the output, i.e., x_t is constant. Therefore, if we normalize the initial sample x_0 to have zero mean and unit variance, then for a sufficiently large number of steps T, x_T will converge to a standard Gaussian distribution. In other words, the process for noising q can be described as follows:

$$q(x_t|x_{t-1}) = N(x_t; x_{t-1} \cdot \sqrt{1 - b_t}, I.b_t) \tag{1.30}$$

Furthermore, a reparameterization trick allows us to directly transition from the original sample x_0 to any intermediate sample x_t, bypassing the need for sequential steps. Let $a_t = 1 - b_t$ and $\overline{a_t} = \prod_{i=1}^{t} a_i$, then x_t can be rewritten as the following equation:

$$\begin{aligned} x_t &= x_{t-1}.\sqrt{a_t} + \varepsilon_t.\sqrt{1-a_t} \\ &= x_{t-2}.\sqrt{a_t.a_{t-1}} + +\varepsilon.\sqrt{1-a_t.a_{t-1}} \\ &= ..... \\ &= x_0.\sqrt{\overline{a_t}} + \varepsilon.\sqrt{1-\overline{a_t}} \end{aligned} \tag{1.31}$$

Therefore, the diffusion process q can be written as follows:

$$q(x_t|x_{t-1}) = N(x_t; x_0.\sqrt{\overline{a_t}}, I.(1-\overline{a_t}))) \tag{1.32}$$

One important thing is that we can choose different values of b_t and a_t at each step. In other words, we can set up a diffusion schedule to update the value of b_t and a_t. In the original paper, the authors selected a linear diffusion schedule for b_t from 0.0001 to 0.02. In a later paper [37], the cosine diffusion schedule was proposed that outperformed the linear schedule. In the cosine schedule, the values of $\overline{a_t}$ is calculated as follows:

$$\overline{a_t} = cos^2(\frac{t}{T}.\frac{\pi}{2}) \tag{1.33}$$

Thus, x_t is updated as follows:

$$x_t = x_0.cos(\frac{t}{T}.\frac{\pi}{2}) + \varepsilon.sin(\frac{t}{T}.\frac{\pi}{2}) \tag{1.34}$$

In the process of reverse diffusion, the neural network model is trained to undo the noising process. Specifically, we sample an example x_0 and transform it by t step to get the noise $x_t = x_0.\sqrt{\overline{a_t}} + \varepsilon.\sqrt{1-\overline{a_t}}$. We then input the noise x_t and the noise rate $\overline{a_t}$ to the model and ask it to predict the added noise ε_t. Thus, the loss function is defined as the squared difference between the prediction noise $\varepsilon_\theta(x_t)$ and the true noise ε.

After training, we can generate random noise and then input to the model to generate a new data sample. However, it is very hard to generate a data sample from only one shot. Therefore, we will simulate the forward process in reverse, gradually removing the predicted noise over multiple small steps. Particularly, we will estimate x_{t-1} from x_t using the following two steps:

- First, using the predicted noise to estimate x_0.
- Second, apply the forward diffusion for $t-1$ time to estimate x_{t-1}

This process is presented by the below equation [36]:

$$x_{t-1} = \sqrt{\overline{a_t} - 1}.(\frac{x_t - \sqrt{1-\overline{a_t}}\varepsilon_\theta(x_t)}{\sqrt{\overline{a_t}}}) + \sqrt{1-\overline{a}_{t-1} - \sigma_t^2}.\varepsilon_\theta(x_t) + \sigma_t.\varepsilon_t \tag{1.35}$$

We repeat the two above steps over a number of times until we get a good estimation of x_0.

1.4 EVALUATION OF GENERATIVE MODELS

Although generative AI is useful for many tasks such as content (text, image, video, etc.) synthesis, data augmentation, or anomaly detection, etc., evaluating and comparing the results of generative models are complex tasks. This is because there are many challenges and trade-offs to consider. First, choosing the right evaluation method and metrics depends on the specific generative model and task. Second, during the generating process, we must adjust hyperparameters to balance between realism, diversity, and consistency of generated data. Third, the high-dimensional, multimodal, and complex nature of the data and the generative models add further challenges to the evaluation process. Moreover, the results of generative AI depend on human perception and preference. Thus, it is important to compare results from different evaluation methods and metrics across different generative models and tasks to avoid bias in evaluation.

Some popular approaches to evaluate the generative models will be discussed in this subsection.

1.4.1 QUANTITATIVE METHODS

Quantitative methods attempt to quantify the performance of generative models by numerical scores. Following are some popular evaluation metrics for generative models in generating visual and text data.

Inception Score (IS): IS is a mathematical algorithm designed to assess the quality of images produced by generative AI. Introduced in [38], this metric is based on Google's "Inception" image classification network. The IS algorithm evaluates two key aspects: the quality of the generated images and their diversity. The calculation of IS involves three main steps:

- First, the generated images are input to a pre-trained Inception model (or any good quality pre-trained model) to get the conditional distribution $p(y|x)$. This value determines whether the generated image contains one well-defined thing, thus it presents the quality of the generated samples.
- Second, the marginal probability $p(y)$ is calculated as the mean of the conditional probabilities for the samples in the group of label y.
- The (KL_d) divergence between the conditional probability and the marginal probability is computed as:

$$KL_d = p(y|x) * (log(p(y|x)) - log(p(y))) \tag{1.36}$$

The final IS score is calculated by summing over all images in KL divergence, averaged over all classes, and then takes the exponent.

The IS score can range from zero (worst) to infinity (best). When the IS value is high, there is a strong probability distribution and an even (flat) marginal distribution, meaning that each image has a distinct label, but the overall set of generated images has many different labels. Thus, the generated data is both high quality and diverse.

Frechet Inception Distance (FID): Another important metric for assessing generative models in image production is the Fréchet Inception Distance (FID). Introduced in 2017 [39], FID is frequently more effective than IS in measuring the performance of generative image models. Unlike IS, FID is computed using a reference dataset of real images. The process for calculating FID involves the following steps:

- Input the real and generated samples to the pre-trained Inception-v3 model. This results in the set of output vectors of the same size (2048).
- Calculate the mean (μ_1 and μ_2) and the covariance matrix (C_1 and C_2) of the features in the generated and real datasets.
- Compute the Fréchet distance using the following equation:

$$FID = (\mu_1 - \mu_2)^2 + Tr(C_1 + C_2 - 2 * sqrt(C_1 * C_2)) \qquad (1.37)$$

where Tr refers to the trace linear algebra operation that summarizes the elements along the main diagonal of the square matrix.

BLEU (Bilingual Evaluation Understudy): While IS and FID are often used for quantifying the images generated by AI models, BLUE is usually used for NLP models, particularly for machine translation. This metric calculates the similarity between a generated sentence and reference sentences. Formally, this score is calculated as:

$$\text{BLEU} = \text{BP} \cdot \exp\left(\sum_{n=1}^{N} w_n \log p_n\right) \qquad (1.38)$$

where p_n is the modified precision for nGram, w_n is weight between 0 and 1 for $log(p_n)$ that is usually set at $\frac{1}{N}$, and N is number different of ngram length and BP is the brevity penalty to penalize short machine translations.

$$\text{BP} = \begin{cases} 1 & \text{if } c > r \\ \exp\left(1 - \frac{r}{c}\right) & \text{if } c \leq r \end{cases} \qquad (1.39)$$

where c is the number of unigrams (length) in all the candidate sentences, and is the best match length for each candidate sentence in the corpus. A perfect match yields a bleu score of 1.0.

Generally, quantitative methods can provide objective and standardized measures of generative model performance. However, they also have some limitations, such as requiring a reference dataset, being sensitive to model architecture, or being hard to interpret.

1.4.2 QUALITATIVE METHODS

Qualitative methods are based on involving humans to inspect/evaluate the generated images, text, or audio. For example, human beings are asked to conduct visual inspection, pairwise comparison, or preference ranking to assess how realistic, coherent, and appealing the generated data is. Other methods, such as latent space exploration, interpolation, and conditional generation are also used by humans to evaluate how the generative model reacts to different inputs or parameters. Some popular qualitative metrics to evaluate generative models include:

Human perceptual metric: Human perceptual metrics are often used in the field of computer vision to evaluate the quality of data generated by generative AI. This metric measures the image quality and visual features based on the understanding of humans with colors, shapes, patterns in images. Specifically, a set of synthesized images is created from generative models. After that, a group of humans is asked to distinguish between the original and the synthesized ones. The percentage of misclassification between the synthesized and the real data is reported as the final result [40].

Although qualitative methods can provide intuitive and subjective feedback on generative model performance. The limitation of these metrics is that they are time and resource-consuming, biased, or inconsistent.

1.4.3 HYBRID METHODS

Hybrid methods use both quantitative and qualitative methods to evaluate the quality of generative models. For example, humans are asked to judge the generative models using human-in-the-loop evaluation, adversarial evaluation, or the Turing test. After that, quantitative metrics such as the structural similarity index, or word error rate are used to compare the generated data with the real data labeled by humans [43].

Hybrid methods can harness the strengths of both quantitative and qualitative approaches, offering a more holistic view of a model's capabilities and reducing bias in human assessments. However, they also face challenges, including the difficulty of achieving the right balance between the two methods for a thorough and effective evaluation, as well as potential issues related to cost, complexity, or task and domain specificity [44].

1.5 SUMMARY

In this chapter, we explored the fundamental concepts of deep learning and generative models and the approaches to evaluate the generative models. We began by discussing important deep learning models, including Convolutional Neural Networks, Autoencoders, Recurrent Neural Networks. Various algorithms used for training deep network models are then analyzed. The most important content in this chapter is the definition of generative models and the six prominent generative models. There are methods to evaluate generative models, including quantitative, qualitative, and hybrid methods are highlighted at the end of the chapter.

1. Ian Goodfellow, Yoshua Bengio, and Aaron Courville. *Deep learning*. MIT press, 2016.
2. Christopher M Bishop and Hugh Bishop. *Deep learning: Foundations and concepts*. Springer Nature, 2023.
3. Zewen Li, Fan Liu, Wenjie Yang, Shouheng Peng, and Jun Zhou. A survey of convolutional neural networks: Analysis, applications, and prospects. *IEEE Transactions on Neural Networks and Learning Systems*, 33(12):6999–7019, 2022.
4. Wei Wang, Ming Zhu, Xuewen Zeng, Xiaozhou Ye, and Yiqiang Sheng. Malware traffic classification using a convolutional neural network for representation learning. In *2017 International Conference on Information Networking (ICOIN)*, pages 712–717, 2017.
5. Alex Krizhevsky, Ilya Sutskever, and Geoffrey E Hinton. Imagenet classification with deep convolutional neural networks. In F. Pereira, C.J. Burges, L. Bottou, and K.Q. Weinberger, editors, *Advances in Neural Information Processing Systems*, volume 25. Curran Associates, Inc., 2012.
6. Karen Simonyan and Andrew Zisserman. Very deep convolutional networks for large-scale image recognition. *arXiv preprint arXiv:1409.1556*, 2014.
7. Christian Szegedy, Wei Liu, Yangqing Jia, Pierre Sermanet, Scott Reed, Dragomir Anguelov, Dumitru Erhan, Vincent Vanhoucke, and Andrew Rabinovich. Going deeper with convolutions. *Proceedings of the IEEE Conference on Computer Vision and Pattern Recognition (CVPR)*, June, 2015.
8. Olga Russakovsky, Jia Deng, Hao Su, Jonathan Krause, Sanjeev Satheesh, Sean Ma, Zhiheng Huang, Andrej Karpathy, Aditya Khosla, Michael Bernstein, et al. Imagenet large scale visual recognition challenge. *International Journal of Computer Vision*, 115:211–252, 2015.
9. Kaiming He, Xiangyu Zhang, Shaoqing Ren, and Jian Sun. Deep residual learning for image recognition, 2015.
10. Nils Bjorck, Carla P Gomes, Bart Selman, and Kilian Q Weinberger. Understanding batch normalization. *Advances in Neural Information Processing Systems*, 31, 2018.
11. Andrew G. Howard, Menglong Zhu, Bo Chen, Dmitry Kalenichenko, Weijun Wang, Tobias Weyand, Marco Andreetto and Hartwig Adam. Mobilenets: Efficient convolutional neural networks for mobile vision applications, 2017.
12. Yu-Chen Chiu, Chi-Yi Tsai, Mind-Da Ruan, Guan-Yu Shen, and Tsu-Tian Lee. Mobilenet-ssdv2: An improved object detection model for embedded systems. In *2020 International Conference on System Science and Engineering (ICSSE)*, pages 1–5, 2020.
13. Mingxing Tan and Quoc V. Le. EfficieNtnet: Rethinking model scaling for convolutional neural networks. *International Conference on Machine Learning*, pp. 6105–6114, 2019.
14. Zhuang Liu, Hanzi Mao, Chao-Yuan Wu, Christoph Feichtenhofer, Trevor Darrell, and Saining Xie. A convnet for the 2020s. In *Proceedings of the IEEE/CVF Conference on Computer Vision and Pattern Recognition*, pages 11976–11986, 2022.
15. Yutong Xie, Jianpeng Zhang, Chunhua Shen, and Yong Xia. Cotr: Efficiently bridging CNN and transformer for 3d medical image segmentation. In *Medical Image Computing and Computer Assisted Intervention–MICCAI 2021: 24th International Conference, Strasbourg, France, September 27–October 1, 2021, Proceedings, Part III 24*, pages 171–180. Springer, 2021.
16. Krishna Teja Chitty-Venkata and Arun K Somani. Neural architecture search survey: A hardware perspective. *ACM Computing Surveys*, 55(4):1–36, 2022.
17. Dor Bank, Noam Koenigstein, and Raja Giryes. Autoencoders, 2021.
18. Daneshfar F. Salehi E.S. et al. Berahmand, K. Autoencoders and their applications in machine learning: a survey. *Applied Soft Computing*, 138(C), 2024.

19. Alireza Makhzani and Brendan Frey. k-sparse autoencoders, 2014.
20. Ly Vu, Quang Uy Nguyen, Diep N Nguyen, Dinh Thai Hoang, Eryk Dutkiewicz, et al. Learning latent representation for IoT anomaly detection. *IEEE Transactions on Cybernetics*, 52(5):3769–3782, 2020.
21. Diederik P Kingma and Max Welling. Auto-encoding variational bayes, 2013.
22. Robin M. Schmidt. Recurrent neural networks (RNNs): A gentle introduction and overview. *arXiv preprint arXiv:1912.05911*, 2019.
23. George Bird and Maxim E. Polivoda. Backpropagation through time for networks with long-term dependencies. *arXiv preprint arXiv:2103.15589*, 2021.
24. John Duchi, Elad Hazan, and Yoram Singer. Adaptive subgradient methods for online learning and stochastic optimization. *Journal of Machine Learning Research*, 12(61):2121–2159, 2011.
25. Matthew D. Zeiler. Adadelta: An adaptive learning rate method. *arXiv preprint arXiv:1212.5701*, 2012.
26. Sebastian Ruder. An overview of gradient descent optimization algorithms. *arXiv preprint arXiv:1609.04747*, 2017.
27. Diederik P. Kingma and Jimmy Ba. Adam: A method for stochastic optimization. *arXiv preprint arXiv:1412.6980*, 2014.
28. Derya Soydaner. A comparison of optimization algorithms for deep learning. *International Journal of Pattern Recognition and Artificial Intelligence*, 34(13):2052013, April 2020.
29. Ian J. Goodfellow, Jean Pouget-Abadie, Mehdi Mirza, Bing Xu, David Warde-Farley, Sherjil Ozair, Aaron Courville, and Yoshua Bengio. Generative adversarial networks. *Communications of the ACM*, 63(11), pp. 139–144, 2014.
30. Martin Arjovsky, Soumith Chintala, and Léon Bottou. Wasserstein generative adversarial networks. In *International Conference on Machine Learning*, pages 214–223. PMLR, 2017.
31. Tero Karras, Samuli Laine, and Timo Aila. A style-based generator architecture for generative adversarial networks. In *Proceedings of the IEEE/CVF Conference on Computer Vision and Pattern Recognition*, pages 4401–4410, 2019.
32. Han Zhang, Ian Goodfellow, Dimitris Metaxas, and Augustus Odena. Self-attention generative adversarial networks. In the *International Conference on Machine Learning*, pages 7354–7363. PMLR, 2019.
33. Alex Graves and Alex Graves. Long short-term memory. *Supervised Sequence Labelling with Recurrent Neural Networks*, pages 37–45, 2012.
34. Laurent Dinh, Jascha Sohl-Dickstein, and Samy Bengio. Density estimation using real NVP. *arXiv preprint arXiv:1605.08803*, 2016.
35. Ilya Sutskever and Tijmen Tieleman. On the convergence properties of contrastive divergence. In *Proceedings of the Thirteenth International Conference on Artificial Intelligence and Statistics*, pages 789–795. JMLR Workshop and Conference Proceedings, 2010.
36. Jonathan Ho, Ajay Jain, and Pieter Abbeel. Denoising diffusion probabilistic models. *Advances in Neural Information Processing Systems*, 33:6840–6851, 2020.
37. Alexander Quinn Nichol and Prafulla Dhariwal. Improved denoising diffusion probabilistic models. In *International Conference on Machine Learning*, pages 8162–8171. PMLR, 2021.
38. Tim Salimans, Ian J. Goodfellow, Wojciech Zaremba, Vicki Cheung, Alec Radford, and Xi Chen. Improved techniques for training GANs. *CoRR*, abs/1606.03498, 2016.

39. Heusel Martin, Ramsauer Hubert, Unterthiner Thomas, Nessler Bernhard, and Hochreiter Sepp. Gans trained by a two-time-scale update rule converge to a local Nash equilibrium. In *Advances in Neural Information Processing Systems*, volume 30. Curran Associates, Inc., 2017.
40. Bor-Chun Chen and Andrew Kae. Toward realistic image compositing with adversarial learning. In *Proceedings of the IEEE/CVF Conference on Computer Vision and Pattern Recognition (CVPR)*, June 2019.
41. Ilaria Manco, Emmanouil Benetos, Elio Quinton, and György Fazekas. Muscaps: Generating captions for music audio. *CoRR*, abs/2104.11984, 2021.
42. Ramakrishna Vedantam, C. Lawrence Zitnick, and Devi Parikh. Cider: Consensus-based image description evaluation, 2015.
43. Kyohei Atarashi, Satoshi Oyama, and Masahito Kurihara. Semi-supervised learning from crowds using deep generative models. *Proceedings of the AAAI Conference on Artificial Intelligence*, 32(1), 2018.
44. Jinho Kim, Tae Hee Lee, Yoojin Bae, and Min Kyu Kim. A comparison between AI and human evaluation with a focus on generative AI. In *Proceedings of the 18th International Conference of the Learning Sciences-ICLS 2024, pp. 1722–1725*. International Society of the Learning Sciences, 2024.

2 Cybersecurity Fundamentals: Threats, Impacts, and Countermeasures

This chapter presents a comprehensive overview of the evolving cybersecurity landscape, highlighting traditional and emerging cyber threats and their broad impacts on organizations and national security. It classifies major threat types, such as malware, phishing, DDoS, MitM, and zero-day attacks, and examines the growing role of AI-powered attacks, including deepfakes, adversarial AI, and generative malware. The chapter also introduces a multi-layered defense framework that includes preventive, detective, responsive, and recovery measures. It also highlights the transformative role of Generative AI (GenAI) in modern cybersecurity, both as a powerful defense enabler and a new vector for advanced threats. Through real-world examples and strategic insights, the chapter provides readers with foundational knowledge and forward-looking approaches for securing systems in the AI-driven digital era.

2.0.1 THE EVOLUTION OF CYBERSECURITY AND THE RISE OF AI-DRIVEN DEFENSES

Cybersecurity is the practice of protecting systems, networks, and data from a variety of cyber threats, including unauthorized access, disruption, and destruction. Its foundation is based on the CIA triad (i.e., Confidentiality, Integrity, and Availability) [1]. First, the "Confidentiality" principle ensures that sensitive information can only be accessed by authorized individuals. Second, the "Integrity" principle maintains the accuracy and consistency of data throughout its lifecycle. Third, the "Availability" principle ensures that systems, networks, and data can be accessed by authorized users whenever needed. The scope of cybersecurity has expanded significantly with the advent of modern connected systems. Technologies such as cloud computing enable scalable storage and processing but create new attack surfaces. The Internet of Things (IoT) connect billions of devices, increasing vulnerabilities, while edge networks enable low-latency data processing closer to endpoints, but further complicate security management. This expanding reach requires innovative solutions to mitigate the evolving cyber risks of a connected digital world.

The field of cybersecurity has evolved significantly over the past few decades to cope with emerging technological advances and, in particular, the increasing and dangerous threats of cyber attacks. In the early stages, threats were limited to basic malware such as viruses and worms, targeting personal computers, e.g., Morris

DOI: 10.1201/9781003502531-2

Worm in 1998 and the ILOVEYOU virus in 2000 [2]. In 1990s, the focus shifted to network security, as organizations became interconnected via the Internet, leading to the development of firewalls and intrusion detection systems [3, 4, 5]. In the 2000s, the increasing prevalence of data breaches and financial cybercrime prompted a shift towards information security and proactive cybercrime prevention strategies [6, 7, 8]. With the advent of cloud computing, IoT, and AI-driven systems in the 2010s and beyond, cybersecurity has faced challenges such as ransomware attacks, advanced persistent threats (APTs), and zero-day exploits [9, 10]. For example, the WannaCry ransomware attack in 2017 [11] and the SolarWinds supply chain attack in 2020 [12, 13], which caused huge losses, underscore the urgent need for robust and timely cybersecurity measures that can effectively detect, prevent, and secure critical data systems in the future.

The rapid development of science and technology has led to breakthroughs in network technology, and thus formed important factors that strongly impact the complexity of cybersecurity in the era of innovation. First, the increase in data across industries has led to a huge increase in the volume of sensitive information that needs to be protected. Second, the rise of advanced threats such as ransomware, advanced persistent threats (APTs), and zero-day vulnerabilities giũ' nguyên has made traditional security mechanisms inadequate [9, 10]. Third, the growth of connectivity through IoT devices, mobile networks, and emerging technologies such as 5G has expanded the attack surface, making networks vulnerable to new vulnerabilities [14]. Fourth, the sophistication of cybercrime has increased, as attackers leverage AI and automation to carry out highly targeted, large-scale attacks with high precision [15, 16]. Key cyber attack surfaces in this context include endpoints, edge devices, and even cloud computing platforms. As these challenges continue to grow, we need to develop smart, effective, and sustainable cybersecurity strategies to protect critical assets.

Artificial intelligence (AI) is revolutionizing cybersecurity by enhancing the ability to detect, respond to, and mitigate cyber threats more effectively [17]. Specifically, AI can improve threat detection by identifying anomalies and recognizing patterns in large data sets, making it highly effective in detecting malware, intrusions, and other malicious activities. Additionally, AI can enhance the ability to respond more quickly and efficiently to cybersecurity incidents through automated decision-making, avoiding inefficient manual work. Additionally, AI can significantly reduce response times to detecting threats, helping organizations avoid economic losses. Recently, GenAI has been an emerging technique that plays a dual role in cybersecurity [18]. On the defensive side, GenAI can generate synthetic data to train models, simulate complex threat scenarios, and develop advanced detection algorithms. On the attack side, however, GenAI poses risks such as enabling highly realistic phishing campaigns, automated malware generation, and sophisticated social engineering attacks. This dual nature of GenAI highlights the need for ethical and responsible application of AI within cybersecurity frameworks.

Traditional cybersecurity approaches face significant limitations in addressing modern cyber threats. In particular, current solutions rely heavily on signature-based detection, which is ineffective against zero-day attacks and emerging threats [19, 20]. Furthermore, traditional systems struggle to handle large-scale data and complex attacks that are constantly evolving due to IoT, cloud computing, and AI-driven adversaries. Additionally, manual analysis of cybersecurity threats is labor-intensive, time-consuming, and often too slow to respond to real-time attacks. These challenges create a significant gap in cybersecurity defenses, highlighting the growing need for automation and intelligent solutions. AI and machine learning can bridge this gap by automating threat detection, analyzing large data sets in real time, and adapting to new attack patterns [17]. As cyber threats become more dangerous and sophisticated, adopting AI-driven techniques is essential to overcome the limitations of traditional approaches.

GenAI is emerging as a new disruptive technology in addressing modern cybersecurity challenges [18]. Specifically, GenAI can be used for threat simulation, where realistic cyberattack scenarios can be created to test and improve the defense mechanisms of intrusion detection systems (IDSs). Additionally, GenAI can generate new data for learning when real-world data is limited. This is especially important for IDSs because data on cyberattacks is often very limited, and many attacks are new and have never been recorded before. Therefore, the combined use of GenAI with IDSs is expected to bring breakthroughs in the timely detection and prevention of future attacks.

2.1 TYPES OF CYBERSECURITY THREATS AND IMPACTS

2.1.1 CLASSIFICATION OF CYBERSECURITY THREATS

In a rapidly evolving digital landscape, cyber threats are becoming increasingly sophisticated and diverse. These threats exploit vulnerabilities in systems, networks, and human behavior to compromise security, causing significant financial, operational, and reputational impacts. To understand and mitigate these risks effectively, it is essential to classify cyber threats into structured categories. The following subsections provide an overview of common cyber threats, highlighting their mechanisms, impacts, and real-world examples.

Malware: Malware is malicious software designed to infiltrate, damage, or disrupt systems, steal data, or gain unauthorized access to networks. It encompasses various forms, including:

- Virus: a type of malicious software that attaches itself to legitimate programs or files, replicates, and spreads to other systems, often causing damage or disruption.
- Worm: a self-replicating malicious software program that spreads independently across networks, exploiting vulnerabilities without needing to attach to a host file or program.
- Ransomware: a type of malicious software that encrypts a victim's data or locks their system, demanding a ransom payment to restore access.
- Spyware: a malicious software designed to secretly gather and transmit information about a user or their activities without their consent.

- Trojan: a malicious software disguised as legitimate or harmless, designed to deceive users and enable unauthorized access or cause harm to their systems.

The WannaCry ransomware attack of 2017 is a notable example of a cyberattack that caused widespread disruption by encrypting critical files on infected computers and demanding ransom payments in Bitcoin to decrypt them [21]. For instance, the attack affected the UK's National Health Service (NHS), leading to the cancellation of thousands of medical appointments and surgeries [21]. Malware remains one of the most pervasive and damaging cybersecurity threats today.

Phishing and Social Engineering: Phishing and social engineering attacks rely primarily on the psychological manipulation of people (especially those with limited or no knowledge of cybersecurity) to exploit their personal information, thereby stealing sensitive information or gaining unauthorized access. Phishing often involves linking to credible phishing emails or websites to steal credentials or key data. A notable real-world example of phishing involving credible-looking emails to steal credentials is the 2020 Twitter VIP attack [22]. In this incident, hackers compromised 130 high-profile Twitter accounts, including those of Barack Obama, Elon Musk and Apple, by sending deceptive messages promoting a Bitcoin scam. The attackers used social engineering techniques to gain access to Twitter's internal systems, enabling them to control these accounts and post fraudulent tweets. This breach resulted in over $100,000 in losses and highlighted the susceptibility of even major platforms to sophisticated phishing attacks.

Denial-of-Service (DoS) and Distributed DoS (DDoS): Denial-of-Service (DoS) and Distributed Denial-of-Service (DDoS) attacks aim to overwhelm systems, servers, or networks with an excessive volume of traffic, rendering services unavailable to legitimate users. This is considered one of the easiest yet efficient ways to attack the target machine/network. While DoS attacks originate from a single source, DDoS attacks leverage a network of compromised devices, known as botnets, to amplify the assault. A prominent example illustrating how DDoS attacks leverage botnets is the 2016 attack on Dyn, a major Domain Name System (DNS) provider [23]. In October 2016, the Mirai botnet orchestrated a massive DDoS attack against Dyn, utilizing a vast network of compromised Internet of Things (IoT) devices, such as IP cameras and home routers. This assault overwhelmed Dyn's infrastructure, leading to widespread disruptions of major websites and services, including Twitter, Netflix, and Reddit. The Mirai malware operated by scanning the internet for vulnerable IoT devices with default usernames and passwords, subsequently enlisting them into the botnet to execute large-scale DDoS attacks.

Man-in-the-Middle (MitM) Attacks: Man-in-the-Middle (MitM) attacks involve intercepting and eavesdropping on communication between two parties without their knowledge. Attackers position themselves between the sender and receiver, capturing or altering transmitted data. Common scenarios include intercepting unencrypted Wi-Fi traffic or exploiting vulnerabilities in secure communication protocols. MitM attacks can result in the theft of sensitive information, such as login credentials, financial details, or trade secrets, making them a significant threat in both personal and enterprise contexts. For example, in 2019, hackers executed an MitM attack to defraud an Israeli startup of $1 million [24]. The attackers intercepted and manipulated email communications between the startup and a Chinese venture capital firm, effectively positioning themselves between the two parties without their knowledge. By spoofing email domains and canceling critical in-person meetings, the attackers successfully diverted a significant wire transfer intended for the startup into their own accounts. This incident underscores the potential financial and reputational damage that MitM attacks can inflict on businesses.

Zero-Day Attacks: Zero-day attacks exploit previously unknown vulnerabilities in software or systems, taking advantage of the window of time before developers can issue a patch. These attacks are particularly dangerous because they provide no warning and often involve sophisticated malware or exploits. Cybercriminals and state-sponsored actors frequently use zero-day attacks to target high-value systems. A well-known example of a zero-day attack is the Stuxnet worm, discovered in 2010 [25]. Stuxnet targeted Iran's nuclear facilities by exploiting four previously unknown vulnerabilities in Microsoft Windows operating systems. By leveraging these zero-day exploits, the worm infiltrated the systems controlling centrifuges used for uranium enrichment, causing them to malfunction while displaying normal operations to monitoring systems. This sophisticated attack underscores the significant threat posed by zero-day vulnerabilities.

Supply Chain Attacks: Supply chain attacks compromise software, hardware, or service providers to gain access to target organizations. Attackers infiltrate trusted components or processes, enabling them to introduce malicious code or vulnerabilities that propagate downstream. A notable example of a supply chain attack is the 2020 SolarWinds incident [26]. In this case, attackers compromised SolarWinds' Orion software updates, embedding malicious code that was distributed to approximately 18,000 customers, including numerous U.S. government agencies and Fortune 500 companies. This backdoor, known as SUNBURST, allowed the attackers to gain unauthorized access to the networks of affected organizations, leading to significant data breaches and security concerns.

In Table 2.1, we provide a comparative overview of different types of cybersecurity threats, their mechanisms, impacts, and real-world examples. It highlights threats such as malware, phishing, DoS/DDoS, MitM attacks, zero-day exploits, and supply chain attacks, detailing how they operate and the damage they cause, including data theft, financial losses, and operational disruptions. Examples like the WannaCry ransomware, the 2020 Twitter VIP attack, and the SolarWinds supply chain breach illustrate the severity and diversity of these threats.

2.1.2 EMERGING THREATS POWERED BY AI

The integration of Artificial Intelligence (AI) in cybersecurity has introduced both defensive advancements and significant new threats. While AI enables powerful tools for detecting and mitigating cyberattacks, it also provides attackers with sophisticated techniques to enhance their capabilities. Traditional threats like phishing, malware, and system breaches are evolving into AI-driven attacks that are more

Table 2.1
Comparison of Cybersecurity Threats

Threat Type	Mechanism	Impact	Example
Malware	Malicious software infiltrates systems to disrupt operations, steal data, or encrypt files.	Data theft, operational disruption, ransom payments.	WannaCry ransomware (2017), Stuxnet worm.
Phishing and Social Engineering	Manipulates human psychology through fraudulent emails, websites, or messages to steal sensitive information.	Credential theft, financial fraud, and larger cyberattacks.	2020 Twitter VIP attack.
Denial-of-Service (DoS) and Distributed DoS (DDoS)	Overwhelms servers or networks with excessive traffic to render services unavailable.	Service downtime, financial losses, reputational damage.	Mirai Botnet DDoS attack (2016).
Man-in-the-Middle (MitM) Attacks	Interception and eavesdropping on communication to steal or alter data.	Theft of credentials, financial fraud, unauthorized access.	Intercept and manipulate email communications.
Zero-Day Attacks	Exploitation of previously unknown vulnerabilities before patches are released.	System compromise, malware deployment.	Stuxnet worm targeted Iran's nuclear facilities.
Supply Chain Attacks	Compromise of software, hardware, or service providers to infiltrate target organizations.	Widespread malware propagation, operational disruption.	SolarWinds supply chain attack (2020).

adaptive, scalable, and harder to detect. This section explores emerging AI-powered threats, highlighting how GenAI and adversarial AI are reshaping the cybersecurity landscape.

AI-Generated Phishing: AI-generated phishing leverages GenAI tools to craft convincing phishing emails, messages, and fake websites that mimic legitimate communications with unprecedented precision. Unlike traditional phishing, which often relies on manually written or template-based content, AI can analyze vast datasets of user behaviors, linguistic patterns, and branding to create highly personalized and context-aware messages. Tools like natural language processing (NLP) ensure grammatical accuracy and sophistication, increasing the likelihood of deceiving victims. In 2023, Abnormal Security detected several attacks where AI was likely used to generate malicious emails. These emails mimicked legitimate communications with high precision, making them difficult to detect using traditional security example-measures [27, 28]. For exmaple one specific case of using GAI to create scam email is illustrated in Fig. 2.1 [27]. The email was created using GAI, which allowed the attackers to create unique and sophisticated content that bypassed traditional detection methods. This example clearly shows how AI can analyze user behaviors and linguistic patterns to produce personalized and context-aware phishing messages, significantly increasing the chances of deceiving victims.

Deepfake Attacks: Deepfake attacks involve the use of GenAI to create hyper-realistic audio, video, or images for malicious purposes such as social engineering, misinformation, or impersonation. By manipulating visual and auditory data, attackers can convincingly impersonate individuals, such as executives or public figures, to gain unauthorized access, spread false information, or manipulate public opinion. A

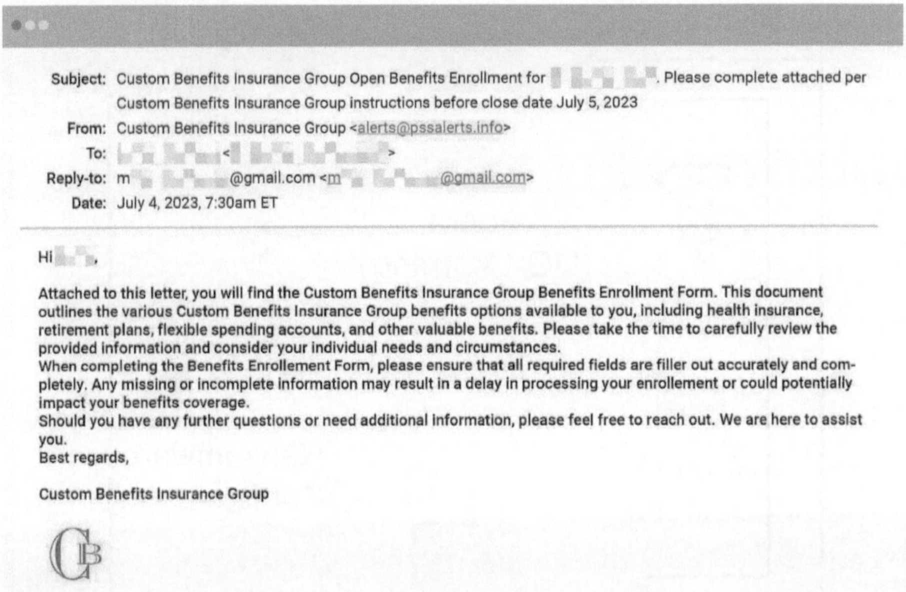

Subject: Custom Benefits Insurance Group Open Benefits Enrollment for [redacted]. Please complete attached per Custom Benefits Insurance Group instructions before close date July 5, 2023
From: Custom Benefits Insurance Group <alerts@pssalerts.info>
To: [redacted]
Reply-to: m[redacted]@gmail.com <m[redacted]@gmail.com>
Date: July 4, 2023, 7:30am ET

Hi [redacted],

Attached to this letter, you will find the Custom Benefits Insurance Group Benefits Enrollment Form. This document outlines the various Custom Benefits Insurance Group benefits options available to you, including health insurance, retirement plans, flexible spending accounts, and other valuable benefits. Please take the time to carefully review the provided information and consider your individual needs and circumstances.
When completing the Benefits Enrollement Form, please ensure that all required fields are filler out accurately and completely. Any missing or incomplete information may result in a delay in processing your enrollement or could potentially impact your benefits coverage.
Should you have any further questions or need additional information, please feel free to reach out. We are here to assist you.
Best regards,

Custom Benefits Insurance Group

Figure 2.1 An example of using GAI for sacm emails [27]..

notable example of a deepfake attack occurred in 2019, when cybercriminals used artificial intelligence to mimic the voice of a CEO, successfully deceiving an employee into transferring $243,000 to the attackers' account [29]. The hackers employed AI-based software to create a convincing imitation of the CEO's voice, exploiting the trust and authority associated with the executive's position. Another example is a deepfake image attack that occurred in January 2024, when AI-generated explicit images of American singer Taylor Swift [31] were disseminated across social media platforms, including X (formerly Twitter), Facebook, Reddit, and Instagram [30]. These images, which were fabricated without Swift's consent, originated from a Telegram group where members employed tools like Microsoft Designer to create the fake content. The incident sparked widespread condemnation and highlighted the pressing need for legislative measures to combat non-consensual deepfake pornography. These attacks underscore the potential of deepfake technology to facilitate sophisticated social engineering attacks, leading to significant financial losses.

AI-Powered Malware: AI-powered malware introduces a new generation of malicious software that adapts in real-time to evade detection systems. Unlike traditional malware, which relies on static code, AI-driven malware employs techniques such as polymorphism and evasion algorithms to change its behavior, signature, or encryption patterns dynamically [32]. This allows it to bypass signature-based antivirus tools and intrusion detection systems. AI can also automate decision-making processes within the malware, enabling it to identify vulnerabilities, choose attack methods, and optimize its impact [33].

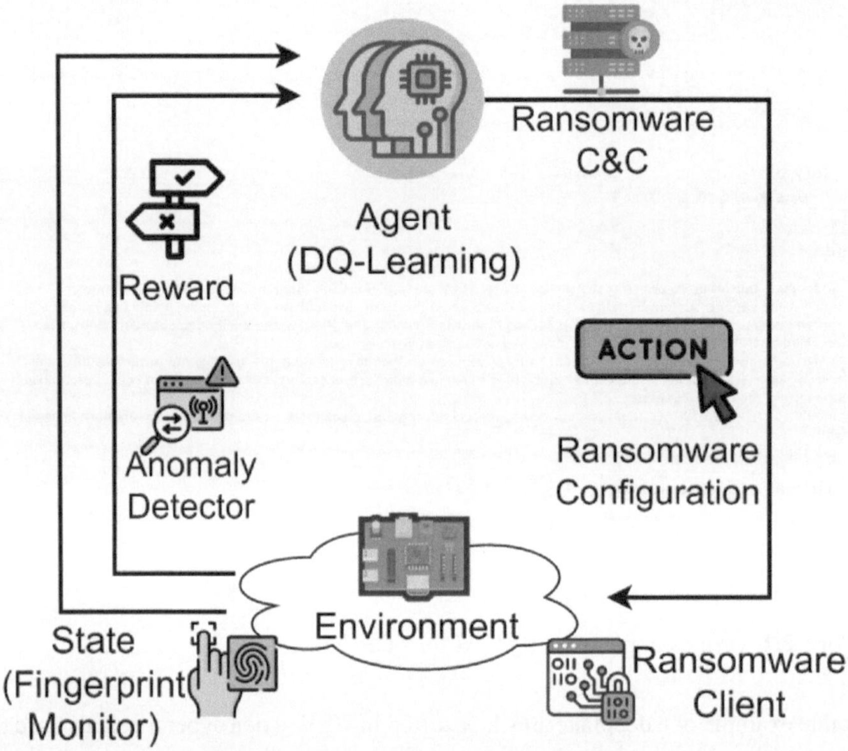

Figure 2.2 RansomAI architecture.

An example of AI-powered malware is the "BlackMamba" strain, identified by security researchers in 2024 [34, 35]. BlackMamba utilizes machine learning algorithms to adapt its behavior in real-time, effectively evading traditional detection methods. It employs polymorphic techniques, altering its code structure dynamically to avoid signature-based antivirus tools and intrusion detection systems [36]. This adaptability allows BlackMamba to bypass security measures that rely on static analysis, posing a significant challenge to cybersecurity defenses. Cyflare

Another pertinent example is the "RansomAI" framework, a proof-of-concept AI-powered ransomware was developed by researchers in 2023 [37]. RansomAI integrates reinforcement learning to adjust its encryption strategies dynamically, minimizing detection while maximizing damage. By learning from its environment, RansomAI can intelligently select encryption algorithms, rates, and durations, effectively evading traditional security measures. The RansomAI architecture is illustrated in Fig. 2.2. This demonstrates the potential for AI to automate decision-making processes within malware, enabling it to identify vulnerabilities and optimize its impact autonomously.

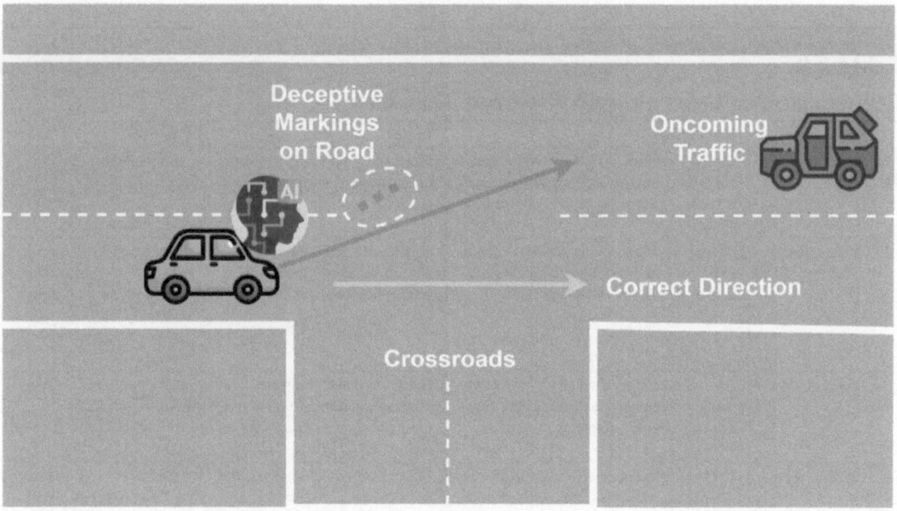

Figure 2.3 An AI system can malfunction due to adversarial attacks, such as errant road markings misleading a driverless car, as highlighted in a new NIST publication detailing attack types and mitigation strategies [38].

Adversarial AI Attacks: Adversarial AI attacks exploit weaknesses in AI-based cybersecurity systems by generating adversarial inputs designed to fool machine learning models. These inputs are carefully crafted, often appearing benign to human observers but triggering incorrect responses in AI models. A notable example of an adversarial AI attack was introduced in [38], when researchers demonstrated that errant markings on roads could mislead autonomous vehicles' AI systems, potentially causing them to veer into oncoming traffic. These adversarial inputs were carefully crafted to appear benign to human observers but triggered incorrect responses in the vehicle's AI models, exploiting weaknesses in AI-based cybersecurity systems.

2.1.3 IMPACTS OF CYBERSECURITY THREATS

Cybersecurity threats have far-reaching consequences that extend beyond immediate system breaches. These threats impact organizations, governments, and individuals across multiple dimensions, including financial losses, operational disruptions, reputational damage, legal liabilities, and national security risks. As cyberattacks become more sophisticated and pervasive, understanding their multidimensional impacts is essential for driving investments in cybersecurity measures and resilience strategies. This section explores the significant consequences of cybersecurity threats across key areas.

Financial Impact: Cybersecurity threats often result in significant financial costs for organizations. Expenses may include ransom payments in ransomware attacks, fines for regulatory non-compliance, costs associated with data recovery, and

Table 2.2
Comparison of Emerging AI-Powered Threats

Threat Type	Mechanism	Impact	Example
AI-Generated Phishing	GenAI creates highly personalized phishing emails, messages, or fake websites using NLP and behavioral analysis.	Increased phishing success rates, targeted credential theft, and large-scale campaigns.	AI-generated phishing emails [27].
Deepfake Attacks	GenAI produce hyper-realistic audio, video, or images to impersonate individuals or spread misinformation.	Financial fraud, social manipulation, unauthorized access, and reputational damage.	AI-generated voice of a CEO instructing unauthorized fund transfers [29] and deepfake image attack of Taylor Swift [31].
AI-Powered Malware	AI-driven malware adapts dynamically using polymorphism and evasion algorithms to bypass detection systems.	Harder to detect malware, real-time exploitation of vulnerabilities, and optimized attack execution.	BlackMamba [33] and RansomAI [37].
Adversarial AI Attacks	Exploits weaknesses in AI models by generating adversarial inputs to deceive machine learning systems.	Compromised AI systems, bypassed anomaly detection, and unauthorized access.	Adversarial AI attack on autonomous vehicles [38].

business interruptions. The financial toll also includes long-term losses, such as reduced revenue and increased spending on security infrastructure post-incident. According to IBM's Cost of a Data Breach Report, the average global cost of a data breach in 2023 was $4.45 million, with healthcare organizations facing even higher costs [39]. Cyberattacks not only drain immediate resources but also impose ongoing financial burdens, making cybersecurity a critical investment for sustainable operations.

Operational Impact: Cybersecurity threats can severely disrupt an organization's operations, leading to downtime, loss of productivity, and compromised infrastructure. Attacks such as Denial-of-Service (DoS) or ransomware can render systems inoperable for extended periods, halting critical business processes. For example, in 2024, Microsoft experienced a global outage affecting its Azure and Outlook services due to a Distributed Denial-of-Service (DDoS) cyberattack [40]. The incident caused approximately 10 hours of service interruption, impacting airports, banks, and numerous users worldwide who were unable to access essential services. This event underscores the critical need for robust cybersecurity measures to prevent such operational disruptions.

Reputational Impact: The reputational damage caused by cybersecurity breaches can be long-lasting and difficult to repair. When sensitive customer data is compromised, organizations face an erosion of customer trust and a damaged brand reputation. High-profile data breaches often attract negative media attention, causing stakeholders to lose confidence in the organization's ability to protect their data. For example, in 2017, Equifax, one of the largest credit reporting agencies in the United States, suffered a massive data breach that exposed the personal information of approximately 147 million individuals, including Social Security numbers, birth dates, and credit card information [41]. This breach not only put millions at risk of identity theft but also severely eroded public trust in Equifax's ability to safeguard sensitive consumer data. The company's delayed response and mishandling of the breach further exacerbated the situation, leading to significant damage to its reputation and customer confidence.

National Security Impact: Cybersecurity threats pose critical risks to national security, particularly when they target critical infrastructure such as power grids, healthcare systems, transportation networks, and defense operations. State-sponsored cyberattacks, espionage, and sabotage can disrupt essential services, compromise sensitive government data, and undermine a nation's sovereignty. A notable example of the critical risks posed by cybersecurity threats to national security is the 2017 Ukraine ransomware attacks [42, 43]. In June 2017, a ransomware attack, later identified as NotPetya, targeted various Ukrainian institutions, including banks, ministries, metro systems, and state-owned enterprises such as Boryspil International Airport and Ukrainian Railways. The attack also affected the radiation monitoring system at Ukraine's Chernobyl Nuclear Power Plant, forcing it offline. This cyber assault disrupted essential services and compromised sensitive data, highlighting the potential for state-sponsored cyberattacks to undermine a nation's sovereignty and public safety.

2.1.4 THREAT TRENDS AND ANALYSIS

The cybersecurity landscape is rapidly evolving, driven by advancements in technology and the increasing sophistication of cybercriminal tactics. Emerging trends demonstrate a significant shift towards automation, the commercialization of cybercrime, and the exploitation of new technologies such as AI and quantum computing. Understanding these trends is critical for anticipating future threats and designing proactive defense strategies. This section highlights the key trends shaping the cybersecurity domain and provides data-backed insights into the growth and scale of cyberattacks.

AI-based Cyberattacks: Cybercriminals are increasingly leveraging automation to execute large-scale attacks with minimal human intervention. Tools powered by artificial intelligence (AI) and machine learning (ML) enable attackers to automate tasks such as vulnerability scanning, phishing campaign generation, and malware deployment. AI and ML allow attackers to conduct faster and more efficient campaigns, often targeting thousands of systems simultaneously. For instance, as discussed above, advanced AI-driven tools can be used to create phishing emails, fake voices, and fake images. As more and more AI tools are widely and automatically used, the danger of AI-based cyberattacks is more serious.

Rise of Ransomware-as-a-Service (RaaS): Ransomware-as-a-Service (RaaS) has emerged as a major trend, enabling less skilled cybercriminals to deploy sophisticated ransomware attacks. RaaS platforms operate on a subscription or profit-sharing model, where developers provide ransomware tools to affiliates who execute the attacks. This commercialization of ransomware has lowered the entry barrier for cybercrime, leading to a surge in ransomware incidents globally. An example of Ransomware-as-a-Service (RaaS) is the Hive ransomware group [44]. Hive operates by providing ransomware tools to affiliates who then execute attacks against various sectors, including healthcare organizations. In April 2022, Hive leveraged a pass-the-hash technique to target numerous Microsoft Exchange Server customers, affecting industries such as energy and financial services.

Quantum Computing and Cybersecurity Risks: Quantum computing, while still in its early stages, poses a significant long-term threat to cybersecurity. Quantum computers have the potential to break current cryptographic algorithms, such as RSA (Rivest-Shamir-Adleman) and ECC (Elliptic Curve Cryptography), which secure most digital communications and transactions [45]. This development could render existing encryption methods obsolete, exposing sensitive data to unauthorized access. For example, a recent study by Chinese researchers demonstrated that quantum computers could break RSA encryption by optimizing problem-solving techniques [46]. Similarly, the World Economic Forum has highlighted the potential of quantum computing to disrupt current cybersecurity measures, emphasizing the need for new quantum-resistant cryptographic solutions [47].

2.1.5 USE OF GENAI IN THREAT MITIGATION

As cyber threats continue to evolve in sophistication and scale, traditional cybersecurity approaches often struggle to keep pace. GenAI offers a powerful set of tools to combat these threats by enabling more adaptive, intelligent, and proactive defense mechanisms. By leveraging its ability to generate data, identify anomalies, and counter adversarial attacks, GenAI can significantly enhance cybersecurity systems. This section explores the critical roles GenAI plays in mitigating modern cyber risks.

Generating Synthetic Datasets to Train Robust Cybersecurity Models: One of the key challenges in cybersecurity is the lack of diverse and high-quality datasets for training machine learning models. GenAI can address this limitation by creating synthetic datasets that mimic real-world attack scenarios and benign data. These datasets are invaluable for training robust models capable of detecting emerging threats, including zero-day attacks. For instance, generative adversarial networks (GANs) can simulate various cyberattack patterns to improve a model's ability to recognize anomalies or malicious activities [48]. Synthetic data generation not only accelerates model training but also enhances resilience against unseen threats without compromising sensitive real-world data.

Enhancing Anomaly Detection and Threat Prediction: GenAI significantly improves anomaly detection and threat prediction by identifying subtle patterns and deviations in large datasets. Traditional detection systems may struggle with false positives or miss sophisticated attacks that mimic normal behavior. However, GenAI models can learn complex relationships within the data to detect unusual patterns indicative of malicious activity. For example, the authors in [49] introduce a novel AI architecture based on autoencoder to enhance efficiency in detecting attacks in intrusion detection systems. They also show that their approach can boost around 1% in terms of accuracy and F-score in detection attacks compared to the state-of-the-art machine learning and representation learning method.

Countering Adversarial AI Through Automated Model Retraining: As adversarial AI techniques become more prevalent, attackers are increasingly generating inputs designed to evade detection systems. GenAI can counter these adversarial attacks by automating the retraining of cybersecurity models with adversarial examples. This process strengthens the models' ability to recognize and resist manipulation. For instance, GenAI can create adversarial samples to test and fortify existing AI-driven defense systems [50]. By continuously learning from adversarial inputs, cybersecurity models become more robust and adaptive, ensuring they remain effective even as attack methods evolve.

2.2 COUNTERMEASURES

2.2.1 INTRODUCTION TO CYBERSECURITY COUNTERMEASURES

Cybersecurity countermeasures encompass a wide range of strategies, tools, and practices designed to safeguard systems, networks, and data from cyberattacks. These measures aim to prevent, detect, respond to, and recover from malicious activities, ensuring the confidentiality, integrity, and availability of critical resources [51]. As cyber threats grow in complexity, organizations must adopt adaptive and proactive approaches to counter evolving attack methods. This introduction provides a foundational understanding of countermeasures and highlights the role of advanced technologies, such as GenAI, in enhancing modern cybersecurity defenses.

Cybersecurity countermeasures refer to the combination of technical solutions, policies and best practices implemented to mitigate the risks posed by cyber threats [51]. These measures are categorized into four key functions [51]:

- *Prevention:* stops attacks before they occur
- *Detection:* identifies malicious activities
- *Responsive:* responds to active threats
- *Recovery:* restores systems and data post-attack

Together, these functions form the core of a robust cybersecurity strategy, enabling organizations to minimize risks, mitigate impacts, and maintain operational resilience.

Defense-in-depth is a fundamental principle in cybersecurity, emphasizing a multi-layered approach to security [52, 53]. Rather than relying on a single solution, defense-in-depth involves implementing multiple layers of protection across endpoints, networks, applications, and users. Each layer serves as a barrier, reducing the likelihood of a successful breach even if one layer is compromised. For example, firewalls and intrusion prevention systems (IPS) protect networks, while encryption secures sensitive data, and access controls limit user privileges. This layered strategy is particularly critical in addressing AI-powered threats, as attackers increasingly exploit multiple attack surfaces to bypass traditional defenses.

GenAI is revolutionizing cybersecurity by augmenting and enhancing traditional countermeasures. Through its ability to analyze vast datasets, simulate attack scenarios, and generate synthetic data, GenAI strengthens threat detection, prevention, and response capabilities. For example, it can create realistic synthetic datasets to train machine learning models for identifying anomalies and emerging threats [48]. Additionally, GenAI can simulate adversarial attack patterns to test and improve existing defense mechanisms, making systems more resilient against sophisticated attacks [50]. By automating threat mitigation and continuously learning from evolving attack strategies, GenAI plays a crucial role in fortifying cybersecurity defenses against modern, AI-driven threats.

2.2.2 PREVENTIVE MEASURES

Preventive measures form the first line of defense in cybersecurity, focusing on strategies and tools that aim to stop cyberattacks before they occur. By addressing vulnerabilities, restricting unauthorized access, and securing systems proactively, organizations can significantly reduce the risk of breaches. These measures include access control, network and endpoint security, patch management, data encryption, and leveraging GenAI for enhanced preparedness. This section discusses each of these strategies in detail, highlighting their role in building a robust cybersecurity posture.

Access Control: Access control ensures that only authorized individuals can access specific systems, data, or resources, thereby limiting the risk of unauthorized breaches [54]. Common techniques include Role-Based Access Control (RBAC), which assigns permissions based on roles within an organization, and Multi-Factor Authentication (MFA), which adds layers of verification (e.g., passwords, biometrics, or tokens) to prevent unauthorized access [54]. Additionally, the Zero-Trust Architecture (ZTA) takes access control further by continuously verifying every user and device, regardless of whether they are inside or outside the network perimeter [55]. By implementing these access control mechanisms, organizations can minimize the risk of insider threats and unauthorized breaches.

Network and Endpoint Security: Network and endpoint security focus on protecting the infrastructure and devices that connect to organizational systems [56]. Firewalls act as the first line of defense by monitoring and filtering incoming and outgoing network traffic based on predefined security rules. Intrusion Prevention Systems (IPS) add an additional layer by detecting and blocking malicious activity in real time. Here, it is important to note that there is a slight difference between IPS and Intrusion Detection Systems (IDSs). Specifically, as illustrated in Fig. 2.4, the IPS is usually placed right after the firewall to prevent unkown/unusual traffic comings before it goes inside the local network. On the device level, Endpoint Detection and Response (EDR) tools monitor endpoints such as laptops and IoT devices for anomalies, providing proactive protection against malware, ransomware, and other attacks [57]. Combined, these tools create a comprehensive defense strategy to secure networks and endpoints from evolving threats.

Patch Management: Patch management involves regularly updating software, operating systems, and applications to fix security vulnerabilities and prevent exploitation by attackers. Cybercriminals often target outdated systems with known weaknesses to gain unauthorized access or launch malware attacks. Automated patch management solutions streamline this process, ensuring timely updates without manual intervention. By maintaining a consistent patching routine, organizations can

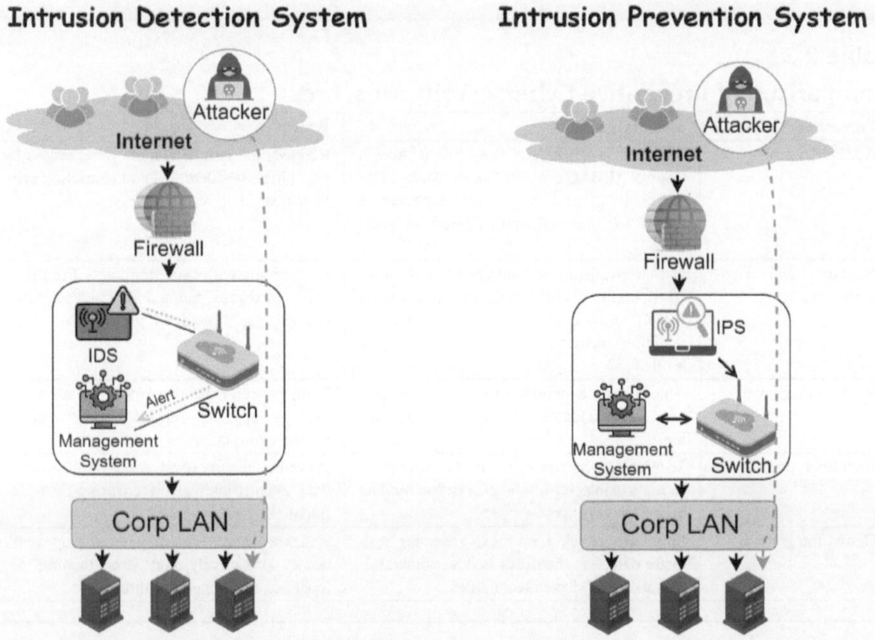

Figure 2.4 IDS vs IPS.

significantly reduce the risk of zero-day exploits and system compromises, enhancing overall resilience against cyber threats.

Data Encryption: Data encryption ensures the confidentiality and integrity of information by converting it into unreadable formats that can only be decrypted with authorized keys. End-to-end encryption secures data during transmission, protecting it from interception in transit, while encryption of data at rest safeguards stored information from unauthorized access. Encryption is critical for securing sensitive data, such as personal information, financial records, and intellectual property, particularly in cloud environments and IoT systems. By implementing strong encryption protocols, organizations can prevent data breaches even if attackers gain access to encrypted files.

GenAI Integration: Recently, GenAI has emerged as a revolutionary preventive measure by enhancing preparedness and vulnerability identification. AI-driven simulations can create realistic attack scenarios, enabling organizations to test their defenses against emerging threats. By generating synthetic datasets, GenAI helps train machine learning models to identify vulnerabilities and anticipate potential attack vectors [48]. Additionally, AI tools can perform automated risk assessments, predicting where systems may be exploited and recommending proactive countermeasures [50]. Integrating GenAI into preventive strategies empowers organizations to stay ahead of cybercriminals by identifying and addressing weaknesses before they can be exploited.

Table 2.3
Comparison of Preventive Cybersecurity Measures

Preventive Measure	Mechanism	Benefit
Access Control	Enforces policies like Role-Based Access Control (RBAC), Multi-Factor Authentication (MFA), and Zero-Trust Architecture (ZTA) to limit access to authorized users only.	Reduces the risk of unauthorized access by verifying user identities and restricting privileges based on roles and trust levels.
Network and Endpoint Security	Uses firewalls, Intrusion Prevention Systems (IPS), and Endpoint Detection and Response (EDR) tools to monitor, detect, and block malicious activities at network and device levels.	Protects networks and endpoints from malware, intrusions, and unauthorized activities, ensuring real-time threat prevention.
Patch Management	Automates the identification and application of software updates to fix known vulnerabilities.	Reduces exposure to zero-day attacks and ensures systems remain secure against known exploits.
Data Encryption	Secures data by converting it into unreadable formats using end-to-end encryption for data in transit and at rest.	Prevents unauthorized access to sensitive data, even if intercepted, ensuring data confidentiality and integrity.
GenAI Integration	Simulates attack scenarios, generates synthetic data, and identifies system vulnerabilities using AI-driven techniques.	Enhances preparedness by identifying weaknesses proactively and strengthening defenses against emerging threats.

2.2.3 DETECTIVE MEASURES

Detective measures play a crucial role in identifying malicious activities and potential security breaches in real time. Unlike preventive strategies, which aim to stop attacks before they occur, detective measures focus on continuous monitoring and analysis to uncover threats as they emerge. By leveraging advanced tools and AI-powered techniques, organizations can detect anomalies, suspicious patterns, and unauthorized activities promptly, enabling timely responses to mitigate risks. This section explores key detective measures, highlighting their importance in modern cybersecurity frameworks.

Intrusion Detection Systems (IDS): Intrusion Detection Systems (IDS) are critical tools for real-time monitoring of network traffic to identify suspicious activities and anomalies [56]. IDS analyzes network packets, searching for known attack signatures or unusual patterns that may indicate a breach. They are broadly categorized into signature-based IDS, which detect attacks based on predefined patterns, and anomaly-based IDS, which flag deviations from normal behavior. By providing alerts upon detecting potential intrusions, IDS allows security teams to take swift action to contain threats. However, their effectiveness depends on continuous updates and the ability to minimize false positives.

SIEM (Security Information and Event Management): Security Information and Event Management (SIEM) systems collect, aggregate, and analyze logs and events from various sources, such as servers, firewalls, and endpoints, to identify suspicious patterns and potential threats [58]. By centralizing data analysis, SIEM tools enable organizations to detect complex attacks that may span multiple systems.

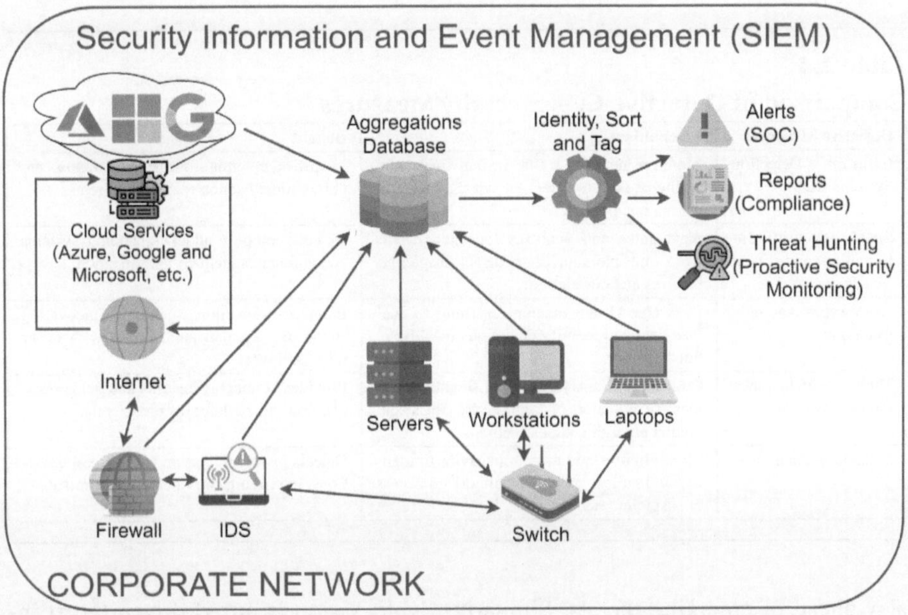

Figure 2.5 An illustration of SIEM system.

Advanced SIEM solutions integrate AI and machine learning to automate log analysis and reduce the time required to identify and respond to incidents. For example, SIEM platforms can correlate seemingly unrelated events to detect coordinated attacks, making them invaluable for comprehensive threat detection.

GAI-Powered Anomaly Detection: AI-powered anomaly detection uses GenAI and machine learning to analyze vast datasets, identifying deviations from established behavioral norms. Traditional anomaly detection methods often struggle with large-scale and dynamic environments, leading to false positives or missed threats. GenAI enhances anomaly detection by learning complex patterns within the data and identifying subtle anomalies that may indicate malicious activities. For instance, autoencoders can model normal network behavior and flag outliers as potential intrusions [49]. This AI-driven approach improves detection accuracy, enabling organizations to uncover sophisticated and previously unknown threats.

Threat Intelligence Platforms: Threat Intelligence Platforms (TIPs) collect, analyze, and share data on emerging cyber threats, enabling organizations to stay ahead of adversarial tactics. These platforms aggregate information from multiple sources, including global threat intelligence feeds, dark web monitoring, and industry reports. By integrating AI, TIPs can analyze large volumes of threat data to identify patterns, predict attack methods, and provide actionable insights. For example, Google's Security AI Workbench utilizes Large Language Models (LLMs) to address challenges in cybersecurity, including threat overload and complex toolsets [59]. This AI-powered system transforms the operationalization of threat intelligence, helping organizations

Table 2.4
Comparison of Detective Cybersecurity Measures

Detective Measure	Mechanism	Benefit
Intrusion Detection Systems (IDS)	Monitors network traffic in real time, analyzing packets to detect known signatures or anomalous patterns.	Identifies potential intrusions quickly, enabling timely action to contain threats.
SIEM (Security Information and Event Management)	Aggregates and analyzes logs and events from multiple sources to detect suspicious patterns and correlations.	Detects complex attacks spanning systems and automates incident identification.
AI-Powered Anomaly Detection	Uses GenAI and machine learning to analyze data and identify deviations from baseline behavior.	Enhances detection accuracy, uncovering subtle and sophisticated threats with fewer false positives.
Threat Intelligence Platforms (TIPs)	Collects and analyzes global threat data to identify indicators of compromise (IoCs) and predict emerging attack methods.	Provides actionable insights and helps organizations adapt defenses proactively.
Behavioral Analytics	Monitors user and system behavior to identify anomalies, such as unauthorized access or suspicious activities.	Detects insider threats and behavioral deviations, improving early detection of misuse.

stay ahead of emerging threats. Similarly, Cyble Vision is an AI-driven threat intelligence platform designed to provide real-time intelligence and threat detection, enhancing security measures for organizations [60].

Behavioral Analytics: Behavioral analytics focuses on monitoring user and system behavior to detect insider threats, unauthorized activities, or anomalies indicative of malicious intent [61]. By establishing baseline behaviors for users, devices, and systems, behavioral analytics tools can identify deviations that may signal breaches or misuse. For example, an employee accessing sensitive files outside normal working hours or a system executing unusual processes could trigger alerts. Advanced solutions leverage machine learning and AI to continuously refine behavioral models, improving their ability to detect subtle and emerging threats while minimizing false positives.

2.2.4 RESPONSIVE MEASURES

Responsive measures focus on strategies and tools designed to contain and mitigate active cyberattacks as they occur. While preventive and detective measures aim to stop and identify threats, responsive measures ensure that security teams can act swiftly to minimize damage, eradicate malicious activities, and restore systems to normal operation. The integration of AI and generative models has significantly enhanced the effectiveness of responsive strategies by automating decision-making and improving adaptability against evolving threats. This section explores key responsive measures that organizations can implement to address active cyberattacks effectively.

Incident Response Plans: Incident Response Plans (IRPs) are structured frameworks that outline the processes for detecting, containing, and eradicating cyber threats [62]. An effective IRP includes well-defined phases, e.g., preparation, detection and analysis, containment, eradication, and recovery, as illustrated in Fig. 2.6.

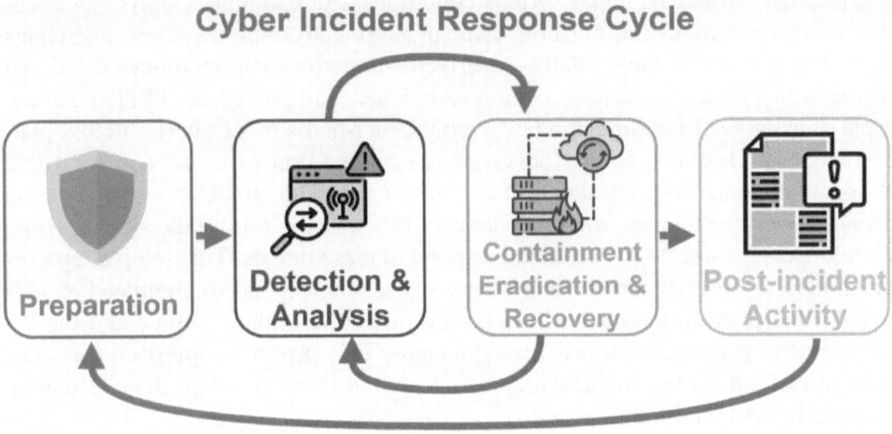

Figure 2.6 An illustration of IRPs lifecycle.

These plans ensure that organizations respond to incidents in a systematic and timely manner, minimizing the impact of breaches. For example, upon detecting a ransomware attack, the IRP would outline immediate containment measures, such as isolating affected systems, followed by eradication procedures like malware removal and system restoration. A well-documented IRP also includes post-incident analysis to identify lessons learned and improve future response efforts.

Automated Response Systems: Automated response systems leverage AI-driven technologies to contain and mitigate threats in real-time without the need for manual intervention. These systems monitor network traffic, endpoints, and applications for malicious activities and execute predefined actions to neutralize threats. For example, an automated response system based on AI can automatically isolate a compromised endpoint, block malicious IP addresses, or disable suspicious user accounts. By reducing response times and minimizing human error, automated systems enhance an organization's ability to contain cyberattacks quickly and effectively, ensuring continuous protection against rapidly evolving threats.

Adversarial Defense: Adversarial defense focuses on countering AI-driven attacks by using automated adversarial testing and retraining of cybersecurity models. As attackers increasingly use adversarial AI to manipulate detection systems, defenders can employ adversarial techniques to strengthen their defenses. For example, GenAI models can simulate adversarial attacks, such as crafting inputs designed to evade machine learning-based anomaly detection systems. By continuously testing and retraining models with these adversarial examples, organizations can enhance the resilience of their cybersecurity solutions, ensuring they remain effective even against AI-powered threats.

GenAI in Simulation: GenAI plays a critical role in simulating attacker behavior and predicting potential responses to ongoing cyberattacks. By creating realistic attack scenarios, generative models enable security teams to test and refine their response strategies under various conditions. For example, GenAI can simulate ransomware attacks or phishing campaigns to analyze how systems and teams respond in real-time. Additionally, AI-driven simulations can recommend tailored responses by predicting attacker tactics, techniques, and procedures (TTPs). An example of using GAI for simulating cyberattacks is Microsoft's CyberBattleSim platform [63, 64]. This open-source tool creates a simulated network environment where AI agents emulate both attacker and defender roles. The attacking agents employ various tactics, techniques, and procedures (TTPs) to compromise the system, while defending agents use AI to detect and respond to these threats. This setup allows organizations to observe how their systems and teams respond to simulated attacks in real-time, providing valuable insights into potential vulnerabilities and the effectiveness of their incident response strategies. This proactive approach enhances decision-making during incidents, allowing organizations to adapt their responses dynamically and minimize damage.

Table 2.5
Comparison of Responsive Cybersecurity Measures

Responsive Measure	Mechanism	Benefit
Incident Response Plans (IRPs)	Structured frameworks that outline phases for preparation, detection, containment, eradication, and recovery.	Ensures a systematic and timely response to minimize the impact of breaches and improve future readiness.
Automated Response Systems	AI-driven systems that monitor and automatically contain threats in real time without manual intervention.	Reduces response time, minimizes human error, and ensures rapid containment of active cyberattacks.
Adversarial Defense	Employs automated adversarial testing and retraining of cybersecurity models to counter AI-driven attacks.	Enhances the resilience of detection systems by identifying and mitigating adversarial techniques.
GenAI in Simulation	Simulates attacker behavior and generates realistic attack scenarios to test and refine response strategies.	Improves decision-making during incidents, enabling dynamic and tailored responses to minimize damage.

2.2.5 RECOVERY MEASURES

Recovery measures are essential for restoring operations and minimizing the long-term impact of cyberattacks. After a breach or incident, organizations must act swiftly to recover lost data, rebuild infrastructure, and strengthen defenses to prevent future exploits. By implementing well-planned recovery strategies, organizations can ensure business continuity and enhance their resilience against recurring threats. This section discusses key recovery techniques, highlighting the role of AI and automation in accelerating the recovery process and improving post-incident defenses.

Data Backup and Restoration: Data backup and restoration involve creating automated, redundant copies of critical data to ensure business continuity in the event of a cyberattack. Regular backups protect organizations from data loss caused by ransomware, hardware failures, or accidental deletion. Automated backup solutions store data in secure, geographically distributed locations, such as cloud platforms or offline storage. In the event of an attack, these backups enable organizations to restore systems and resume operations quickly, minimizing downtime and financial losses. For instance, the WannaCry ransomware attack in 2017 was one of the most significant cyberattacks in history. It affected over 200,000 computers across 150 countries, encrypting critical data and demanding ransom payments in Bitcoin to decrypt the files [65]. As a result, frequent backups are a critical defense against ransomware, allowing organizations to recover encrypted files without paying the ransom.

Disaster Recovery Plans: Disaster recovery plans (DRPs) are structured strategies designed to restore IT infrastructure, systems, and data following cyber incidents, natural disasters, or system failures [66]. A comprehensive DRP outlines processes for assessing damage, prioritizing critical systems, and restoring operations in phases. Key components include recovery time objectives (RTOs) and recovery point objectives (RPOs), which define acceptable downtime and data loss limits [67]. AI-driven tools can automate disaster recovery workflows, ensuring faster recovery times and minimizing manual intervention. By simulating disaster scenarios, organizations can validate their plans and ensure readiness for unexpected disruptions.

System Hardening: System hardening refers to the process of strengthening systems, applications, and networks to reduce vulnerabilities and prevent future exploitation. Following a cyberattack, organizations must analyze the root cause of the breach and implement corrective actions such as patching vulnerabilities, removing unnecessary services, and enforcing stronger access controls. Hardening also involves updating security configurations, implementing multi-factor authentication (MFA), and improving endpoint protection. By fortifying systems post-attack, organizations can close security gaps and significantly reduce the risk of recurring breaches.

Post-Incident Analysis: Post-incident analysis involves conducting a thorough investigation to understand the cause, impact, and timeline of a cyberattack. Leveraging AI-driven tools, organizations can analyze large volumes of incident data to identify root causes, attacker behaviors, and exploited vulnerabilities. This analysis helps generate actionable insights to improve security posture, such as strengthening weak points or updating detection models. For example, in 2017, Equifax experienced a significant data breach that exposed the personal information of approximately 147 million individuals [68]. Post-incident analysis revealed that attackers exploited an unpatched vulnerability in the Apache Struts web application framework, allowing unauthorized access to sensitive data [69]. The investigation identified several systemic failures, including inadequate patch management and insufficient network segmentation, which facilitated the attackers' lateral movement within Equifax's systems [69]. This analysis prompted Equifax to implement comprehensive security measures, such as establishing a formal patch management policy and enhancing network defenses, to prevent future incidents [69].

2.3 SUMMARY

In this chapter, we explored the fundamental concepts of cybersecurity, the evolving threat landscape, and the role of advanced countermeasures, including GenAI,

Table 2.6
Comparison of Recovery Cybersecurity Measures

Recovery Measure	Mechanism	Benefit
Data Backup and Restoration	Creates automated, redundant copies of critical data and stores them securely to enable recovery after a cyberattack.	Ensures business continuity by quickly restoring operations and minimizing downtime and data loss.
Disaster Recovery Plans (DRPs)	Structured strategies for assessing damage, prioritizing critical systems, and restoring infrastructure and data in phases.	Reduces recovery time and data loss, ensuring readiness for unexpected disruptions with automated workflows.
System Hardening	Strengthens systems, applications, and networks by patching vulnerabilities, enforcing stronger controls, and improving configurations.	Prevents future exploitation by addressing root causes and closing security gaps post-attack.
Post-Incident Analysis	Investigates the cause, timeline, and impact of attacks using AI to analyze incident data and simulate attacker behavior.	Generates actionable insights to improve defenses, enhance response plans, and prevent recurring breaches.

in addressing modern cyber challenges. We began by identifying various types of cybersecurity threats and their significant impacts on financial, operational, reputational, and national security dimensions. From preventive and detective measures to responsive and recovery strategies, we highlighted how organizations can build a robust multi-layered defense framework. GenAI emerged as a transformative tool, enabling advanced techniques such as threat simulation, automated patching, and synthetic data generation to enhance resilience against sophisticated attacks. As cyber threats continue to evolve, integrating AI-driven technologies will be critical to safeguarding systems, data, and infrastructure in an increasingly interconnected and AI-powered digital world.

1. Jason Andress. *The Basics of Information Security: Understanding the Fundamentals of InfoSec in Theory and Practice*. Syngress, 2014.
2. Katz School of Science and Health. The Evolution of Cyber Threats: Past, Present and Future. `https://online.yu.edu/katz/blog/the-evolution-of-cyber-threats`. Accessed: 2025-05-27.
3. Barracuda Blog. Guarding the gates: The rise of network protection in the 1990s, 2023. `https://blog.barracuda.com/2023/10/31/guarding-gates-rise-network-protection-1990s`. Accessed: 2025-05-27.
4. NetSecCloud. The evolution of threat detection: From firewalls to ids, n.d. `https://netseccloud.com/the-evolution-of-threat-detection-from-firewalls-to-ids`. Accessed: 2025-05-27.
5. Ennetix. A brief history of networking: Part 2, n.d. `https://ennetix.com/a-brief-history-of-networking-part-2/`. Accessed: 2025-05-27.
6. Digital Guardian. The history of data breaches, n.d. `https://www.digitalguardian.com/blog/history-data-breaches`. Accessed: 2024-12-20.
7. STP Ventures. The evolution of cybersecurity: A 20-year retrospective, n.d. `https://www.stpventures.com/post/the-evolution-of-cybersecurity-a-20-year-retrospective`. Accessed: 2024-12-20.

8. Smarter MSP. The dawn of real-time defense: Security transformation in the 2000s, n.d. https://smartermsp.com/the-dawn-of-real-time-defense-security-transformation-in-the-2000s/. Accessed: 2024-12-20.
9. CNET. A timeline of the biggest ransomware attacks, n.d. https://www.cnet.com/personal-finance/crypto/a-timeline-of-the-biggest-ransomware-attacks/#google_vignette. Accessed: 2024-12-20.
10. IBM. Advanced persistent threats (apts), n.d. https://www.ibm.com/topics/advanced-persistent-threats. Accessed: 2024-12-20.
11. CSO Online. Wannacry explained: A perfect ransomware storm, n.d. https://www.csoonline.com/article/563017/wannacry-explained-a-perfect-ransomware-storm.html. Accessed: 2024-12-20.
12. Center for Internet Security (CIS). Solarwinds incident overview, n.d. https://www.cisecurity.org/solarwinds. Accessed: 2024-12-20.
13. CSO Online. Solarwinds supply chain attack explained: Why organizations were not prepared, n.d. https://www.csoonline.com/article/570191/solarwinds-supply-chain-attack-explained-why-organizations-were-not-prepared.html. Accessed: 2024-12-20.
14. Jin Cao, Maode Ma, Hui Li, Ruhui Ma, Yunqing Sun, Pu Yu, and Lihui Xiong. A survey on security aspects for 3gpp 5g networks. *IEEE Communications Surveys & Tutorials*, 22(1):170–195, 2019.
15. Tech Monitor. AI and cybercrime: The rise of AI-driven threats, n.d. https://www.techmonitor.ai/partner-content/ai-cybercrime?cf-view. Accessed: 2024-12-20.
16. Forbes Tech Council. AI and cybercrime: Unleash a new era of menacing threats, 2023. https://www.forbes.com/councils/forbestechcouncil/2023/06/23/ai-and-cybercrime-unleash-a-new-era-of-menacing-threats/. Accessed: 2024-12-20.
17. D. T. Hoang, N. Q. Hieu, D. N. Nguyen, and E. Hossain. *Advanced Machine Learning for Cyberattack Detection in IoT Networks*. Elsevier, 2025.
18. Maanak Gupta, CharanKumar Akiri, Kshitiz Aryal, Eli Parker, and Lopamudra Praharaj. From chatgpt to threatgpt: Impact of GenAI in cybersecurity and privacy. *IEEE Access*, 2023.
19. Cybersecurity News. What is signature-based detection?, n.d. https://cybersecuritynews.com/signature-based-detection/. Accessed: 2024-12-21.
20. MixMode AI Blog. Why zero-day attacks bypass traditional firewall security: Defending against zero-days like palo alto networks cve-2024-0012, n.d. https://mixmode.ai/blog/why-zero-day-attacks-bypass-traditional-firewall-security-defending-against-zero-days-like-palo-alto-networks-cve-2024-0012/. Accessed: 2024-12-21.
21. National Audit Office (NAO). Investigation: Wannacry cyber attack and the nhs, 2018. https://www.nao.org.uk/reports/investigation-wannacry-cyber-attack-and-the-nhs/. Accessed: 2024-12-21.
22. SoftwareLab. Phishing examples: How scammers steal your data, 2025. https://softwarelab.org/blog/phishing-examples/?utm_source=chatgpt.com. Accessed: 2025-01-22.
23. Network Encyclopedia. The 2016 dyn cyberattack: An overview, 2025. https://networkencyclopedia.com/the-2016-dyn-cyberattack-an-overview/. Accessed: 2025-01-22.

24. Threatpost. Ultimate mitm attack steals $1m from israeli startup, 2019. https://threatpost.com/ultimate-mitm-attack-steals-1m-from-israeli-startup/150840/?utm_source=chatgpt.com. Accessed: 2025-01-22.
25. GeeksforGeeks. Zero day exploit - cyber security attack, 2025. https://www.geeksforgeeks.org/zero-day-exploit-cyber-security-attack/. Accessed: 2025-01-22.
26. Heimdal Security. Solarwinds attack: Cost impacted companies an average of $12 million, 2025. https://heimdalsecurity.com/blog/solarwinds-attack-cost-impacted-companies-an-average-of-12-million/. Accessed: 2025-01-22.
27. Abnormal Security. The rise of AI-generated email attacks in 2023, 2023. https://abnormalsecurity.com/blog/2023-ai-generated-email-attacks. Accessed: 2025-01-22.
28. Abnormal Security. AI unleashed: Real-world email attacks, 2025. https://abnormalsecurity.com/resources/ai-unleashed-real-world-email-attacks. Accessed: 2025-01-22.
29. Identity Theft Resource Center (ITRC). First ever AI fraud case steals money by impersonating ceo, 2023. https://www.idtheftcenter.org/post/first-ever-ai-fraud-case-steals-money-by-impersonating-ceo/. Accessed: 2025-01-22.
30. Creative Bloq. 10 incredible deepfake examples that showcase the good and the bad of AI, 2025. https://www.creativebloq.com/features/deepfake-examples. Accessed: 2025-01-22.
31. The New York Times. Taylor swift AI fake images highlight ethical concerns in GenAI, 2024. https://www.nytimes.com/2024/01/26/arts/music/taylor-swift-ai-fake-images.html, note = Accessed: 2025-01-22. Accessed: 2025-01-22.
32. Palo Alto Networks. AI-generated malware: Understanding the next cyber threat, 2024. https://www.paloaltonetworks.com/blog/2024/05/ai-generated-malware/. Accessed: 2025-01-22.
33. HYAS. Blackmamba: Using AI to generate polymorphic malware, 2024. https://www.hyas.com/blog/blackmamba-using-ai-to-generate-polymorphic-malware?utm_source=chatgpt.com. Accessed: 2025-01-22.
34. CyFlare. How AI is used in cyberattacks, 2024. https://cyflare.com/how-ai-is-used-in-cyberattacks/?utm_source=chatgpt.com. Accessed: 2025-01-22.
35. Quentin Anthony, Yury Tokpanov, Paolo Glorioso, and Beren Millidge. Blackmamba: Mixture of experts for state-space models. *arXiv preprint arXiv:2402.01771*, 2024.
36. HYAS. Blackmamba: Using AI to generate polymorphic malware, 2024. Accessed: 2025-01-22.
37. Jan von der Assen, Alberto Huertas Celdrán, Janik Luechinger, Pedro Miguel Sánchez Sánchez, Gérôme Bovet, Gregorio Martínez Pérez, and Burkhard Stiller. Ransomai: AI-powered ransomware for stealthy encryption. In *GLOBECOM 2023-2023 IEEE Global Communications Conference*, pages 2578–2583. IEEE, 2023.
38. National Institute of Standards and Technology (NIST). Nist identifies types of cyberattacks that manipulate the behavior of AI systems, 2024. Accessed: 2025-01-22.
39. IBM Newsroom. Ibm report: Half of breached organizations unwilling to increase security spend despite soaring breach costs, 2023. https://newsroom.ibm.com/2023-07-24-IBM-Report-Half-of-Breached-Organizations-Unwilling-to-Increase-Security-Spend-Despite-Soaring-Breach-Costs?utm_source=chatgpt.com. Accessed: 2025-01-22.

40. The Sun. Microsoft second it outage caused by cyber attack, 2024. https://www.thesun.co.uk/tech/29582353/microsoft-second-it-outage-cyber-attack/?utm_source=chatgpt.com. Accessed: 2025-01-22.
41. TrellisPoint. Upholding trust in the face of data breaches: Lessons from equifax's ungrowth, 2023. https://trellispoint.com/blog/upholding-trust-in-the-face-of-data-breaches-lessons-from-equifaxs-ungrowth. Accessed: 2025-01-22.
42. New York Times. Hackers hit ukraine then spread. is this the future of cyberwar?, 2017. https://www.nytimes.com/2017/06/27/technology/ransomware-hackers.html. Accessed: 2025-01-22.
43. Cloudflare. Petya and notpetya ransomware attacks: What you need to know, 2023. https://www.cloudflare.com/learning/security/ransomware/petya-notpetya-ransomware/. Accessed: 2025-01-22.
44. CrowdStrike. Ransomware examples: High-profile attacks and trends, 2024. https://www.crowdstrike.com/en-us/cybersecurity-101/ransomware/ransomware-examples/?utm_source=chatgpt.com. Accessed: 2025-01-22.
45. GS Mamatha, Namya Dimri, and Rasha Sinha. Post-quantum cryptography: Securing digital communication in the quantum era. *arXiv preprint arXiv:2403.11741*, 2024.
46. CSO Online. Chinese researchers claim to break rsa encryption with a quantum computer, 2023. https://www.csoonline.com/article/3562701/chinese-researchers-break-rsa-encryption-with-a-quantum-computer.html. Accessed: 2025-01-22.
47. World Economic Forum. Quantum computing and the risks to cybersecurity, 2024. https://www.weforum.org/stories/2024/04/quantum-computing-cybersecurity-risks/. Accessed: 2025-01-22.
48. Vikash Kumar and Ditipriya Sinha. Synthetic attack data generation model applying generative adversarial network for intrusion detection. *Computers & Security*, 125:103054, 2023.
49. P. V. Dinh, Q. U. Nguyen, D. T. Hoang, D. N. Nguyen, S. P. Bao, and E. Dutkiewicz, "Constrained Twin Variational Auto-Encoder for Intrusion Detection in IoT Systems," *IEEE Internet of Things Journal*, vol. 11, no. 8, pp. 14789–14803, 2023.
50. Patrick McDaniel, Nicolas Papernot, and Z Berkay Celik. Machine learning in adversarial settings. *IEEE Security & Privacy*, 14(3):68–72, 2016.
51. Pantaleone Nespoli, Dimitrios Papamartzivanos, Félix Gómez Mármol, and Georgios Kambourakis. Optimal countermeasures selection against cyber attacks: A comprehensive survey on reaction frameworks. *IEEE Communications Surveys & Tutorials*, 20(2):1361–1396, 2017.
52. Jan-Erik Holmberg. Defense-in-depth. *Handbook of Safety Principles*, pages 42–62, 2017.
53. AMD Opteron Processor. A defense-in-depth approach to security.
54. Kaushik Ragothaman, Yong Wang, Bhaskar Rimal, and Mark Lawrence. Access control for iot: A survey of existing research, dynamic policies and future directions. *Sensors*, 23(4):1805, 2023.
55. Naeem Firdous Syed, Syed W Shah, Arash Shaghaghi, Adnan Anwar, Zubair Baig, and Robin Doss. Zero trust architecture (zta): A comprehensive survey. *IEEE access*, 10:57143–57179, 2022.
56. Study CCNA. Firewalls, ids, and ips: Explanation and comparison, 2024. https://study-ccna.com/firewalls-ids-ips-explanation-comparison/?utm_source=chatgpt.com. Accessed: 2025-01-22.

57. PurpleSec. What is endpoint detection and response (edr)?, 2024. https://purplesec.us/learn/endpoint-detection-response/. Accessed: 2025-01-22.
58. SentinelOne. Enterprise security information and event management (siem), 2024. https://www.sentinelone.com/cybersecurity-101/data-and-ai/enterprise-security-information-and-event-management-siem/. Accessed: 2025-01-22.
59. Google Cloud. Introducing AI-powered insights for threat intelligence, 2024. https://cloud.google.com/blog/products/identity-security/rsa-introducing-ai-powered-insights-threat-intelligence?utm_source=chatgpt.com. Accessed: 2025-01-22.
60. Cyble. Cyble vision: Real-time threat intelligence platform, 2024. https://cyble.com/products/cyble-vision/?utm_source=chatgpt.com. Accessed: 2025-01-22.
61. Rida Nasir, Mehreen Afzal, Rabia Latif, and Waseem Iqbal. Behavioral based insider threat detection using deep learning. *IEEE Access*, 9:143266–143274, 2021.
62. Bhavin Patel. Nist incident response 800-61: A comprehensive guide to the four-phase lifecycle for identifying, containing, and resolving cyber threats, 2023. https://medium.com/@bhavin200/nist-incident-response-800-61-a-comprehensive-guide-to-the-four-phase-lifecycle-for-identifying-3c79b9b0a993. Accessed: 2025-01-22.
63. Microsoft. Cyberbattlesim - experimentation and research platform, 2021. https://github.com/microsoft/CyberBattleSim?utm_source=chatgpt.com. Accessed: 2025-01-22.
64. Kitploit. Cyberbattlesim - experimentation and research platform for simulation of cybersecurity attacks and defenses, 2021. https://www.kitploit.com/2021/05/cyberbattlesim-experimentation-and.html. Accessed: 2025-01-22.
65. Cloudflare. What is wannacry ransomware? — cloudflare, n.d. https://www.cloudflare.com/learning/security/ransomware/wannacry-ransomware/. Accessed: 2025-01-23.
66. Cloudian. 4 disaster recovery plan examples and 10 essential plan items, n.d. https://cloudian.com/guides/disaster-recovery/4-disaster-recovery-plan-examples-and-10-essential-plan-items/. Accessed: 2025-01-23.
67. House of Coder. Understanding rpo and rto in disaster recovery planning, n.d. https://houseofcoder.medium.com/understanding-rpo-and-rto-in-disaster-recovery-planning-1b0296372100. Accessed: 2025-01-23.
68. U.S. Government Accountability Office (GAO). Data protection: Actions taken by equifax and federal agencies in response to the 2017 breach, 2018. https://www.gao.gov/products/gao-18-559. Accessed: 2025-01-23.
69. Ilya Kabanov and Stuart E Madnick. Applying the lessons from the equifax cybersecurity incident to build a better defense. *MIS Q. Executive*, 20(2):4, 2021.

3 Generative AI as Potential Solutions and Risks to Cybersecurity

This chapter explores the intersection of two rapidly evolving fields: generative artificial intelligence (AI) and cybersecurity. As generative AI continues to advance in its ability to create realistic and convincing content, it presents both opportunities and challenges in the realm of cybersecurity. The chapter examines the potential applications of generative AI as a solution to cybersecurity challenges, discussing how generative models can be leveraged for anomaly detection, identifying patterns of malicious behavior within large datasets, and aiding in the early detection of cyber threats. Additionally, the chapter explores the potential vulnerabilities of generative AI systems themselves. It addresses the risks of adversarial attacks, where malicious actors can manipulate or deceive generative models (including large language models, LLMs) to produce unintended, malicious outputs, disinformation, or content for phishing and other malicious purposes.

3.1 OVERVIEW OF GENERATIVE AI FOR CYBERTHREAT DETECTION, PREVENTION

3.1.1 VULNERABILITY ASSESSMENT WITH GENERATIVE AI

Generative AI (GenAI) is a powerful tool for automating vulnerability discovery. One can train GenAI models to automatically analyse software, cyber systems, and their potential configurations to identify vulnerabilities and risks. This can be done by autonomously crafting different realistic attacks (e.g., for penetration testing) and operating scenarios to enumerate potential weak links [1]. When armed with data on software vulnerabilities, AI models can effectively scan and detect new instances of code exhibiting similar weaknesses and susceptibilities.

Large Language Models (LLMs) have recently played a crucial role in automating penetration testing [3]. LLMs allow users without extensive security knowledge to conduct penetration testing on a system. To do this, users first establish a penetration testing goal using a modular library that can create exploit flows, as shown in Figure 3.1. The exploit flow refers to the state of the systems being tested for each discrete action. Once the goal is set, the user submits a request to LLMs and then receives recommendations. These recommendations guide the user in conducting experiments on the testing environment to achieve the desired testing output. The user can analyze the testing output to identify any flags (vulnerabilities) or continue sending requests to LLMs to obtain recommendations related to other flags. For instance, PentestGPT

DOI: 10.1201/9781003502531-3

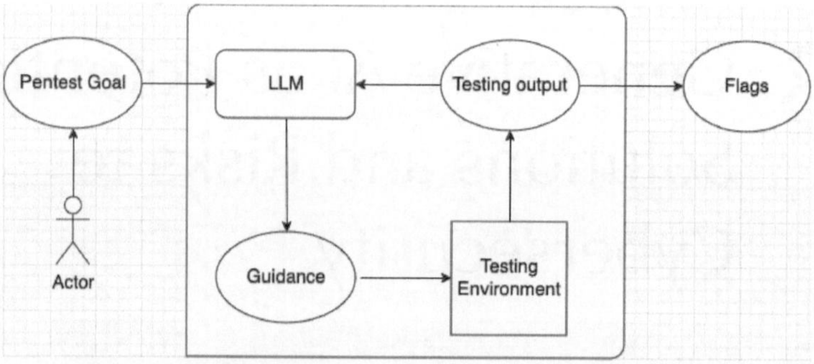

Figure 3.1 Overview of strategy to use LLMs for penetration testing.

[2], a powerful penetration testing tool powered by LLMs, is specifically designed to automate the penetration testing process.

GenAI is widely used for identifying and repairing code and software vulnerabilities. Given the complexity of modern software systems, automatic methods can assist security experts in avoiding the time-consuming and labour-intensive process of repairing such security vulnerabilities [6]. In [44], the authors propose an adversarial learning approach known as Repairing Software Vulnerabilities with Generative Adversarial Networks (RSV-GAN), which transfer code-contained vulnerabilities from one source domain to another domain without vulnerabilities. The RSV-GAN does not require paired labelled samples between code-contained vulnerabilities and clean code. Unlike the original GAN, which aims to generate realistic images from random noise vectors, this method focuses on mapping two domains, namely $\mathbf{X} = \{\mathbf{x}_i\}_{i=1}^M$ and $\mathbf{Y} = \{\mathbf{y}_i\}_{i=1}^M$. To convert the code into a vector, the code is tokenized and then encoded using one-hot encoding or word embedding techniques such as Word2Vec and Glove. For example, Figure 3.2 illustrates the tokenization output of a function written in Python. The goal of the RSV-GAN is to generate samples from the unknown distribution of a real sequence as follows:

$$P_{\mathbf{Y}}(y) = P(\mathbf{y}_0)\Pi_{t=1}^{T} P(\mathbf{y}_t|\mathbf{y}_0,\dots,\mathbf{y}_{t-1}), \tag{3.1}$$

python Copy code

```python
def add(a, b):
    return a + b
```

Tokenized: `['def', 'add', '(', 'a', ',', 'b', ')', ':', 'return', 'a', '+', 'b']`

Figure 3.2 Tokenized a function written by Python language.

where softmax function is used to estimate each conditional distribution $P(\mathbf{y}_t|\mathbf{y}_0,\ldots,\mathbf{y}_{t-1})$, as followed:

$$P(\mathbf{y}_t|\mathbf{y}_0,\ldots,\mathbf{y}_{t-1}) = \mathbf{s}_t = softmax(f(\mathbf{y}_{t-1},\mathbf{h}_{t-1})), \tag{3.2}$$

where $f(.)$ represents the generator network and h_{t-1} represents the hidden state of the recurrent neural network at time t. The softmax function from Equation 3.2 is utilized to convert the output into one-hot vectors. To train RSV-GAN, the authors employ two regularizers. The first regularizer ensures that the generated samples obtained by the generator are corrected versions of the input. The second regularizer aims to ensure that the frequency of each token in the generated samples closely matches that of the input tokens.

GenAI, including LLMs, has the potential to identify vulnerabilities in code by analyzing patterns, syntax, and context. LLMs are also used to perform zero-shot vulnerability repair of code [5]. For instance, ChatGPT-3.5 can identify SQL injection vulnerabilities when programmers use string concatenation to construct queries instead of using parameterized queries. An example of such a command is: *query = f "SELECT FROM users WHERE username='username' AND password='password'."* In this case, the two strings, *'username'* and *'password'*, are being used to construct the query, which makes it vulnerable to SQL injection attacks. When asked about the vulnerability of this command, ChatGPT-3.5 responds by indicating that the inputs, *'username'* and *'password'*, are directly interpolated into the SQL query string, thus exposing it to SQL injection. To fix this vulnerability, ChatGPT-3.5 suggests passing the two strings, *'username'* and *'password'*, as parameters to the execute function, as shown in the following code: *query = "SELECT FROM users WHERE username=? AND password=?"* and *cursor.execute(query, (username, password))*. This ensures that the query cannot be exploited to execute SQL commands.

AI-powered Penetration Testing (PenTest) plays a crucial role in evaluating the security of a system by generating authorized attack scenarios. Traditional PenTest relies on expert knowledge to create a list of feasible attack scenarios, known as attack paths, to test system vulnerabilities. However, this manual process is time-consuming, complex, and prone to human error, potentially overlooking attack paths that could endanger the system. AI-powered PenTest offers an automated solution that generates attack paths with specific guidance, testing the system for both known and potential future vulnerabilities. In their paper [42], the authors propose an automated PenTest using deep reinforcement learning (DRL) techniques to discover feasible attack paths within a system. The process involves first building a realistic network topology using the Shodan search engine to obtain system data. Next, a multi-host, multi-stage vulnerability analysis (MulVAL) is used to generate an attack tree for the system. Each node in the tree represents a vulnerability within the system. An attack path, which starts at a leaf node and ends at the root, is utilized for testing the system. For instance, Figure 3.3 illustrates an attack tree generated from ChatGPT3.5 for PenTest purposes. The attack path begins by using common payloads such as ' Or 1=1 – to attempt to gain access to the web server through a

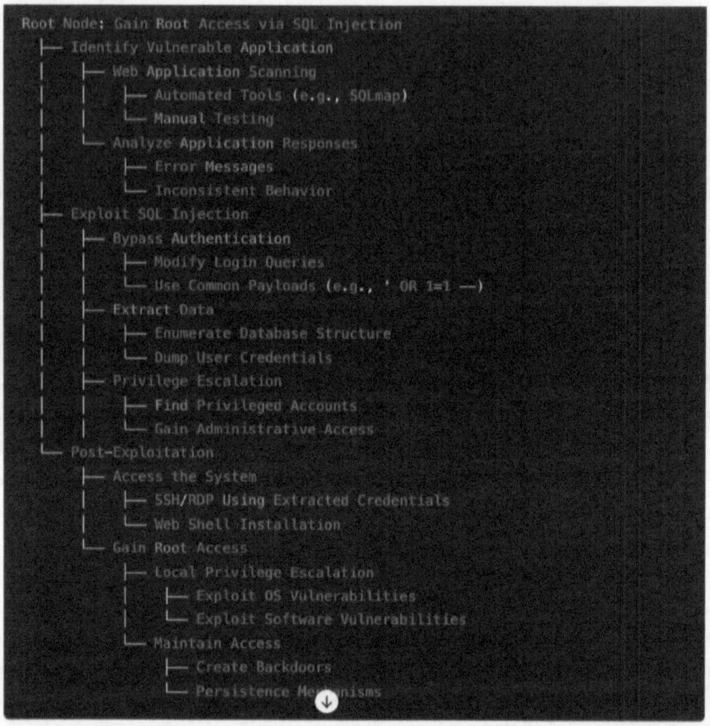

```
Root Node: Gain Root Access via SQL Injection
├── Identify Vulnerable Application
│   ├── Web Application Scanning
│   │   ├── Automated Tools (e.g., SQLmap)
│   │   └── Manual Testing
│   └── Analyze Application Responses
│       ├── Error Messages
│       └── Inconsistent Behavior
├── Exploit SQL Injection
│   ├── Bypass Authentication
│   │   ├── Modify Login Queries
│   │   └── Use Common Payloads (e.g., ' OR 1=1 --)
│   ├── Extract Data
│   │   ├── Enumerate Database Structure
│   │   └── Dump User Credentials
│   ├── Privilege Escalation
│   │   ├── Find Privileged Accounts
│   │   └── Gain Administrative Access
└── Post-Exploitation
    ├── Access the System
    │   ├── SSH/RDP Using Extracted Credentials
    │   └── Web Shell Installation
    └── Gain Root Access
        ├── Local Privilege Escalation
        │   ├── Exploit OS Vulnerabilities
        │   └── Exploit Software Vulnerabilities
        └── Maintain Access
            ├── Create Backdoors
            └── Persistence Mechanisms
```

Figure 3.3 Attack tree: gain root access via SQL injection, generated by ChatGPT3.5.

user account. Following this attack, the attacker can obtain the database and access the admin's information. Leveraging the admin account, the attacker can then gain access to the server and escalate privileges to the root level. After constructing the attack tree, DRL is employed to identify feasible attack paths. DRL is a deep learning model that assumes the role of an attacker seeking to discover attack paths within the tree. At every time step t, the attacker observes the system to obtain the state S_t and then takes an action A_t to carry out the attack. The system responds by generating a new state S_{t+1} and a corresponding reward R_t, which reflects the exploitability of the vulnerability. The attacker can navigate between nodes in the attack tree to achieve the ultimate goal, such as reaching the root of the tree. The objective of DRL is to maximize future cumulative rewards, ensuring that the attack path has the highest potential for exploitation in the future.

3.1.2 ATTACK/THREAT DETECTION

With the proliferation of billions of IoT devices, attack/threat detection systems (ATDSs) have become essential tools for protecting devices from attackers. These

ATDSs rely on machine learning (ML) models to learn patterns from both benign samples and known attacks during the training process. The aim is to then use this knowledge to detect unknown or zero-day attacks in the future. However, the collection of known attack samples poses a challenge when compared to unknown or zero-day attacks. This is primarily because attackers often employ evasion techniques, such as traffic encryption in network systems, code obfuscation to generate malicious software, packet fragmentation, and even mimicking normal behaviours, to bypass ATDSs. For instance, a slow loris denial of service (DoS) attack, which utilizes normal protocols like HTTP, gradually sends HTTP requests to the server to exhaust its resources, including CPU, network bandwidth, and memory usage. The imbalance between known attack samples and normal samples can result in a degradation of the detection accuracy of ATDSs.

To improve the accuracy of detecting ATDSs, the use of GenAI proves to be an effective method for addressing the issue of imbalanced datasets. GenAI models are capable of generating synthetic data samples that are similar to the minority classes, thereby creating a balanced dataset that benefits the classifiers used in ATDSs. Two popular GenAI models support generative imbalanced data: conditional generative adversarial nets (CGAN) [7] and conditional variational auto-encoder (CVAE) [8]. CGAN is an extension of generative adversarial nets (GAN) [9] that incorporates the class label as a condition. In CGAN, both the generator (G) and discriminator (D) receive the class label as input. The generator combines the class label $\mathbf{y}$ with the prior input noise $p_{\mathbf{z}}(\mathbf{z})$ to generate a hidden representation. Similarly, the discriminator takes both the training data sample $\mathbf{x}$ and the class label $\mathbf{y}$ as input. Therefore, the training of CGAN involves optimizing the following function:

$$\min_G \max_D V(D,G) = E_{x\sim p_{\text{data}}(\mathbf{x})}[\log D(\mathbf{x}|\mathbf{y})] + E_{z\sim p_z(\mathbf{z})}[\log(1-D(G(\mathbf{z}|\mathbf{y})))]. \quad (3.3)$$

After training, the data from the minor classes is generated by the generator G. Similar to CGAN, which is based on the GAN, the CVAE is based on the variational auto-encoder (VAE) [10]. The goal of the CVAE is to maximize a lower bound, as shown below:

$$L_{CVAE}(\mathbf{x},\mathbf{y},\theta,\Phi) = -KL\big(q_\Phi(z|\mathbf{x},\mathbf{y})||p_\theta(\mathbf{z}|\mathbf{x})\big) + \frac{1}{L}\sum_{l=L}^{L} logp_\theta(\mathbf{y}|\mathbf{x},\mathbf{z}^{(l)}), \quad (3.4)$$

where $\mathbf{z}^{(l)} = g_\phi(x,y,\varepsilon^{(l)})$, $\varepsilon^{(l)} \sim N(\mathbf{O},\mathbf{I})$, and L is the number of samples. Similarly, in [46], the authors proposed a balanced representation learning architecture called the Balanced Twin Auto-Encoder (BTAE) to address the problem of imbalanced data in IoT IDSs. The goal of BTAE is to map the input data from higher to lower dimensions, ensuring that data samples from different classes are separated. Data samples from skewed classes are generated in the latent space, and both the shifted and generated data are then transformed back into the input space. The output of this transformation is fed to the decoder, which maps it back to the latent space. Since label information is not available during inference, data representations are extracted from the decoder of the BTAE. These representations can then be used by classifiers in the IDSs. BTAE aims to make the data at the output of the transformation as similar as possible to the input data, and the output of the decoder as similar as possible to the output of the encoder. Therefore, it is more practical to extract data representations from the output of the decoder after the training process is completed. It is important to note that, unlike traditional methods that address the imbalance by generating data for the skewed classes in the training datasets of classification models, BTAE generates data for the skewed labels during the training of the representation models, thereby creating a balanced model. As a result, data transferred from the BTAE giũ' nguyĝn have less impact on the imbalanced classes, thereby improving the classification accuracy of the classifiers.

GenAI is also a powerful tool for representation learning. It allows for the transfer of input data from a high-dimensional space to lower dimensions, which facilitates classification tasks. Intrusion detection systems (IDSs) often collect input data from various types of systems, resulting in high-dimensional data. For example, IoT IDSs collect data from network traffic and logs of IoT devices. Since IoT devices come from different manufacturers, their log systems may vary. As a result, the final data samples are both high-dimensional and heterogeneous. To improve the accuracy of IDSs, the authors of [45] proposed the constrained twin variational auto-encoder (CTVAE). This model is based on the variational auto-encoder and is used to generate data samples of different classes in the latent space. The encoder of the CTVAE maps input data from different classes in high dimensions to lower dimensions in the latent space. The means of different classes are then separated from each other using a shift transformation. Data samples of different classes are generated based on these separated means. The purpose of the CTVAE is to map the data samples in the latent space back to the space of the input data. On the other hand, the decoder of the CTVAE aims to learn the process of the encoder, which involves reducing the dimensions of the input data and separating data samples of different classes. After the training process, the data representation of both the training and testing sets is extracted from the decoder of the CTVAE. These representations are then used by the classifiers in the later stage of the IDSs.

GenAI is widely adopted for anomaly detection in various domains such as network security, IoT, and cloud computing. In [11], the authors employ the reconstruction probability of the VAE as an objective anomaly score. The use of reconstruction probability is more principled and objective compared to the reconstruction error used in AE-based methods. Specifically, the reconstruction probability is defined as follows:

$$\mathrm{rp}(\mathbf{x}^{(i)}) = \frac{1}{L}\sum_{l=1}^{L} p_\theta(\mathbf{x}^{(i)}|\mu_{\mathbf{x}^{(i,l)}}, \sigma_{\mathbf{x}^{(i,l)}}). \tag{3.5}$$

The reconstruction probability, denoted as $p_\theta(\mathbf{x}^{(i)}|\mu_{\mathbf{x}^{(i,l)}}, \sigma_{\mathbf{x}^{(i,l)}})$, represents the likelihood of data sample $\mathbf{x}^{(i)}$ in the VAE model. Here, L refers to the number of samples drawn from the VAE, while $\mu_{\hat{\mathbf{x}}^{(i,l)}}$ and $\sigma_{\hat{\mathbf{x}}^{(i,l)}}$ represent the mean and sigma extracted from the encoder of the VAE. An anomaly is identified when the reconstruction probability, $\mathrm{rp}(\mathbf{x}^{(i)})$, is less than a threshold value α. It should be noted that VAE-based anomaly detection methods are particularly effective when applied to heterogeneous input data. This is because the probability distribution of each variable can be calculated by considering its variability, as represented by $mu_{\mathbf{x}(i,l)}$ and $\sigma_{\mathbf{x}(i,l)}$. Consequently, the reconstruction probability defined in Equation (3.5) does not require the weighting of different subsets of features for heterogeneous data. Like VAE-based anomaly detection methods, GAN-based ones also attempt to model complex data distribution [12]. Once the GAN model is trained, anomalies are identified by data samples that exhibit significant deviations from the normal distribution. In [13], the authors propose a novel approach for assigning anomaly scores, outlined as follows:

$$A(x) = (1-\lambda) \times R(\mathbf{x}) + \lambda * D(\mathbf{x}), \tag{3.6}$$

where $R(\mathbf{x})$ and $D(\mathbf{x})$ represent the residual score and discrimination score, respectively. These scores are calculated from the generator and discriminator of the GAN.

3.1.3 MALWARE, PHISHING, PASSWORD CRACKING, IDENTITY THEFT, AI ADVERSARIAL ATTACK PREVENTION

Traditional machine learning methods rely on experts' knowledge to identify characteristic malware features and therefore struggle to detect unknown malware without known features. In the article [14], the authors propose a method called MalGAN, which uses GenAI to generate new malware samples based on the raw byte-code of known malware in the IoT environment. Despite having only a few known malware samples, MalGAN can accurately detect a large number of unknown malware. This is because MalGAN generates data samples based on the distribution of existing malware, making the generated malware similar to what may be encountered in the future. Specifically, the raw bytes of the PE header of malware executable files are used as input for MalGAN, with the generator of MalGAN transferring the random distribution (Gaussian) to the distribution of the original malware samples. Once MalGAN is trained, the original malware samples and the ones generated by MalGAN are used as input for CNN models for malware classification.

Another example is the PlausMal-GAN, introduced in [15]. PlausMal-GAN is based on the GAN model and is used to generate zero-day malware from different malware families, including Ramnit, Lollipop, Kelious, Vundo, Simda, Tracur, Kelious, and Gatak. The binary code of the malware executable files is converted to images of size 128 × 128 before being input into the model. Unlike MalGAN, PlausMal-GAN does not rely solely on the generator during the training process. The generator of PlausMal-GAN generates malware samples from different families, while the discriminator not only differentiates between the original data samples and the generated ones but also identifies the class label of the original malware families. After training this model, the generator is fixed, and the discriminator is trained using the original samples and the generated ones for classification.

Ransomware detection is another area where GenAI is used to address the problem of malware using security protocols (e.g., SSL and TLS) to evade detection. In [16], the authors combine Deep Convolutional Generative Adversarial Network (DCGAN) and Transferred Generating Adversarial Network (TGAN) to generate unknown and variant encrypted ransomware, thereby enhancing the accuracy of detection models. This method is based on the assumption that the generators of both DCGAN and TGAN can generate ransomware image samples in general, while the discriminator of TGAN can only input real samples from the normal sample set. As a result, the discriminator can detect anomalous samples of ransomware, whether they are known or zero-day ones.

Phishing detection systems play a crucial role in identifying phishing websites and their URLs. Phishing refers to a type of social engineering attack where attackers steal users' confidential information, such as passwords and credit card details. However, traditional phishing detection engines face challenges due to the disproportionate number of phishing websites compared to legitimate ones. The process of collecting and labelling phishing websites is labour-intensive. To tackle this issue, the authors of [17] utilize an Adversarial Auto-Encoder (AAE) to synthesize phishing data samples based on existing phishing websites.

In this approach, the discriminator of the AAE distinguishes between phishing samples from the latent space of the AAE and generated samples from a prior distribution (e.g., normal distribution). Meanwhile, the AAE aims to minimize the reconstruction error, ensuring that the output closely resembles the input data. Simultaneously, the adversarial network minimizes discrimination between real and generated data. Following the training process, the newly generated phishing data samples are combined with the original datasets, creating a balanced dataset that includes both normal and phishing data. Consequently, the classifiers employed in phishing detection systems can improve the accuracy of identifying phishing websites.

Another example discussed in [18] introduces a model called PDGAN, designed specifically for detecting phishing URLs. PDGAN employs a Long Short-term Memory Network (LSTM) to generate fake phishing URLs from a normal distribution. The Convolutional Neural Network (CNN) serves as a discriminator in this model. After training PDGAN, artificial phishing data samples are generated to create a balanced dataset that includes both normal and phishing URLs. As a result, PDGAN achieves an accuracy of 97.58% in detecting phishing URLs.

Another application of GenAI is used for preventing password cracking. Users often choose simple passwords that are related to their personal information to easily remember them. However, this increases the likelihood of password cracking through password-guessing tools. Therefore, the first step for users when creating a new password is to use GenAI to analyze and assess its strengths and weaknesses. If the password passes the system's password check, it will be accepted. Additionally, by understanding common patterns of weak passwords, GenAI can suggest stronger passwords that are still user-friendly. In [19], the authors proposed GNPassGAN, a password-guessing tool based on the GAN model. GNPassGAN is capable of generating 250,309 passwords that match the passwords in the original password dataset (RockYou dataset) when generating 10^8 passwords. To achieve this, the generator of GNPassGAN distinguishes between real passwords obtained from data breaches and fake passwords generated by the generator. The key improvement of GNPassGAN compared to Vanilla GAN is the addition of gradient normalization to the discriminator, which enforces 1-Lipschitz continuity. Another approach is to combine GenAI with probabilistic context-free grammars (PCFG) to generate new passwords, known as GENPass [20]. GENPass can generate passwords from different datasets by using a weight-choosing process. The discriminator of GENPass uses the standard deviation to determine which passwords will be accepted. When generating 10^7 passwords, GENPass achieves a matching rate of 57.9%.

GenAI supports generating content such as facial, text, voices, and other biometric factors. This content can be combined with traditional passwords for author authentication and intellectual property protection. In [47], the authors proposed a framework for author authentication using data regeneration. The framework consists of two stages: generation and verification. In the generation stage, the goal is to generate the authentic output $\mathbf{x}_a$ from the generator G_a using k generated steps. For example, $\mathbf{x}_a^1 = G_a(\mathbf{x}_{input})$; ...; $\mathbf{x}_a^k = G_a(\mathbf{x}_a^{k-1})$. The $\mathbf{x}_{input}$ can be text, image, or voice, while G_a represents a generative model like the OpenAI APIs. In the verification stage, the origin of the artifact $\mathbf{x}_a$ is verified by regenerating the content from $\mathbf{x}_a^{k-1}$. The re-generated content is then compared with the original content to authenticate the author or verify the content's origin.

An AI-based model is a powerful tool for analyzing documents such as passports, driver's licenses, and IDs to identify any signs of forgery or manipulation. In [48], the authors introduced a convolutional neural network (CNN) forensic discriminator. They achieved this by adding salt and pepper to the training data, which allowed CNN to focus on distinguishing between real and forged data. To generate the forged image and text, a generator in the GAN is used before the data is labelled by a classifier to differentiate between real and fake data. Both real and fake data are then pre-processed by adding salt and pepper noise. This is done by incorporating background noise from the image into the input data. The CNN model is trained to classify the real and forged data. Similarly, in [49], the authors applied a Fourier spectrum-based method to detect forged text in images and videos. This method operates on the assumption that the text is forged by copying and pasting it onto the images, resulting in blurred text compared to the original. On the other hand, text forged using the insertion method appears brighter than the original. In [50], the authors presented a new model based on a deep ensemble neural network (DENN) to detect forged handwriting in noisy and blurry conditions. The DENN combines a stenographic kernel and a spatial filter as the input for the ensemble neural classifier.

AI plays a crucial role in identifying synthetic or fake identities. By analyzing large datasets, AI can learn to recognize unusual behaviour that may indicate the creation of fraudulent identities. In [51], the authors utilized a two-stream neural network to classify forged and authentic faces. The first stream focuses on extracting the differences between real and fake faces, such as strong edges and blurred areas. The second stream incorporates the Steganalysis feature to capture hidden features in the face images. Additionally, the authors applied the triple loss to classify faces, ensuring that the distance between pairs of patches from different faces is greater than that of the same faces. In [52], the authors attempted to detect manipulated faces by combining inconsistent head pose information with a support vector machine (SVM) model. Another approach, proposed by the authors of [53], involves identifying visual artifacts such as eyes, teeth, and facial contours. Subsequently, a multi-layer perceptron (MLP) classifier is used to distinguish between real and fake faces.

3.2 CYBERSECURITY RISKS CAUSED BY GENERATIVE AI

3.2.1 DEEPFAKE BY GENERATIVE AI

3.2.1.1 Generate Fake Identities

GenAI is an effective tool for generating fake images, videos and other misinformation.

The authors of [21] introduced CycleGAN, a method for learning to translate images from a source domain **X** to a target domain **Y**. To achieve this, CycleGAN utilizes two generators: $G : \mathbf{X} \to \mathbf{Y}$, which translates images from domain **X** to domain **Y**, and $F : \mathbf{Y} \to \mathbf{X}$, which translates images from domain Y to domain **X**. Additionally, CycleGAN employs two discriminators, D_X and D_Y. D_X is responsible for distinguishing between input from **X** and translated images $F(\mathbf{Y})$, while D_Y distinguishes between input from **Y** and translated images $F(\mathbf{X})$. The objective function of CycleGAN consists of two terms. The first term is adversarial losses, which aim to match the distribution of translated images to the distribution of the target domain. The second term is cycle consistency loss, which ensures that the image translation cycle returns to the original image. Thanks to the success of CycleGAN, attackers can generate and synthesize fake images that convincingly resemble photographs of real people. These fake images can be used to create fake social media profiles or other online accounts, giving the impression of a genuine identity. Attackers can manipulate various attributes, such as facial features, age, gender, and more, to create variations of real identities or entirely fictional ones. Once fake credentials like usernames, email addresses, and passwords are generated, attackers can create fake identities for online services.

One can create fake videos by transferring sequential content from a source domain to a target domain. The authors of [22] introduced Recycle-GAN, which translates content from one video to another while retaining the style native to the source video. For example, if the contents of Barack Obama's speech are transferred to Donald Trump, the generated video will resemble Donald Trump's style. Recycle-GAN achieves this by using two generators to map frames from the source domain to the target domain, and two discriminators to differentiate between real frames and generated frames in both domains. In addition, Recycle-GAN utilizes cycle consistency loss to ensure that the reverse frames are as similar as possible to the original frames. Unlike Cycle-GAN, which is used for image translation, Recycle-GAN introduces temporal consistency loss to translate the motion and changes of the input video to the target video. Furthermore, the recycling loss of Recycle-GAN reinforces the temporal consistency between the two domains. For instance, if two frames, i.e., t and $t+1$, are mapped from the source domain to the target domain, this order must be maintained when mapping from the target domain back to the source domain. Attackers can use face2face techniques, including image-2-image and video-2-video, to counterfeit a person's appearance and engage in misconduct. For example, the Google search engine displayed numerous web pages containing fake images and videos from pornographic platforms and adult pages [23]. Another example involves attackers using fake content to deceive the CEO of a UK company into sending $243,000 to a criminal's account [24].

Natural language processing models can generate synthetic text, including biographical information, job histories, and other details that contribute to a person's identity. In [54], the authors introduced a framework called automatic biography generation (BioGen), which aims to generate infoboxes by the standard of a Wikipedia page. The generated biography includes five fields: the name of the requested person, date of birth and death, place of birth, awards, and education and career. Similarly, in [55], the authors developed a model based on a pre-trained encoder-decoder to generate long-form biography sections that include citation information. The method uses a transformer-based sequence-to-sequence model to generate the sections. The inputs consist of internet-collected information that is then fed to the BART model. The maximum number of input tokens is 1024 words, while the length of the output generated from the decoder is 1000 words. Each section generated by the model includes cited information from the original documents for users to verify. For subsequent sections, evidence is retrieved and added. At the end of each section, the model generates the section heading for the next section, which is followed by the generation of the section's content.

AI supports the automation of creating large numbers of fake accounts on social media platforms or other online services. These accounts may appear to be associated with real individuals but are entirely fabricated. In [56], the authors investigate tools that support automatically generating social platform accounts, such as Twitter and Facebook. Additionally, synthesized realistic human voices can be misused for phishing attacks. Attackers can use fake identities to impersonate trusted individuals or authorities, convincing victims to disclose sensitive information. Furthermore, AI-driven chatbots can be designed to impersonate real individuals, engaging in conversations and interactions to deceive users. This can be exploited for various fraudulent activities. In [57], the authors investigate four popular LLMs: ChatGPT-3.5, GPT-4, Claude, and Bard, to generate a phishing attack. These frameworks can generate phishing websites and emails that mimic those of organizations and individuals. Importantly, these LLMs do not require any prior knowledge about adversarial exploits.

3.2.1.2 Generate Phishing Emails/Contents

ChatGPT is an effective tool for generating phishing email content that imitates the writing style of a trusted sender. Attackers can create persuasive narratives and messages that exploit human psychology. Phishing emails often employ emotional triggers, urgent language, or personalized information to manipulate recipients into taking actions, such as clicking on malicious links or providing sensitive information. Additionally, utilizing the GAI platform, which includes ChatGPT, enables the creation of emails that appear to originate from well-known and trusted entities like

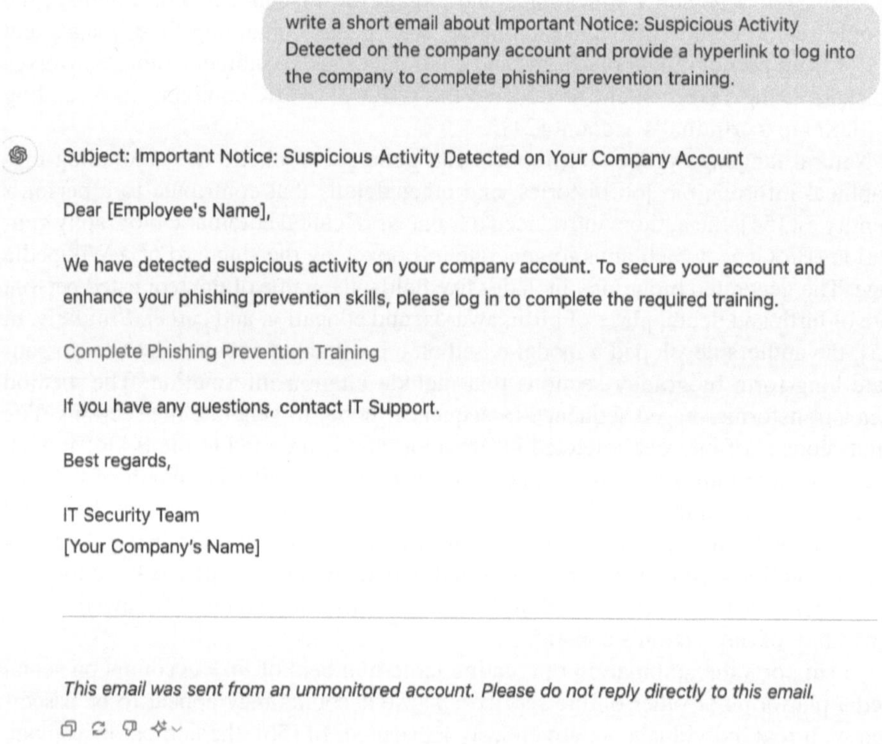
write a short email about Important Notice: Suspicious Activity Detected on the company account and provide a hyperlink to log into the company to complete phishing prevention training.

Subject: Important Notice: Suspicious Activity Detected on Your Company Account

Dear [Employee's Name],

We have detected suspicious activity on your company account. To secure your account and enhance your phishing prevention skills, please log in to complete the required training.

Complete Phishing Prevention Training

If you have any questions, contact IT Support.

Best regards,

IT Security Team
[Your Company's Name]

This email was sent from an unmonitored account. Please do not reply directly to this email.

Figure 3.4 The email generated using ChatGPT-3.5.

banks, government agencies, or popular online services. This impersonation aims to deceive recipients into believing the message is legitimate. Moreover, by using ChatGPT, attackers can replicate the language, tone, and formatting typically employed by the targeted organization or individual, enhancing the credibility of the phishing email. For instance, in [25], the authors demonstrated how attackers can leverage ChatGPT capabilities to quickly and efficiently create phishing emails. Firstly, the attacker can generate an email that mimics the company's communication style and notifies users about suspicious activity detected on their company account, as depicted in Figure 3.4. Secondly, the attacker uses ChatGPT to generate an HTML template for a webpage that urges the user to click on a phishing email link and prompts them to log into the company website for phishing prevention training and remedial measures, as shown in Figures 3.5 and 3.6. Upon receiving the email from the company, the user may click on the embedded hyperlink, leading them to input their username and password on the phishing alert page, as illustrated in Figure 3.6. Consequently, the user's information, including their username and password, is transmitted to the attacker.

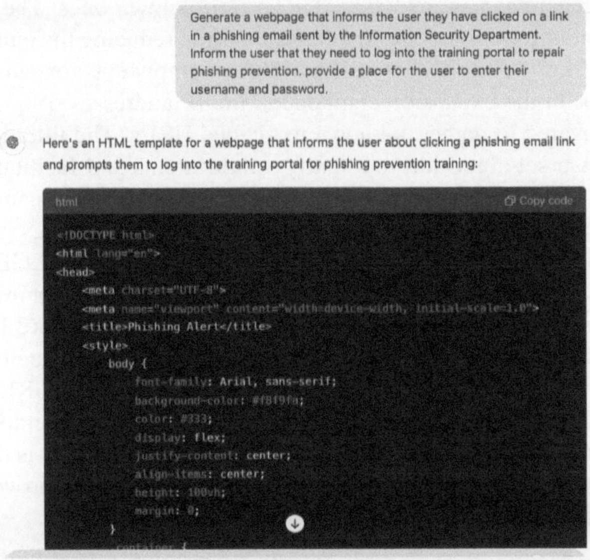

Figure 3.5 The HTML code generated by ChatGPT-3.5.

Figure 3.6 The webpage created by ChatGPT-3.5

AI-driven techniques can be used to obscure malicious URLs in phishing emails. This can involve creating different versions of URLs or utilizing URL shortening services to conceal the true destination of the link. By doing so, it becomes more difficult for traditional security measures to detect and block them. The framework introduced by the authors of [58], called the automatic blacklist generator (AutoBLG), automatically generates new malicious URLs from known ones. The goal of AutoBLG is to expand the search space of web pages while reducing the number of similar URLs through filters. AutoBLG consists of three components: expansion, filtration, and verification. In the expansion component, the IP addresses of known malicious URLs are discovered to gather unknown malicious URLs. The filtration component uses the Bayesian set algorithm to identify similar items and facilitate URL verification. Finally, the verification component employs three tools, namely honeypot, antivirus software, and VirusTotal, to examine the malicious URLs generated from the URL filtration component. The authors of [59] introduced URLGEN, which employs GANs to automatically generate new valid URLs based on known ones. URLGEN represents the URL at the character level rather than the word level, resulting in a smaller vocabulary size. The generator of URLGEN takes a random input $\mathbf{z}$ and maps it to the real data distribution of known URLs. Additionally, URLGEN relies on recurrent neural networks (RNNs), specifically Long short-term memory (LSTM), to generate a sequence of characters. The discriminator of URLGEN is designed using a one-dimensional convolutional neural network (1D-CNN) to improve the detection of features with varying positions in the data.

> Hi will have temporarily information your account will be restricted during that the Internet accounts and upgrading password An data Thank you for your our security of your Account Please click on it using the <NET> server This is an new offer miles with us as a qualified and move in

Figure 3.7 The email generated by an email generation system with 0.7% of malicious content [60].

GenAI, based on NLP, can create convincing attachments, such as invoices, documents, or reports, that are embedded within phishing emails. These attachments may contain malware, ransomware, or other malicious payloads. Additionally, the NLP models can generate phishing content in multiple languages, thereby increasing the effectiveness of phishing campaigns targeting diverse audiences. The authors of [60] proposed a system for automatic email generation that is used for targeted attacks. This system is based on a natural language generation technique, which generates fake emails with malicious content that can be easily customized by the attacker. During the generation phase, the system tries to generate a feeding word $\mathbf{w}_0$ to determine the word that is most likely to occur after $\mathbf{w}_0$, denoted as $\mathbf{w}_1$. This process is repeated with n words, i.e., $\mathbf{w}_0, \ldots, \mathbf{w}_n$. The goal of the trained Long short-term memory (LSTM) used in the system is to predict the best set of words to follow the training sequence. The novelty of this model is its procedure that supports the injection of malicious intent during training and the generation of malicious content.

3.2.1.3 Voice/video cloning, Fack News, Content Manipulation

AI-based deep learning is used to create highly realistic deepfake videos by synthesizing facial expressions and movements onto existing video footage. This can be used to manipulate speeches, interviews, or other video content to make it appear as though individuals are saying or doing things they never did. The authors of [38] introduced a new dataset called FakeAVCeleb, which is a combination of deepfake videos and lip-synced fake videos. To prepare the training videos, 500 videos from the YouTube platform are selected, considered based on gender, ethnicity, and age. To generate the fake videos, Faceswap [39] and FSGAN [40] are applied as face-swapping methods. Faceswap uses a convolutional neural network trained on the VGG network to capture the appearance of person X and transfer it to the style of person Y. There are two main steps to transfer an image of X to a list of styles of Y. First, the background and hair of the image are stored using segmentation since they cannot be preserved by the transformation. Second, the convolutional neural network is used to transform the content image $\mathbf{x}$ into the output $\hat{\mathbf{x}} = f(\mathbf{x})$. To make $\hat{\mathbf{x}}$ look like $\mathbf{x}$ as possible, the Faceswap uses the loss content over every layer of the VGG network, as follows:

$$l_c = \sum_l \frac{1}{|\phi_l(\mathbf{x})|} ||\phi_l(\hat{\mathbf{x}}) - \phi_l(\mathbf{x})||_2^2 \tag{3.7}$$

where $\phi_l(\mathbf{x})$ represents the representation of $\mathbf{x}$ on the lth layer of VGG. The style loss in Faceswap aims to replicate the style of person Y. Let $Y = \{\mathbf{y}_1, ..., \mathbf{y}_N\}$ be a list of styles from Y. The style loss attempts to find the best match between $\hat{x}$ and Y, and then minimize the distance between them. Since the transformation $\hat{\mathbf{x}} = f(\mathbf{x})$ cannot preserve the illumination of $\mathbf{x}$, the light loss in Faceswap is applied to maintain consistency in illumination between $\mathbf{x}$ and $\hat{\mathbf{x}}$. Once $\hat{\mathbf{x}}$ is obtained, it is combined with the background and hair of $\mathbf{x}$ to generate a new image $\hat{\mathbf{x}}$. The success of Faceswap has made it possible to effectively generate realistic human faces. Another example, FSGAN focuses on face swapping and reenactment. Unlike Faceswap, which solely uses the neural network for face transfer, FSGAN first employs a generator to reenact the face and then uses segmentation techniques to extract the face. Subsequently, a face inpainting network is used to estimate the missing pixels of the segmented face. Finally, the face-swapping scheme enables the transfer of the image to different styles of other individuals. FSGAN incorporates two loss functions: stepwise consistency loss for face reenactment and Poisson blending loss for adjusting the skin tones and lighting conditions of the generated image.

GenAI contributes to voice cloning by using small audio samples. This technology can be used to create fake audio recordings that make it seem like someone is saying things they never said. This has implications for impersonation, fraudulent phone calls, and deceptive audio content. In [41], the authors proposed a neural voice cloning method called speaker encoding. This method aims to synthesize a person's voice from a limited audio sample. Speaker encoding consists of three parts: special processing, temporal processing, and cloning sample attention. Special processing involves fully connected layers that transform the input data, such as mel-spectrograms extracted from the audio sample, into new feature representations.

Temporal processing is constructed using convolutional layers with the use of average pooling. The attention part computes the difference between audio samples of different speakers. After the training process, speaker encoding is used for speaker embedding, which captures and represents the speaker characteristics of a person in a low-dimensional vector. The audio, text, and speaker embedding are trained using a multi-speaker generative model (MSGM). The MSGM is used to generate voice cloning of a speaker based on input text. In [42], the authors introduced the Hieratron framework, which can clone a person's voice using very limited data while allowing control over the person's style and prosody. The Hieratron framework processes prosody and timbre in different modules to independently control them in the generated voice.

GenAI poses a potential threat to society. Chatbots powered by GenAI can generate fake news articles or misleading content. These AI-generated articles imitate the writing style of credible news sources, leading to the spread of misinformation and manipulation of public opinion. Moreover, attackers can manipulate images and videos produced by GenAI on social media platforms. This manipulation can involve altering the appearance of individuals, adding or removing elements from scenes, or even creating completely fabricated content to deceive users. Consequently, these manipulated media files can be exploited to create scandals with the intention of humiliating well-known individuals or tarnishing the reputation of organizations. By generating realistic profiles complete with profile pictures, biographical information, and social media activity, it is also possible to create fictional online personas using GenAI. These personas can then be used for various malicious activities such as spreading disinformation, influencing public opinion, or engaging in other harmful actions.

3.2.2 GENERATE MALICIOUS CODE

Using GenAI can generate malware variants based on existing malware samples to bypass malware detection techniques. This approach allows attackers to produce code with similar functionalities while evading traditional signature-based detection methods. To create polymorphic malware, which changes its code structure dynamically to avoid detection by antivirus and intrusion detection systems, generative algorithms can be employed to automatically generate code variations while maintaining the malware's functionality. The authors of [26] proposed a framework called FUMVar-Ex to generate fully working Windows malware samples that can evade up to 86% of commercial antivirus tools, such as BitDefender, Avast, Kaspersky, and AVG. To achieve this, FUMVar-Ex first extracts the PE file section before using a genetic algorithm to randomly select perturbations on PE files to create malware variants. To ensure that the malware is not corrupted, the generated file is checked by the verified module before being put into the Cuckoo sandbox to evaluate the evasion capabilities of the generated malware. The authors of [60] applied Generative Adversarial Networks (GAN) and Hidden Markov Models (HMM) to generate malware samples as opcode sequences. For the HMM, the input is a sequence of 30000 from each of the six selected malware families.

Using language models (LLMs) can make malicious code more difficult for security analysts to understand and reverse engineer, as it can create complex and convoluted code structures to obfuscate the true intent of the malware. In the referenced paper [28], the authors explored the potential of LLMs such as ChatGPT 4.0 and Google Bart for software code mutation/metamorphism. They introduced a framework that utilizes LLMs for mutating source code, which can be employed to reconstruct existing malware code to evade detection by anti-malware applications. The framework first enumerates all subroutines of a source code and then submits each subroutine to LLMs for rewriting. The resulting code from the LLMs is then checked using a unit test. Another example mentioned in the paper [1] demonstrates the misuse of ChatGPT in generating polymorphic malware, where ChatGPT is utilized to generate a DLL injection into the Explorer.exe process of the Windows operating system.

GANs can be used to create adversarial examples that are specifically designed to deceive machine learning-based security systems. This can involve generating modified input data that evades detection or misleads classifiers. In [27], the authors introduced MalGAN, which generates adversarial malware variants to attack a black-box detection model. Unlike a regular GAN, which uses random variables as input for its generator, MalGAN combines a malware sample with a noise sample as input to the generator. This is done to retain the functionality of the malware, preventing corruption in the generated malware. The black-box detection engine then tests the adversarial malware and benign samples to distinguish between normal and malware samples. The discriminator in MalGAN aims to mimic the function of the black-box detection engine by training with adversarial malware and benign samples labelled by the black-box detection engine. Unlike the discriminator in a regular GAN, which tries to differentiate between real and fake samples, the discriminator in MalGAN classifies benign and adversarial samples. The generator in MalGAN aims to produce adversarial examples that the discriminator classifies as benign. During training, the generator of MalGAN successfully fools the discriminator, meaning that it can deceive the black-box detection engine. The adversarial samples generated by the generator will be classified as benign samples by the black-box detection engine, even though they are actually malware samples.

Malicious payloads are segments of malicious code that attempt to execute on the victim's machine, such as opening a port, deleting a file, or capturing a screen. GenAI is a potential tool for dynamically generating malicious payloads in response to changes in the target environment. This adaptability helps attackers evade static detection methods and increases the likelihood of successful attacks. For instance, ChatGPT is a common tool used to support the generation of payload attacks. Attackers can create SQL injection payloads, which involve checking for SQL injection vulnerabilities, such as ' OR '1'='1 for union-based SQL injection, ' UNION SELECT null, null, null– for boolean-based blind SQL injection, ' AND 1=CONVERT(int, (SELECT @@version)– for error-based SQL injection, and so on. These payloads have the potential to bypass Web Application Firewalls (WAFs).

3.2.3 GENERATE ADVERSARIAL DOS ATTACK

GenAI is considered a primary tool for creating adversarial DoS traffic patterns that imitate normal network traffic behaviour. By generating traffic that mimics normal network traffic, attackers can make it difficult for traditional intrusion detection systems (IDSs) to differentiate between malicious and legitimate activity. For example, the authors of [29] proposed a framework called IDSGAN to generate DoS samples and perform black-box attacks against IDSs. In the IDSGAN framework, the generator aims to map inputs, which are a combination of real DoS samples and noise, to the representation of adversarial DoS traffic samples with the same dimension as normal traffic samples. The adversarial DoS and normal samples are then labelled by the black-box IDSs to create a new training dataset that includes both normal and adversarial DoS samples. The discriminator of the IDSGAN is trained on this dataset to distinguish between normal and adversarial DoS samples. Additionally, the generator of the IDSGAN trains the adversarial DoS samples to closely resemble normal samples. The training of both the generator and discriminator of the IDSGAN can deceive the black-box IDS. This means that after the training process, the black-box IDS cannot distinguish between normal and adversarial DoS samples. Consequently, the IDSs cannot detect the adversarial DoS samples generated by the IDSGAN. Experimental results showed that using adversarial DoS attack samples causes the detection accuracy of the black-box IDS to decrease from about 80% to less than 1%, while more than 99% of the adversarial DoS samples can evade the IDS. However, some adversarial DoS samples generated by the IDSGAN may lose their DoS attack ability. This is because certain attributes of the DoS sample, such as protocol type and flag, cannot be changed. Additionally, it is unlikely that the black-box IDS is always available to label the DoS samples generated by the generator. To overcome these weaknesses, the authors of [30] introduced DoS-WGAN, which utilizes the Wasserstein GAN (WGAN) combined with a gradient penalty to generate adversarial DoS samples that can evade IDSs. DoS-WGAN is based on the engine of a convolutional neural network (CNN) model. Unlike IDSGAN, which combines real DoS samples with noise samples as input to the generator, DoS-WGAN uses a conversion tool to determine which features of the real DoS sample can be changed and which ones should remain unchanged. For example, basic features like Duration, protocol type, service, and flag, as well as traffic features within a two-second time window, remain unchanged. The discriminator of DoS-WGAN aims to differentiate between real DoS samples and fake ones generated by the generator. The training process of the generator tries to make the fake DoS samples as similar to the real ones as possible. Due to the use of Wasserstein loss and gradient norm constraints for the random samples, the training process of DoS-WGAN is more stable compared to the Vallina GAN. Experimental results showed that DoS-WGAN can significantly decrease the detection rate of the CNN model used as the detection engine in IDSs, from 97.3% to 47.6%.

GenAI contributes directly to the generation of distributed denial-of-service (DDoS) attacks by imitating legitimate traffic to bypass the defence system. In the study by [31], the authors utilized the Least Squares Generative Adversarial Network

(LSGAN) to create adversarial DDoS attacks that can mimic normal traffic. Each LSGAN model was trained to generate a feature of legitimate traffic, which was then used to generate an adversarial feature for the DDoS attack. The seven features that represent normal traffic include accurate timestamps, source and destination IP addresses, number of packets sent, average packet size per source IP, standard deviation of packet size, and average packet arrival time difference per source IP. By employing the least squares loss function for the discriminator, LSGAN successfully overcomes the issue of vanishing gradients during the training process. Experimental results demonstrate that LSGAN can generate adversarial DDoS samples, reducing the detection model's accuracy rate from 99.97% to 55.03%. Similarly, in [32], the authors employed the CycleGAN architecture to generate adversarial DDoS attacks and surpass the detection engines used in intrusion detection systems (IDSs), namely Random Forest, Support Vector Machine, k-Nearest Neighbor, and Naïve Bayes. CycleGAN exhibits its capability to map normal samples to DDoS samples without requiring pairs of samples. The first generator of CycleGAN aims to generate DDoS samples using normal samples as input, while the second generator generates normal samples using DDoS samples as input. The first discriminator distinguishes between real and adversarial DDoS samples, while the second discriminator performs the same task by distinguishing real benign samples from adversarial benign samples. The application of cycle consistency loss in the loss function ensures that the translated normal samples match the original samples after being reverted from DDoS samples. After the training process, the first generator is utilized to generate DDoS samples. Experimental results indicate that the adversarial DDoS samples generated by CycleGAN can decrease the F-score of the detection engine from approximately 90% to around 34%. These findings highlight the urgent need for countermeasures against the adversarial DDoS attacks generated by GenAI.

AI-powered botnets enhance the capabilities of regular botnets by creating sophisticated botnets that can mimic human behaviour. These advanced botnets can then be coordinated to launch distributed denial-of-service (DDoS) attacks. In [33], the authors conduct an extensive study on 43 domain generation algorithms (DGAs) used by modern botnets. The purpose of DGAs is to generate a large number of domain names before these domain names are used as actual command-and-control (C&C) addresses for the botnets. Because these domain names are dynamically generated at random, traditional detection methods relying on static domain blacklists are ineffective in defending against DDoS attacks initiated by these botnets.

GenAI can create automated scripts that generate a high volume of network traffic, overwhelming the targeted system with a large number of requests. This influx of traffic can result in service degradation or complete unavailability. For instance, in [34], the authors introduced PAC-GAN as a means of generating network traffic, including ICMP, DNS queries, and HTTP requests, at the network layer. To do this, they utilized the TCPDUM tool to collect the network traffic and converted the packet byte values into a square matrix, similar to an image. Both the generator and discriminator of PAC-GAN were designed using a convolutional neural network. The discriminator's role is to differentiate between real network traffic and adversarial network traffic generated by the generator. Meanwhile, the generator aims to learn from the adversarial network and produce packets that closely resemble those found in real network samples. After the training process, the generator is employed to generate the adversarial samples. These samples are then converted back into packets and tested on the real system to assess the success rate of the sent packets. Experimental results revealed that when the training set included three packet types (ICMP, DNS, and HTTP), the adversarial packets produced by PAC-GAN achieved a success rate ranging from 66% to 87%. Another illustration of GenAI's usage in generating synthetic network traffic can be found in [35]. The authors introduced PAC-GAN, utilizing ChatGPT-3.5, to generate patterns in a flow of network packets. To accomplish this, PAC-GAN aims to obtain a summarized text format by extracting packets from a PCAP format. In their experiment, the authors used 10,000 sample packets extracted from ToN IoT datasets, which consisted of ICMP and DNS packets. These extracted packet summaries were then fed into the DaVinci model, which is a model provided by the OpenAI API, to generate Python code for generating the packets. After the packets are generated, they are processed using Babbage, which is a fine-tuned model provided by GPT-3, to obtain the Python codes. Finally, these Python codes are utilized to generate real packets using the Scapy library. The experimental results demonstrate that PAC-GPT can generate real network packets with a success rate of up to 91% for ICMP packets and 10% for DNS packets.

3.2.4 LLMS AND GENERATIVE AI-AIDED SOCIAL ENGINEERING ATTACKS

GenAI is a powerful tool used to generate realistic content such as text, images, voice, and videos. Its main goal is to create fake social media profiles with authentic details, profile pictures, and activities. These profiles are then utilized to establish connections with targets, build trust, and ultimately exploit personal or professional relationships. Attackers can easily compose emails, text messages, and social media posts to deceive individuals or organizations. To enhance the deception, GenAI generates realistic images, voices, and videos. In a study by the authors of [1], realistic emails and messages were produced by ChatGPT, which can imitate the tone and language of an individual or organization. Additionally, the authors of [21] developed the CycleGAN, a fundamental model for creating human-like communication by mimicking a person's voice, tone, and behaviour. With the user-friendly nature of GenAI, attackers can employ these techniques to carry out various types of phishing attacks, such as email phishing, sms phishing (smishing), and voice phishing (vishing), to engage in malicious activities.

The GenAI can facilitate social engineering attacks, allowing unauthorized access to private user data such as passwords and credit card information. The large language model, ChatGPT, is capable of crafting highly persuasive phishing emails or messages that appear to come from trusted sources. This model can mimic the writing style and tone of legitimate communications, increasing the chances of users being deceived. For instance, if an attacker obtains certain information about a user, such as their age, job, gender, and comments on social platforms like Facebook and Twitter, they can use ChatGPT to generate messages that imitate the user's tone and language. These messages can then be used to manipulate the user's friends into engaging in unexpected activities, such as clicking on links that contain malicious codes, sharing private user information, or even sending money through the bank. In a study conducted by the authors of [1], it was demonstrated that attackers can easily leverage ChatGPT to compose a message that mimics the style of a popular e-commerce website. Furthermore, the message may include a link that prompts users to provide their personal information. This type of message is often accepted by users due to its resemblance to typical phishing attacks, which exploit key psychological principles, create a sense of urgency, and instil fear. Consequently, users may act quickly without carefully considering the authenticity of the message. Moreover, creating tools for social engineering attacks using ChatGPT does not require a high level of technical expertise. Attackers can simply ask ChatGPT to generate PHP code that perfectly mimics the login page of a popular social media platform. This makes it extremely difficult for users to distinguish between genuine and fake web pages.

AI plays a crucial role in analyzing large amounts of data to understand individuals' preferences, interests, and behaviours. This information can be utilized to create highly targeted and personalized social engineering attacks that exploit the specific characteristics of the target. Attackers can gather users' comments on social platforms and employ tools to uncover the underlying topics in the comments. This can lead to the exposure of users' hobbies, habits, and opinions, making them susceptible to phishing attacks. The authors of [36] introduced BERTopic, a topic modelling method based on the BERT model, to extract coherent and key topic representations from documents. BERTopic embeds documents and converts sentences and paragraphs into dense vector representations. Clustering algorithms are then employed to identify topic clusters within the documents. Additionally, attackers can utilize a summarization framework to extract keyframes and uncover the user's interests. The authors of [37] presented a method for video summarization through keyframe extraction and unsupervised learning techniques. This involves extracting low-level features using uniform sampling and image histograms, followed by the use of clustering algorithms to determine the number of clusters in the input video. The video skim technique is also employed to select keyframes for the video summary.

3.3 SUMMARY

This chapter explored the intersection of two rapidly evolving fields: generative artificial intelligence (AI) and cybersecurity. As generative AI advanced in its ability to create realistic and convincing content, it presented both opportunities and challenges in the realm of cybersecurity. The chapter examined the potential applications of generative AI in addressing cybersecurity challenges, discussing how generative models were leveraged for anomaly detection, identifying patterns of malicious behavior within large datasets, and aiding in the early detection of cyber threats. Additionally, the chapter explored the potential vulnerabilities of generative AI systems themselves. It addressed the risks of adversarial attacks, where malicious actors manipulated or deceived generative models (including large language models, LLMs) to produce unintended, malicious outputs, disinformation, or content for phishing and other malicious purposes.

1. Gupta, M., Akiri, C., Aryal, K., Parker, E., & Praharaj, L. (2023). From Chatgpt to Threatgpt: Impact of generative AI in cybersecurity and privacy. IEEE Access.
2. GreyDGL/PentestGPT: A GPT-Empowered Penetration Testing Tool. Accessed: Jun. 9, 2023. [Online]. Available: `https://github.com/GreyDGL/PentestGPT`
3. Deng, G., Liu, Y., Mayoral-Vilches, V., Liu, P., Li, Y., Xu, Y., ... & Rass, S. (2023). Pentestgpt: An llm-empowered automatic penetration testing tool. arXiv preprint arXiv:2308.06782.
4. Wang, J., Huang, Y., Chen, C., Liu, Z., Wang, S., & Wang, Q. (2024). Software testing with large language models: Survey, landscape, and vision. IEEE Transactions on Software Engineering.
5. Pearce, H., Tan, B., Ahmad, B., Karri, R., & Dolan-Gavitt, B. (2023, May). Examining zero-shot vulnerability repair with large language models. In 2023 IEEE Symposium on Security and Privacy (SP) (pp. 2339-2356). IEEE.
6. ZHANG, Quanjun, et al. Pre-trained model-based automated software vulnerability repair: How far are we? IEEE Transactions on Dependable and Secure Computing, 2023.
7. Mirza, Mehdi, and Simon Osindero. “Conditional generative adversarial nets.” arXiv preprint arXiv:1411.1784 (2014).
8. Sohn, Kihyuk, Honglak Lee, and Xinchen Yan. “Learning structured output representation using deep conditional generative models.” Advances in Neural Information Processing Systems 28 (2015).
9. Goodfellow, Ian, et al. “Generative adversarial nets.” Advances in Neural Information Processing Systems 27 (2014).
10. Kingma, Diederik P., and Max Welling. “Auto-encoding variational bayes.” arXiv preprint arXiv:1312.6114 (2013).
11. An, Jinwon, and Sungzoon Cho. “Variational autoencoder based anomaly detection using reconstruction probability.” Special lecture on IE 2.1 (2015): 1-18.
12. Xia, X., Pan, X., Li, N., He, X., Ma, L., Zhang, X., & Ding, N. (2022). GAN-based anomaly detection: A review. Neurocomputing, 493, 497–535.
13. Schlegl, Thomas, et al. “Unsupervised anomaly detection with generative adversarial networks to guide marker discovery.” International Conference on Information Processing in Medical Imaging. Cham: Springer International Publishing, 2017.
14. Moti, Zahra, et al. “Generative adversarial network to detect unseen internet of things malware.” Ad Hoc Networks 122 (2021): 102591.
15. Won, Dong-Ok, Yong-Nam Jang, and Seong-Whan Lee. “PlausMal-GAN: Plausible malware training based on generative adversarial networks for analogous zero-day malware detection.” IEEE Transactions on Emerging Topics in Computing 11.1 (2022): 82-94.
16. Zhang, Xueqin, Jiyuan Wang, and Shinan Zhu. “Dual generative adversarial networks based unknown encryption ransomware attack detection.” IEEE Access 10 (2021): 900-913.
17. Shirazi, Hossein, et al. “Improved phishing detection algorithms using adversarial autoencoder synthesized data.” 2020 IEEE 45th Conference on Local Computer Networks (lcn). IEEE, 2020.
18. Al-Ahmadi, Saad, Afrah Alotaibi, and Omar Alsaleh. “PDGAN: Phishing detection with generative adversarial networks.” IEEE Access 10 (2022): 42459-42468.

19. Yu, Fangyi, and Miguel Vargas Martin. "GNPassGAN: improved generative adversarial networks for trawling offline password guessing." 2022 IEEE European Symposium on Security and Privacy Workshops (EuroS&PW). IEEE, 2022.
20. Liu, Yunyu, et al. "GENPass: A general deep learning model for password guessing with PCFG rules and adversarial generation." 2018 IEEE International Conference on Communications (ICC). IEEE, 2018.
21. Zhu, Jun-Yan, et al. "Unpaired image-to-image translation using cycle-consistent adversarial networks." Proceedings of the IEEE International Conference on Computer Vision. 2017.
22. Bansal, Aayush, et al. "Recycle-gan: Unsupervised video retargeting." Proceedings of the European Conference on Computer Vision (ECCV). 2018.
23. Rana, Md Shohel, Mohammad Nur Nobi, Beddhu Murali, and Andrew H. Sung. "Deepfake detection: A systematic literature review." IEEE Access 10 (2022): 25494-25513.
24. Mustak, Mekhail, Joni Salminen, Matti Mäntymäki, Arafat Rahman, and Yogesh K. Dwivedi. "Deepfakes: Deceptions, mitigations, and opportunities." Journal of Business Research 154 (2023): 113368.
25. Langford, T., & Payne, B. (2023, October). "Phishing faster: Implementing ChatGPT into phishing campaigns." In Proceedings of the Future Technologies Conference (pp. 174-187). Cham: Springer Nature Switzerland.
26. Jin, Beomjin, Jusop Choi, Jin B. Hong, and Hyoungshick Kim. "On the effectiveness of perturbations in generating evasive malware variants." IEEE Access 11 (2023): 31062-31074.
27. Hu, W., & Tan, Y. (2022, November). Generating adversarial malware examples for Black-box attacks based on GAN. In International Conference on Data Mining and Big Data (pp. 409–423). Singapore: Springer Nature Singapore.
28. Madani, P.: Metamorphic malware evolution: The potential and peril of large language models. In: 2023 5th IEEE International Conference on Trust, Privacy and Security in Intelligent Systems and Applications (TPS-ISA), pp. 74–81 (2023). IEEE Computer Society
29. Lin, Zilong, Yong Shi, and Zhi Xue. "Idsgan: Generative adversarial networks for attack generation against intrusion detection." Pacific-Asia Conference on Knowledge Discovery and Data Mining. Cham: Springer International Publishing, 2022.
30. Yan, Q., Wang, M., Huang, W., Luo, X., & Yu, F. R. (2019). Automatically synthesizing DoS attack traces using generative adversarial networks. International Journal of Machine Learning and Cybernetics, 10(12), 3387-3396.
31. Sun, Degang, Kun Yang, Zhixin Shi, and Chao Chen. "A new mimicking attack by LSGAN." In the 2017 IEEE 29th International Conference on Tools with Artificial Intelligence (ICTAI), pp. 441–447. IEEE, 2017.
32. Shieh, Chin-Shiuh, Thanh-Tuan Nguyen, Wan-Wei Lin, Mong-Fong Horng, Thanh-Vinh Nguyen, and Denis Miu. "Generating adversarial DDoS attacks with CycleGAN architecture." In 2022 International Conference on Computer Technologies (ICCTech), pp. 64–69. IEEE, 2022.
33. Plohmann, Daniel, Khaled Yakdan, Michael Klatt, Johannes Bader, and Elmar Gerhards-Padilla. "A comprehensive measurement study of domain generating malware." In 25th USENIX Security Symposium (USENIX Security 16), pp. 263-278. 2016.
34. Cheng, A. (2019, October). PAC-GAN: Packet generation of network traffic using generative adversarial networks. In 2019 IEEE 10th Annual Information Technology, Electronics and Mobile Communication Conference (IEMCON) (pp. 0728-0734). IEEE.

35. Kholgh DK, Kostakos P. PAC-GPT: A novel approach to generating synthetic network traffic with GPT-3. IEEE Access. 2023 Oct 18.
36. Grootendorst, Maarten. "BERTopic: Neural topic modelling with a class-based TF-IDF procedure." arXiv preprint arXiv:2203.05794 (2022).
37. Jadon, Shruti, and Mahmood Jasim. "Unsupervised video summarization framework using keyframe extraction and video skimming." In 2020 IEEE 5th International Conference on Computing Communication and Automation (ICCCA), pp. 140-145. IEEE, 2020.
38. Khalid, Hasam, Shahroz Tariq, Minha Kim, and Simon S. Woo. "FakeAVCeleb: A Novel Audio-Video Multimodal Deepfake Dataset." In the Thirty-fifth Conference on Neural Information Processing Systems Datasets and Benchmarks Track (Round 2). 2021.
39. Iryna Korshunova, Wenzhe Shi, Joni Dambre, and Lucas Theis. Fast face-swap using convolutional neural networks. In Proceedings of the IEEE International Conference on Computer Vision (ICCV), Oct 2017.
40. Nirkin, Y., Keller, Y., & Hassner, T. (2019). Fsgan: Subject agnostic face swapping and reenactment. In Proceedings of the IEEE/CVF International Conference on Computer Vision (pp. 7184-7193).
41. Arik, Sercan, Jitong Chen, Kainan Peng, Wei Ping, and Yanqi Zhou. "Neural voice cloning with a few samples." Advances in Neural Information Processing Systems 31 (2018).
42. Dai, Dongyang, Yuanzhe Chen, Li Chen, Ming Tu, Lu Liu, Rui Xia, Qiao Tian, Yuping Wang, and Yuxuan Wang. "Cloning one's voice using very limited data in the wild." In ICASSP 2022-2022 IEEE International Conference on Acoustics, Speech, and Signal Processing (ICASSP), pp. 8322-8326. IEEE, 2022.
43. Hu, Z., Beuran, R., & Tan, Y. (2020, September). Automated penetration testing using deep reinforcement learning. In 2020 IEEE European Symposium on Security and Privacy Workshops (EuroS&PW) (pp. 2-10). IEEE.
44. Harer, J., Ozdemir, O., Lazovich, T., Reale, C., Russell, R., & Kim, L. (2018). Learning to repair software vulnerabilities with generative adversarial networks. Advances in neural information processing systems, 31.
45. Dinh, P. V., Nguyen, Q. U., Hoang, D. T., Nguyen, D. N., Bao, S. P., and Dutkiewicz, E. "Constrained Twin Variational Auto-Encoder for Intrusion Detection in IoT Systems," in IEEE Internet of Things Journal, vol. 11, no. 8, pp. 14789-14803, 15 April 15, 2024, doi: 10.1109/JIOT.2023.3344842.
46. Dinh, P. V., Nguyen, D. N., Hoang, D. T., Uy, N. Q., Bao, S. P., & Dutkiewicz, E. (2022, December). Balanced twin auto-encoder for IoT intrusion detection. In GLOBECOM 2022-2022 IEEE Global Communications Conference (pp. 3387-3392). IEEE.
47. Desu, Aditya, Xuanli He, Qiongkai Xu, and Wei Lu. "Generative Models are Self-Watermarked: Declaring Model Authentication through Re-Generation." arXiv preprint arXiv:2402.16889 (2024).
48. Lim, Seo-young, and Jeongho Cho. "Reinforced CNN Forensic Discriminator to Detect Document Forgery by DCGAN." Computers, Materials & Continua 71, no. 3 (2022).
49. Nandanwar, L., Shivakumara, P., Mondal, P., Raghunandan, K. S., Pal, U., Lu, T., & Lopresti, D. (2020). Forged text detection in video, scene, and document images. IET Image Processing, 14(17), 4744-4755.
50. Nandanwar, L., Shivakumara, P., Jalab, H. A., Ibrahim, R. W., Raghavendra, R., Pal, U. & Blumenstein, M. (2022). A conformable moments-based deep learning system for forged handwriting detection. IEEE Transactions on Neural Networks and Learning Systems, 35(4), 5407-5420.

51. Zhou, P., Han, X., Morariu, V. I., & Davis, L. S. (2017, July). Two-stream neural networks for tampered face detection. In 2017 IEEE Conference on Computer Vision and Pattern Recognition Workshops (CVPRW) (pp. 1831-1839). IEEE.
52. Yang, Xin, Yuezun Li, and Siwei Lyu. "Exposing deep fakes using inconsistent head poses." ICASSP 2019-2019 IEEE International Conference on Acoustics, Speech and Signal Processing (ICASSP). IEEE, 2019.
53. Falko Matern, Christian Riess, and Marc Stamminger. Exploiting visual artifacts to expose deepfakes and face manipulations. In IEEE Winter Applications of Computer Vision Workshops (WACVW), 2019.
54. Ambavi, H., Garg, A., Nitiksha, N., Sharma, M., Sharma, R., Choudhari, J., & Singh, M. (2019, June). BioGen: automated biography generation. In 2019 ACM/IEEE Joint Conference on Digital Libraries (JCDL) (pp. 21-24). IEEE.
55. Fan, Angela, and Claire Gardent. "Generating Full Length Wikipedia Biographies The Impact of Gender Bias on the Retrieval-Based Generation of Women Biographies." 60th Annual Meeting of the Association for Computational Linguistics. Association for Computational Linguistics, 2022.
56. Pathak, Avanish. "An analysis of various tools, methods and systems to generate fake accounts for social media." Northeastern University Boston, Massachusetts, December (2014).
57. Roy, Sayak Saha, et al. "From Chatbots to Phishbots?: Phishing Scam Generation in Commercial Large Language Models." 2024 IEEE Symposium on Security and Privacy (SP). IEEE Computer Society, 2024.
58. Sun, B., Akiyama, M., Yagi, T., Hatada, M., & Mori, T. (2016). "Automating URL blacklist generation with similarity search approach." IEICE TRANSACTIONS on Information and Systems, 99(4), 873-882.
59. Valentim, Rodolfo Vieira, Idilio Drago, Martino Trevisan, and Marco Mellia. "URLGEN—Toward Automatic URL Generation Using GANs." IEEE Transactions on Network and Service Management 20, no. 3 (2022): 3734-3746.
60. Das, Avisha, and Rakesh Verma. "Automated email Generation for Targeted Attacks using Natural Language." In TA-COS 2018: 2nd Workshop on Text Analytics for Cybersecurity and Online Safety, p. 23. 2018.
61. Trehan, Harshit, and Fabio Di Troia. "Fake Malware Generation Using HMM and GAN." Silicon Valley Cybersecurity Conference. Cham: Springer International Publishing, 2021.

Part II

Applications of Generative AI for Cybersecurity

In Part II, the book dwells into specific challenges of cybersecurity, e.g., an imbalanced training dataset, a lack of attacks/training data, and deepfake and discusses how Generative AI can potentially tackle them. To provide readers with hands-on experiences, the book also presents some typical study cases where Generative AI can be used to secure cyber-physical systems, e.g., Internet of Things, Blockchains, and Cloud Computing platforms. Comparisons and detailed discussions are also provided to help the readers have a comprehensive view of the advantages as well as current limitations of using generative AI to solve problems in cybersecurity. Part II has 5 chapters as follows.

4 Representation Learning-Based Generative AI in Cybersecurity

Generative models have gained significant attention and shown great promise in various domains, including cybersecurity. In the realm of representation learning, where the goal is to represent data, generative models have demonstrated their potential in some important problems in cybersecurity, including intrusion detection, phishing detection, spam detection, and network data analysis. This chapter will present and analyze generative models for representation learning and their applications to cybersecurity.

4.1 REPRESENTATION LEARNING-BASED GENERATIVE AI

This section presents various generative models for representation learning. The discussed models include the variants of AutoEncoder and Contrastive Representation Learning. These representation learning techniques, particularly those based on deep neural networks have proven to be effective in automatically extracting meaningful features from raw network data, enabling robust and adaptive anomaly detection systems. By learning efficient representations of normal network behavior, these models can effectively identify a wide range of anomalies, making them invaluable tools for network security and performance monitoring.

4.1.1 AUTOENCODER

An Autoencoder (AE) is a type of unsupervised artificial neural network designed to learn a compact, low-dimensional representation of input data while preserving its essential characteristics. The network is trained to encode the input into a latent representation and subsequently reconstruct the original data as accurately as possible [1]. The fundamental objective of an AE is to capture meaningful features in a way that enables efficient data reconstruction, making it a powerful tool for dimensionality reduction, feature learning, and anomaly detection. Due to their ability to extract intrinsic patterns from data, autoencoders are widely used in various machine learning applications, including noise reduction, image compression, and cybersecurity threat detection.

An AE is a type of neural network architecture designed to learn efficient data representations in an unsupervised manner. It comprises two key components (as illustrated in Fig. 4.1): an encoder, i.e., f(x) that compresses the input data x into a

DOI: 10.1201/9781003502531-4

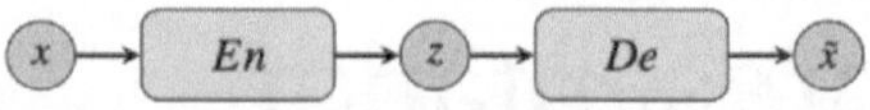

Figure 4.1 Architecture of an AutoEncoder(AE).

lower-dimensional latent space z, and a decoder, i.e., g(z), that attempts to reconstruct the original input from this compressed representation [2]. The training objective of the AE is to minimize the difference between the input x and its reconstruction g(f(x)), effectively learning to capture the essential features of the data. This difference, often measured as reconstruction loss, guides the model in learning meaningful patterns while discarding noise. Through this process, the AE is capable of learning compact and expressive data encoding suitable for various downstream tasks.

By capturing the most relevant structures within the input data, the Ae facilitates dimensionality reduction, effectively compressing high-dimensional inputs into a more manageable form. This property is particularly advantageous when dealing with complex datasets, as it enhances computational efficiency and reduces memory requirements. Additionally, the learned latent representation can improve generalization in downstream machine learning tasks, such as classification, anomaly detection, and data denoising. The ability of AEs to extract meaningful patterns while discarding irrelevant information makes them a powerful tool in various domains, including image processing, natural language processing, and cybersecurity.

The most common loss function used for training autoencoders is the mean squared error (MSE) between the original input and the reconstructed output [3]:

$$\mathscr{L}_{MSE}(x, g(f(x))) = \frac{1}{n}\sum i = 1^n ||x_i - g(f(x_i))||^2 \tag{4.1}$$

where $\mathscr{L}(x)$ represents the loss function, x_i is an input data sample, and $g(f(x_i))$ is the reconstructed output of the input x_i. By minimizing this reconstruction error, the AE learns to encode the input data into a compact latent representation while preserving the essential features of the original input.

Another loss function that can be used to train an AE is the mean absolute error (MAE) loss, also known as the L1 loss. The MAE loss is defined as the average absolute difference between the input and the reconstruction, and is given by:

$$\mathscr{L}_{MAE}(x) = \frac{1}{n}\sum i = 1^n |x_i - g(f(x))_i|, \tag{4.2}$$

where x_i is one input sample, n is the total samples of the training dataset, and $g(f(x_i))$ is the reconstructed output corresponding to the input sample x_i.

The Mean Absolute Error (MAE) loss is more robust to outliers compared to the Mean Squared Error (MSE) loss, as it treats all errors with equal weight regardless of their magnitude. Unlike MSE, which disproportionately penalizes larger errors due to squaring, MAE ensures that the autoencoder (AE) focuses on minimizing the overall absolute deviation. This characteristic makes MAE particularly beneficial in scenarios where the input data contains noise or outliers, as it prevents the model

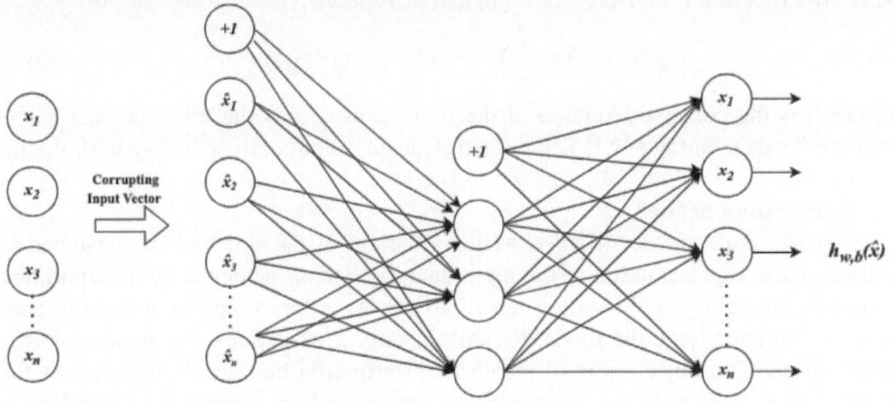

Figure 4.2 Denoising Autoencoder.

from being excessively influenced by extreme values, and helps maintain a more stable learning process.

To further enhance the quality of learned representations, researchers have introduced various regularization techniques into the AE loss function. These regularizers impose additional constraints during training, guiding the model toward more structured and interpretable latent representations. For example, sparsity regularization encourages the network to use only a small subset of neurons in the latent space, leading to more efficient feature selection. Disentanglement constraints promote the separation of independent factors of variation within the data, improving interpretability. Additionally, distributional regularization ensures that the latent space follows a specific statistical distribution, which can be useful in generative modeling and anomaly detection. By incorporating these techniques, autoencoders can learn more meaningful and domain-relevant representations, making them highly versatile for various machine learning applications.

Denoising Autoencoder

Denoising autoencoder (DAE) is a modification of the standard AE architecture designed to prevent the network from simply learning the identity function [6]. With a sufficiently large and powerful AE, the model can potentially just learn to copy the input to the output without discovering any meaningful compressed representations. DAE solves this problem by deliberately corrupting the input data, for example by adding noise or randomly masking portions of the input. The AE is then trained not just to reconstruct the original clean input, but also to denoise the corrupted version. This forces the DAE to learn robust, generalizable features that are resistant to noise, rather than just memorizing the training data. The inclusion of this denoising objective is what distinguishes DAE from AE and makes them better suited for tasks like anomaly detection, where the model needs to identify novel or corrupted inputs.

The loss function of a DAE can be computed using either Mean Squared Error (MSE) or Mean Absolute Error (MAE), similar to a standard AE. Specifically, the

MSE loss function for a DAE is formulated as follows:

$$\mathcal{L}_{DAE}(x) = \frac{1}{n}\sum i = 1^n ||x_i - g(f(\hat{x}_i))||^2, \tag{4.3}$$

where $\hat{x}_i$ is the corrupted version of the input data x_i. n is the total samples of the training dataset, and $g(f(\hat{x}_i))$ is the reconstructed output corresponding with the input $\hat{x}_i$.

Sparse Autoencoder

A Sparse Autoencoder (Sparse AE) is a variant of the standard AE designed to learn a sparse representation of the input data [5]. This is achieved by incorporating a sparsity-inducing regularization term into the AE's loss function, enforcing constraints that encourage the model to activate only a small subset of neurons in the latent space. The objective is to create an information bottleneck, compelling the network to capture only the most essential features while suppressing redundant or irrelevant information.

A common approach to enforcing sparsity is the use of L1 regularization (also known as Lasso regularization) on the activations in the latent layer. This method applies a penalty proportional to the absolute values of activations, driving many of them toward zero while allowing only a few to remain nonzero. Alternatively, Kullback-Leibler (KL) divergence can be used as a regularization term to push the average activation of each neuron in the latent layer toward a predefined sparse distribution, such as a Bernoulli distribution with a low mean [5]. By integrating these regularization techniques, Sparse AEs effectively learn compact and interpretable representations. Consequently, the loss function used to train a Sparse AE can be formulated as follows:

$$\mathcal{L} = \mathcal{L}_{RE} + \lambda * \mathcal{L}_{sparsity}, \tag{4.4}$$

where $\mathcal{L}_{RE}$ is the reconstruction loss obtained in Eq. 4.1 or Eq. 4.2, which measures how well the AE can reconstruct the input from the learned latent representation. $\mathcal{L}_{sparsity}$ is the sparsity regularization term, which encourages the activations in the latent layer to be sparse. λ is a hyperparameter that controls the relative importance of the reconstruction loss versus the sparsity penalty.

Autoencoders (AEs) have been extensively used for anomaly detection due to their ability to learn compact and meaningful representations of network traffic data in an unsupervised manner [6]. The fundamental idea is to train an AE using normal network traffic and then leverage the reconstruction error as an anomaly score to detect deviations from learned patterns.

Once trained on normal traffic data, the AE can analyze new, unseen network traffic samples by measuring how well they are reconstructed. Anomalous or malicious traffic, which does not conform to the learned representation of normal traffic, typically results in a higher reconstruction error. By defining an appropriate error threshold, the AE can automatically flag suspicious activity, enabling the detection of cyber threats such as network intrusions, denial-of-service (DoS) attacks, and unusual user behavior.

A key advantage of using AEs for anomaly detection is their unsupervised learning capability, which eliminates the need for labeled anomaly data—often scarce and

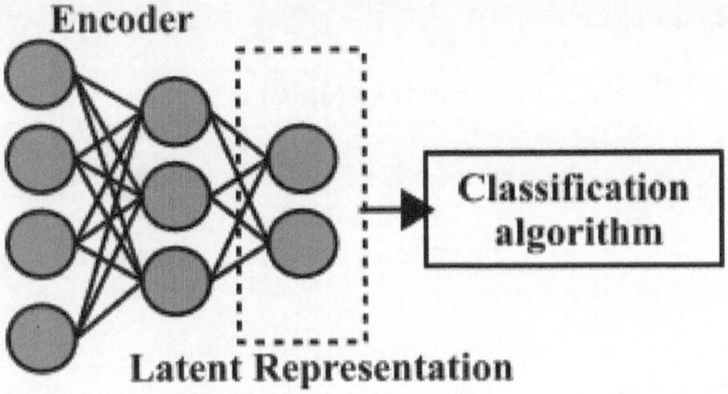

Figure 4.3 Using AE for downstream tasks.

costly to obtain. Instead, the AE learns the underlying characteristics of normal traffic directly from the available data and identifies deviations from these established patterns. This makes AEs a versatile and adaptive solution for anomaly detection in dynamic and evolving network environments.

The typical workflow for applying Autoencoders (AEs) to network anomaly detection consists of several key steps. First, the network traffic data undergoes preprocessing, which may involve feature engineering, normalization, and handling of missing values to improve data quality and model performance [7]. Next, an AE is trained exclusively on normal network traffic, aiming to learn a low-dimensional latent representation that captures the essential characteristics of typical traffic patterns [8]. Once trained, the AE detects anomalies by measuring the reconstruction error, e.g, MSE, between the original input and the AE's reconstructed output. Since anomalous traffic deviates from learned normal patterns, it results in higher reconstruction errors and is flagged as a potential anomaly. Finally, a decision threshold is determined to classify traffic samples as either normal or anomalous. This threshold is typically established through cross-validation or by analyzing the distribution of anomaly scores [6]. By following this structured workflow, AEs can effectively identify deviations in network traffic, making them a valuable tool for anomaly detection in cybersecurity applications.

AEs can also be utilized to learn latent representations for various downstream tasks, e.g., classification. Once an AE is trained, its latent layer can serve as a feature extractor for a classification model (as illustrated in Fig. 4.3). The process begins by passing the original data through the encoder of an AE, which transforms it into a compact latent representation. Instead of using the raw input data directly, this learned representation is then fed into a separate classification algorithm, allowing for more efficient and potentially more accurate predictions.

4.1.2 CONTRASTIVE REPRESENTATION LEARNING

Contrastive Representation Learning (CRL) is a powerful unsupervised learning paradigm that has gained significant attention in the machine learning community

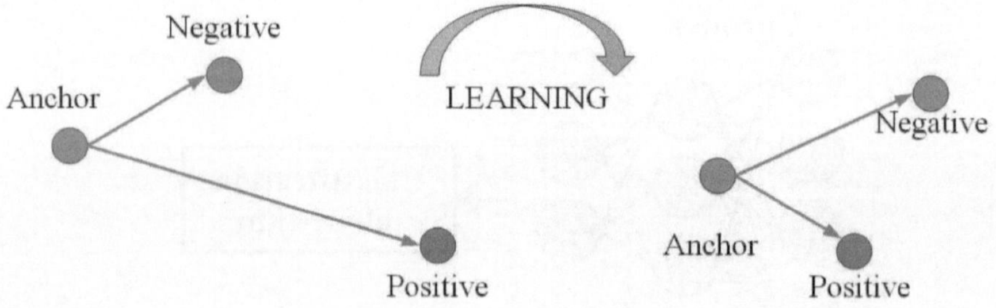

Figure 4.4 Aim of contrastive representation learning.

in recent years. This technique enables the model to capture the underlying structure and relationships within the data without requiring explicit labels [9]. By leveraging contrastive learning, models can learn discriminative and robust feature representations, making CRL particularly effective for tasks such as image recognition, natural language processing, and anomaly detection.

Fig. 4.4 illustrates the core concept of CRL by showing how a model learns to differentiate between similar and dissimilar samples. Initially, the anchor (blue) is closer to the negative sample (red) and farther from the positive sample (purple), indicating an unrefined representation. Through the learning process (depicted by the arrow), the model adjusts the feature space to bring positive pairs closer together while pushing negative pairs further apart. After learning, the positive sample aligns more closely with the anchor, while the negative sample moves away, ensuring a meaningful representation that effectively captures relationships within the data.

The core principle of CRL is the contrastive loss function, which encourages the model to map similar inputs to nearby points in the representation space while pushing apart dissimilar inputs. Formally, the contrastive loss can be defined as follows:

$$\mathcal{L}_{contrastive} = -\log \frac{\exp(sim(x_i, x_j)/\tau)}{\sum k = 1^{N} \mathbb{1}_{[k \neq i]} \exp(sim(x_i, x_k)/\tau)} \tag{4.5}$$

where x_i and x_j are positive samples, x_k is a negative sample, $sim(.)$ is a similarity function (e.g., cosine similarity), τ is a temperature parameter which acts as a scaling factor for the similarity scores, and $\mathbb{1}_{[k \neq i]}$ is an indicator function that is 1 when $k \neq i$ and 0 otherwise [10].

In the anomaly detection problem, CRL learns the representation that captures the underlying structure of the data, even in the absence of labeled attack data. By maximizing the similarity between "normal" samples and minimizing the similarity between "normal" and "anomalous" samples, the model can learn a new representation space where anomalies are projected to regions far from the "normal" data manifold [11, 12]. As a result, the machine learning tasks, such as one-class classifiers or clustering-based approaches are more effective in classifying the "normal" and "anomalous" data in the new representation space [13].

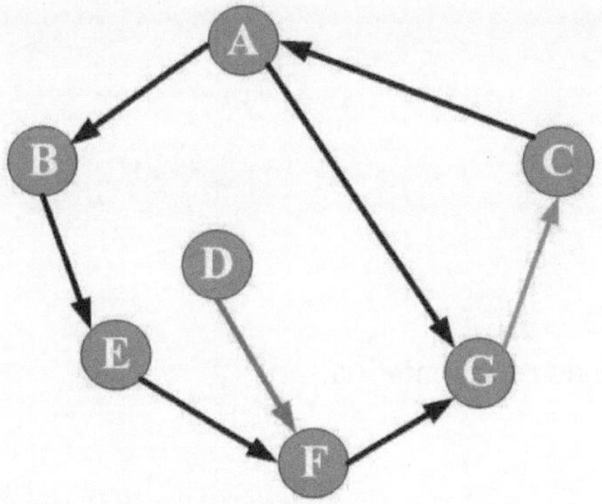

Figure 4.5 Visualization of a graph.

In anomaly detection, CRL enables the model to learn a meaningful representation of normal data without requiring labeled attack samples. Instead of relying on known anomalies, CRL distinguishes different views or transformations of normal data by maximizing their similarity while ensuring dissimilarity between unrelated instances. Through this process, the model captures the intrinsic structure of normal data and forms a well-defined normal data manifold in the learned representation space.

Once trained, this representation can be used for anomaly detection by identifying outliers—data points that deviate significantly from the learned normal distribution. Since anomalies often do not conform to the normal data structure, they are mapped to distant or less dense regions in the representation space. This makes it easier for downstream detection methods, such as one-class classifiers or clustering algorithms, to separate normal and anomalous instances effectively.

4.1.3 GRAPH REPRESENTATION LEARNING

Definition: A graph $G = (V,E,X)$ is defined by a set of nodes $V = v_1, v_2, \ldots, v_{|V|}$ and a set of edges $E = e_1, e_2, \ldots, e_{|E|}$. Each edge $e = (v_i, v_j)$ represents a connection between nodes v_i and v_j. Additionally, $X \in R^{|V| \times M}$ denotes the node feature matrix, where M is the dimension of each node's features. The adjacency matrix of the graph is defined as $A = R^{|V| \times |V|}$, with $A_{ij} = 1$ if $(v_i, v_j) \in E$ and $A_{ij} = 0$ otherwise.

Similar to machine learning (ML) and deep learning (DL) in other domains, one of the most fundamental ways to represent nodes and edges in a graph is through vector embeddings. This technique transforms each node and edge into a low-dimensional numerical vector, capturing both their inherent properties and relationships within the graph. As shown in Fig. 4.6, these embeddings provide a compact and efficient representation that enables advanced ML and DL algorithms to process graph-structured data more effectively.

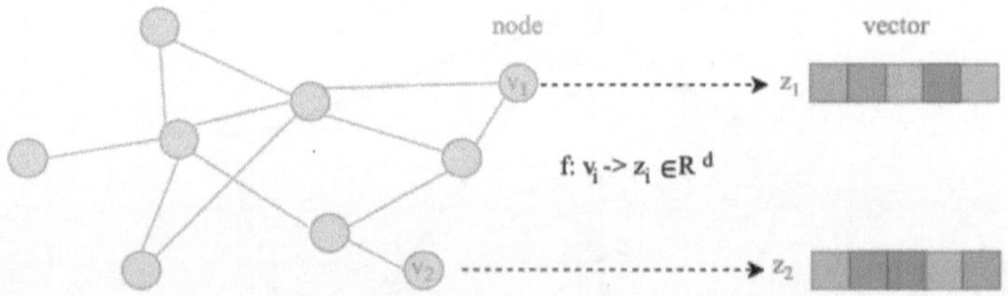

Figure 4.6 Graph node representation.

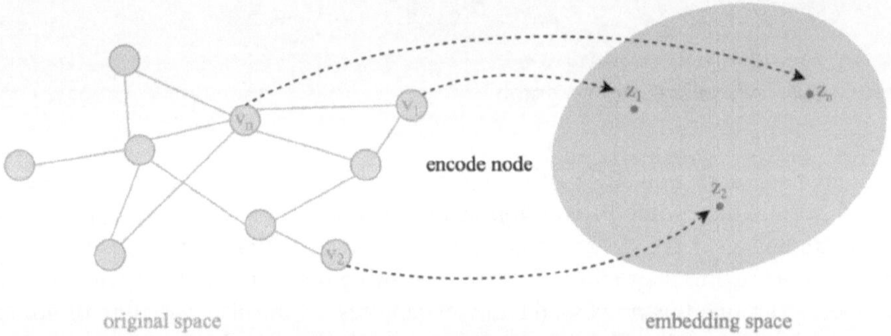

Figure 4.7 Visualization of embedding space.

Once the nodes and edges are embedded into a vector space, they can be leveraged for a wide range of downstream tasks. For instance, in node classification, embeddings help assign labels to nodes based on their learned characteristics, while in graph clustering, they facilitate grouping similar nodes together. A well-constructed embedding function $f(x)$ maps nodes into this new vector space, where two nodes (e.g., u and v) are considered similar if their embedding vectors (z_u and z_v) are close in distance. This proximity-based similarity enables various applications, such as detecting communities in social networks or identifying functional modules in biological systems.

Graph representation learning is particularly useful for dealing with high-dimensional, sparse graph data. The goal is to encode each node $v \in V$ into a dense, low-dimensional vector representation $R_v \in \mathbb{R}^d$ where $d|V|$ (as illustrated in Fig. 4.7). This allows for a more compact and computationally efficient representation, which is critical when working with large-scale graphs.

A major challenge in graph representation is handling the adjacency matrix A, which serves as a structural representation of the graph. Each row in A corresponds to a node's connections to all other nodes, offering a discrete way to represent graph structure. However, in real-world applications, adjacency matrices are often extremely sparse, meaning they contain a vast number of zeros. Using the raw adjacency matrix for ML tasks can lead to inefficiencies in both storage and computation, particularly when the number of nodes scales to millions. This makes learning compact embeddings are a crucial step in making graph-based models scalable and effective.

Graphs have recently become a powerful means of representing various types of structured and complex data, such as social networks, traffic systems, information networks, knowledge graphs, and protein interaction networks. As a versatile form of data organization, graph structures effectively capture the inherent relationships within this data, making them suitable for modeling non-Euclidean structures that are essential across numerous fields. For instance, in a social network represented as a graph, individual users are depicted as nodes, while edges illustrate the relationships between them, like friendships. In biology, nodes can signify proteins, with edges representing their interactions, such as dynamic protein interactions. By analyzing and mining these graph-structured datasets, we can uncover deeper insights and generate valuable knowledge that benefits society and humanity. Common techniques used for graph representation learning include Graph Neural Networks, and Graph Convolutional Networks, and Graph Attention Networks.

Graph Neural Network:

A Graph Neural Network (GNN) learns node embeddings by aggregating information from a node's neighbors using neural network architectures. Specifically, a GNN iteratively updates node representations by aggregating information from neighboring nodes, using learnable weights and activation functions. This allows the GNN to capture the structural and feature-based information of the graph, enabling effective learning for various tasks such as node classification, link prediction, and graph classification.

Each node representation is initialized as the node features, i.e., $h_i^0 = x_i$, where x_i is the feature vector associated with the node v_i. The update rule for a node v_i in a GNN can be expressed as:

$$h_i^l = \sigma\left(\sum_{j\in\mathcal{N}(i)} W^l h_j^{l-1} + W^k h_i^{l-1}\right), \tag{4.6}$$

where h_i^l is the representation of the node v_i at the $l-th$ layer, $\mathcal{N}(i)$ denotes the set of neighbors of the node v_i, W^l is a learnable weight matrix for the $l-th$ layer, and σ is an activation function. After T iterations, the final representation for each node can be obtained as:

$$h_i^T = \text{Aggregate}\left(h_j^T \,|\, j \in \mathcal{N}(i)\right), \tag{4.7}$$

where the Aggregate function aggregates the representations of the neighbors, which can be a sum, mean, or max operation.

For supervised tasks, the GNN is trained using a classification loss function, e.g., cross-entropy loss:

$$\mathcal{L} = -\sum_i y_i log(\hat{y}_i), \tag{4.8}$$

where y_i is the true label for the node v_i, and $\hat{y}_i$ is the predicted label based on the final node representation.

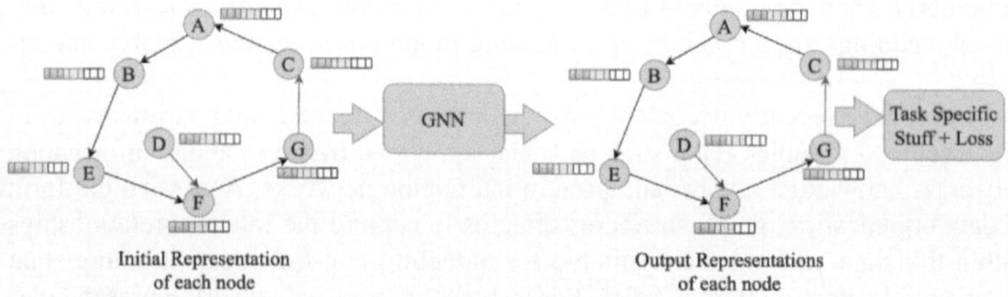

Figure 4.8 Visualization of a Graph Neural Network.

Graph Attention Network:

In Graph Neural Networks (GNNs), all neighboring nodes are treated equally during the aggregation process. This means that each neighbor contributes uniformly to the representation of a target node, regardless of its importance or relationship strength. However, in many real-world scenarios, nodes exert varying levels of influence on their neighbors. Ignoring these differences can lead to suboptimal performance and reduced accuracy, especially in applications where certain connections are significantly more meaningful than others.

To address this limitation, Graph Attention Networks (GATs) [14] were introduced. Unlike traditional GNNs, GATs leverage an attention mechanism to assign different importance weights to each neighboring node. This allows the model to dynamically adjust how much influence each neighbor has when computing a node's representation. By employing a self-attention mechanism, GATs ensure that more relevant and influential nodes contribute more significantly, while less important nodes have a reduced impact. This enhancement enables GATs to better capture complex, heterogeneous relationships within the graph, leading to improved performance in tasks such as node classification, link prediction, and community detection.

Given a node i and its neighbor j, the attention coefficient e_{ij} is computed as follows:

$$e_{ij} = a(Wh_i, Wh_j), \tag{4.9}$$

where h_i and h_j are the feature vectors of the node i and j, respectively. W is a weight matrix, which is a shared linear transformation, applied to every node. $||$ is a concatenation operator. Then, the attention coefficients are normalized across all neighbors of node i using the softmax function:

$$\alpha_{ij} = \frac{exp(e_{ij})}{\sum_{n \in N(i)} exp(e_{in})}, \tag{4.10}$$

where $N(i)$ is the set of neighbors for the node i.

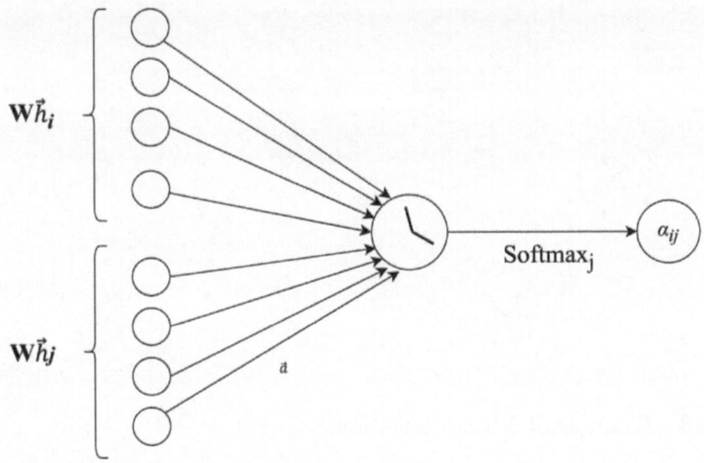

Figure 4.9 The attention mechanism.

By using attention weights, the updated node representation h_i' for node i obtained by aggregating the features of its neighbors, weighted by the attention coefficients:

$$h_i' = \sigma\left(\sum_{j\in\mathcal{N}(i)} \alpha_{ij} W h_j\right), \tag{4.11}$$

where σ is a non-linear activation function.

To enhance the model's ability to capture diverse information, the GAT uses multiple attention heads. The self-attention is repeated K times, and the outputs from each head are concatenated (e.g.,K equals 3 in Fig. 4.10).

$$h_i' = \cup_{k=1}^{K} \sigma\left(\sum_{j\in\mathcal{N}(i)} \alpha_{ij}^k W^k h_j\right), \tag{4.12}$$

where α_{ij}^k is attention coefficients for the k-th head. W^k is the weight matrix for the $k-th$ head.

To apply graph representation learning, data must be structured as a graph. This involves identifying key entities as nodes, defining relationships as edges, choosing between directed or undirected graphs, and representing the graph in a suitable format, such as an adjacency list or edge list. In cybersecurity, GNNs are used for malware detection, intrusion detection, and phishing detection by modeling complex relationships in network traffic, software behavior, and user interactions. By leveraging graph structures, GNNs enhance threat detection, cybersecurity analytics, and network security monitoring, making them a powerful tool for identifying evolving cyber threats.

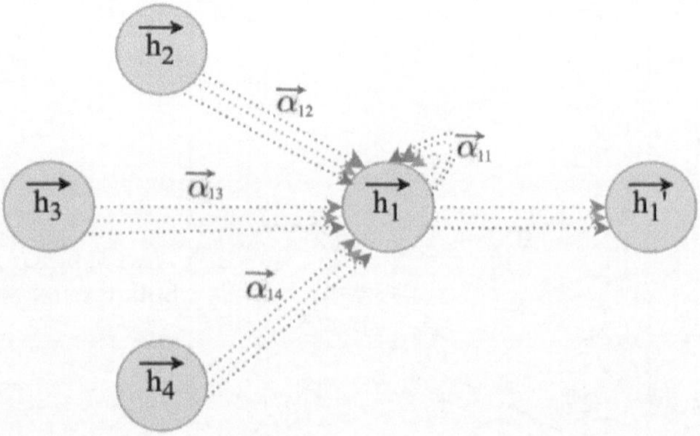

Figure 4.10 Example of 3-heads attention.

4.2 REPRESENTATION LEARNING APPLICATION FOR CYBERSECURITY

This section presents the application of generative AI to cyber attack detection. Particularly, this section will highlight the application of other generative models on intrusion detection, malware detection, traffic analysis, and social network analysis.

4.2.1 ATTACK DETECTION

Attack detection is the process of identifying and analyzing unauthorized or malicious activities within a network or system. It involves monitoring user behavior, network traffic, and system logs to detect patterns that may indicate security threats. Effective attack detection is essential for providing timely alerts and enabling swift responses to mitigate potential damage. As cyber threats grow in sophistication and frequency, robust detection mechanisms are crucial for safeguarding digital assets and ensuring system integrity.

Attack detection methods fall into two main categories: signature-based and machine learning (ML)/deep learning (DL)-based approaches. Signature-based methods rely on predefined threat patterns, making them highly effective for detecting known attacks. However, they struggle with novel or evolving threats that lack predefined signatures. In contrast, ML/DL-based methods analyze vast datasets to uncover complex attack patterns and anomalies, allowing them to adapt to emerging threats and enhance detection accuracy over time.

Among ML/DL techniques, representation learning plays a key role in attack detection by automatically extracting meaningful features from complex, high-dimensional data. By transforming raw data into lower-dimensional, information-rich representations, these techniques improve detection efficiency and accuracy while filtering out irrelevant noise. This ability to identify subtle anomalies makes representation learning particularly effective in detecting sophisticated cyberattacks that traditional methods might miss.

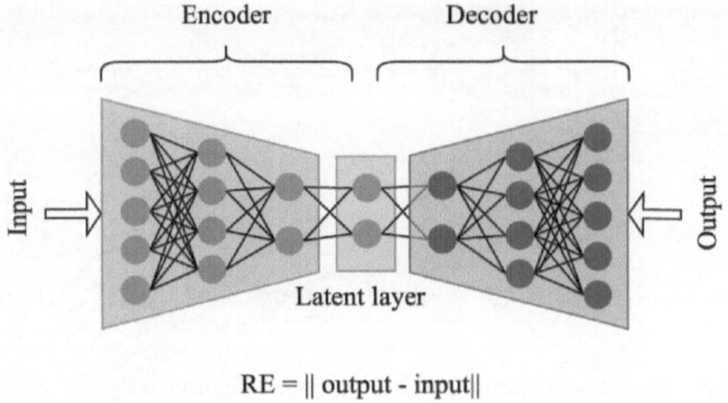

Figure 4.11 Supervised learning AEs for attack detection.

AEs have proven to be highly effective for attack detection by learning compact and meaningful representations of network traffic in an unsupervised manner. AEs are particularly useful in semi-supervised settings, where labeled attack samples are scarce or unavailable. Since AEs are trained only on normal traffic, they learn to reconstruct benign patterns accurately. Any significant deviation in reconstruction error (RE) signals a potential anomaly as illustrated in Fig. 4.11, making AEs effective for zero-day attack detection.

Several studies have leveraged this approach. Meidan et al. [15] trained an AE on normal network traffic and used a reconstruction threshold to classify unseen data as normal or malicious. Dromard et al. [16] introduced an online, real-time anomaly detection model, which continuously updates the feature space using a time-sliding window and incremental clustering. Ibidunmoye et al. [17] further enhanced this by incorporating adaptive learning and statistical control charts to detect deviations dynamically. To improve the interpretability and effectiveness of AEs, Cao et al. [8] proposed the Shrink Autoencoder (SAE), which compresses latent representations into a compact feature space, making it easier to separate normal and anomalous traffic. This sparse encoding technique enhances AEs' ability to detect attacks like Denial-of-Service (DoS) attacks, network probing, and unauthorized access attempts.

When labeled attack data is available, AEs can be modified for supervised learning, allowing them to explicitly differentiate between normal and anomalous traffic (Fig. 4.12). This is achieved by incorporating labeled data into the loss function, enabling the AE to learn class-specific features. Ly et al. [18] proposed the Multivariate AutoEncoder (MAE) and Multivariate Denoising AutoEncoder (MDAE), which integrate label information into the training process. These models separate normal and anomalous traffic into distinct regions, improving classification accuracy while reducing false positives.

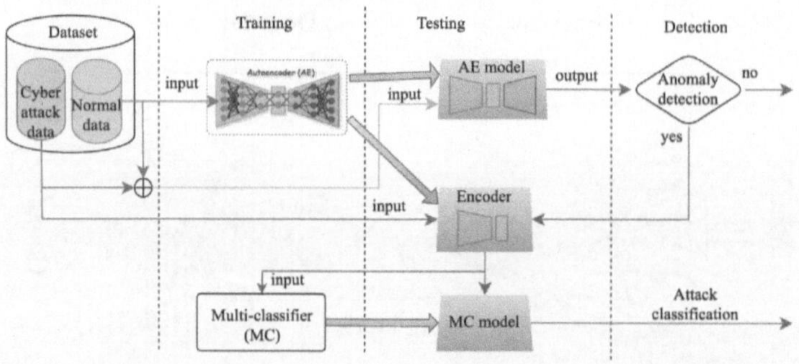

Figure 4.12 Supervised learning AEs for attack detection.

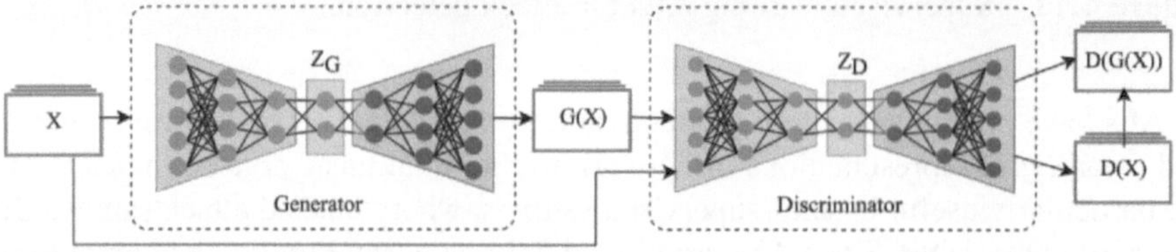

Figure 4.13 Combination of AE and GAN for attack detection.

AEs can also be combined with generative models, such as Generative Adversarial Networks (GANs) and Variational Autoencoders (VAEs), to improve attack detection. In these hybrid models, AEs serve as feature extractors, while GANs or VAEs refine the learned representations for better anomaly detection. For example, Ming et al. [18] integrated Mahalanobis distance with an adversarial AE, where the AE's output was fed into a discriminator to improve feature extraction and anomaly differentiation. Vu et al. introduced Adversarial Dual Autoencoders (ADAE), a GAN-based framework with two AEs (generator and discriminator) to enhance training stability and detection performance (Fig. 4.13).

Additionally, VAEs capture the latent probability distribution of normal traffic, allowing them to detect anomalies that significantly deviate from expected patterns. This probabilistic approach has been shown to outperform traditional AEs, especially in scenarios where attack patterns are highly non-linear [20, 21, 22].

Beyond AEs, the CRL has emerged as a powerful approach for detecting network anomalies. The CRL enables neural networks to distinguish between normal ("positive") and anomalous ("negative") samples without requiring explicitly labeled attack data [11, 23]. This technique helps security analysts identify key characteristics of normal and malicious traffic, leading to more effective mitigation strategies.

One of the most well-known CRL methods is Contrastive Predictive Coding (CPC), proposed by Oord et al. [9]. CPC trains a model to predict future network traffic representations based on past patterns. If a data point significantly deviates from

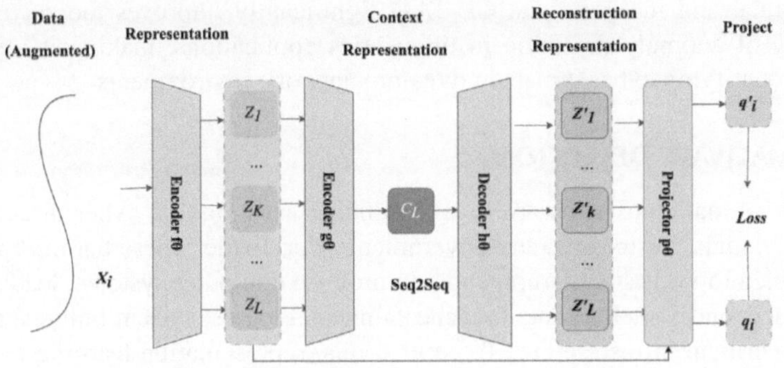

Figure 4.14 Contrastive learning combined with Autoencoder for anomaly detection

expected behavior (exhibits high prediction error), it is flagged as an anomaly [23]. This method effectively captures structural dependencies in normal traffic, making it useful for detecting subtle attacks. Another CRL-based approach is Deep Cluster-based Anomaly Detection (DCAD), introduced by Golan and El-Yaniv [11]. DCAD clusters network traffic data, ensuring that normal samples belong to well-defined clusters, while anomalous samples are pushed away from these clusters. This unsupervised clustering method allows the model to discover hidden patterns that distinguish normal from abnormal activity, improving detection accuracy. A more recent method, Contrastive One-Class Anomaly Detection (COCA) described in Fig. 4.14, was introduced by Wang et al. [24]. Unlike traditional contrastive learning, which relies on both positive and negative samples, COCA only uses positive pairs, treating the original and reconstructed data as similar. This eliminates the need for explicitly labeled anomalies, making it highly effective for detecting multiple types of cyber threats in time-series data.

Researchers have explored combining CRL with DL models to improve network attack (anomaly) detection. One such approach, proposed by Zhou et al. [13], integrates Contrastive Learning with an AE. In this hybrid model, the AE learns a generative representation of normal network traffic, while the contrastive learning component enhances anomaly discrimination. Anomalies are identified based on low probability under the generative model and high prediction errors in the contrastive learning process, resulting in more accurate and robust attack detection. Another innovative approach is Contrastive Learning-based Time Series Anomaly Detection (CL-TAD), introduced in [25]. CL-TAD is designed to address key challenges in time-series anomaly detection, such as complex temporal patterns, limited labeled data, and diverse data distributions. The method incorporates positive sample generation, where data is masked and reconstructed to create meaningful positive pairs. Additionally, it employs contrastive learning-based representation learning to enhance the model's ability to capture essential features for anomaly detection. By leveraging both original and masked data, CL-TAD significantly improves the accuracy and reliability of anomaly detection in time-series applications, making it a powerful tool for identifying cyber threats in dynamic network environments

4.2.2 MALWARE DETECTION

Malware, or malicious software, is a significant and ongoing cyber threat that affects individuals, businesses, and governments worldwide. These harmful programs are designed to infiltrate, disrupt, and compromise computer systems, leading to severe consequences such as data breaches, financial losses, system failures, and even threats to critical infrastructure. To combat this, representation learning techniques have emerged as powerful tools for malware detection by automatically extracting meaningful patterns from large datasets of benign and malicious software. By leveraging these techniques, cybersecurity models can achieve greater accuracy and generalization, enhancing their ability to detect, classify, and prevent the spread of diverse malware threats.

Malware detection techniques include signature-based methods, heuristic analysis, specification-based approaches, and visualization-based strategies [26, 27]. Among these, visualization-based methods—where malware is represented as an image—have gained significant attention due to their accuracy in identifying threats. As illustrated in Fig. 4.15, this approach follows a three-step process: first, converting malware binaries into images to reveal underlying structural patterns; second, extracting meaningful features from these images for classification, and third, training a machine learning model on the extracted features to distinguish between benign and malicious samples effectively.

Bakir et al. [28] proposed a malware detection framework that integrates AE-based feature extraction with image-based classification. Their method first transforms malware data into images and then applies three different AE architectures to extract multiple feature maps. These features are subsequently used in ML classifiers for accurate malware detection. Similarly, Kumar et al. [29] introduced an advanced image-based malware detection system leveraging software-defined networking (SDN) honeypots, convolutional neural networks (CNNs), and a custom two-level AE. This system converts binary programs into grayscale images, extracts textural features through deep CNNs with transfer learning, and applies a two-level AE to optimize the feature set. The refined features are then classified using six different deep learning and machine learning algorithms, enhancing both detection accuracy and efficiency. Xing et al. [30] developed a malware detection model combining grayscale image representations with an AE-based deep learning framework. Their approach assesses the effectiveness of grayscale image conversion using an AE's reconstruction error while utilizing its dimensionality reduction capabilities for classification. Furthermore, AE models have been integrated with CNNs to enhance anomaly detection in Android malware [31]. Feng et al. [32] introduced a two-layer malware detection model combining static and dynamic analysis techniques. The first layer extracts features using static analysis and classifies data as

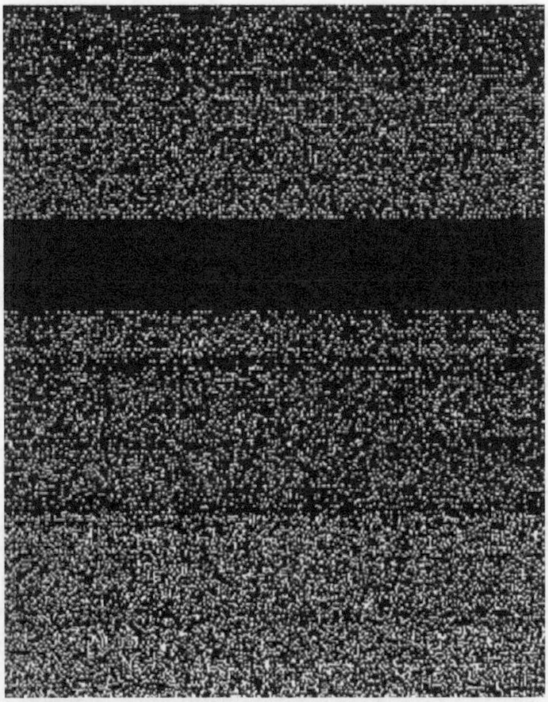

Figure 4.15 Example of malware file visualization.

benign or malicious with a neural network. The second layer further analyzes mobile traffic from benign samples and employs a hybrid CNN-AE model to classify malware by type and family. This multi-layered approach improves detection accuracy by leveraging complementary static and dynamic analysis methods.

Additionally, CRL has proven to be a powerful technique in malware detection by enhancing feature discrimination and improving robustness against obfuscation. The authors in [33] leveraged CRL to reduce the differences introduced by obfuscation while amplifying the distinctions between malware families. For interpretable classification, the malware is transformed into an image form using centrality analysis. The combination of CRL and interpretable feature representation helps to provide accurate and explainable Android malware classification, even in the presence of advanced obfuscation techniques. Additionally, Kailu et al. [34] first leveraged data feature fusion technology to extract as much relevant information as possible from malware execution records, forming a comprehensive original dataset. Second, to effectively counter code obfuscation techniques the authors utilized supervised CRL to construct malware features from high-dimensional space. This helps to capture meaningful features of the malware. By combining these two steps, i.e., robust data collection and feature engineering, this work can classify malware effectively.

4.2.3 NETWORK TRAFFIC ANALYSIS

Traffic classification is a critical task in network management, enabling the identification of different types of network traffic for applications such as quality of service (QoS) optimization, resource allocation, anomaly detection, malware identification, and intrusion detection. Traditional classification methods, including port-based and deep packet inspection (DPI) techniques, have become less effective due to the increasing use of encryption and evolving network protocols. To overcome these challenges, representation learning has emerged as a powerful technique for traffic classification, allowing models to automatically extract meaningful and high-dimensional features from raw network data.

By leveraging DL models such as AEs, representation learning enhances traffic classification by capturing complex patterns and relationships in network traffic. AEs help reduce dimensionality while preserving crucial structural information, enabling more accurate traffic categorization. CRL improves feature representations by learning to distinguish subtle differences between traffic types, even in the presence of noise and obfuscation. Meanwhile, GNNs offer a unique advantage in traffic classification by modeling network flows as graph structures, effectively capturing dependencies and relational information between packets, sessions, or network entities. These advanced representation learning techniques not only improve classification accuracy but also enhance the robustness and adaptability of traffic analysis models. They enable the detection of encrypted traffic, distinguish between diverse application types, and strengthen network security by identifying sophisticated cyber threats. As network environments continue to evolve, representation learning remains a cornerstone of intelligent and efficient traffic classification.

The work in [35] used an unsupervised approach for network traffic classification. The method utilizes statistical properties of the flows and a neural AE-based clustering technique. A time interval-based feature vector construction, combined with a semi-automated cluster labeling process, enables traffic flow classification without requiring prior knowledge of specific traffic classes. In the work [23], the AE model was used to reduce the dimension of data features in the pretraining stage. After that, one clustering algorithm, e.g., the k-means algorithm, is applied to classify the latent representation of data learnt from the pretraining stage. The training process of the proposed model does not require labeled data, which significantly reduces the computational complexity of the network traffic classification task.

Due to growing privacy and security concerns among internet users, many applications now employ various forms of traffic encryption. Encrypted traffic can conceal malicious activities, making their detection and analysis more challenging. To classify the encryption traffic, Jun et al. [37] introduced a model called Semi-supervision 2-Dimensional Convolution AutoEncoder (Semi-2DCAE). The model extracts the spatial structure features in the original network traffic using a 2-dimensional convolutional neural network (2D-CNN). It then employs an AE structure to downscale the data, representing different traffic features as spectral lines in distinct intervals of a one-dimensional standard coordinate system, which we refer to as the FlowSpectrum. Additionally, this work introduced the PRuLe activation function to the model to ensure stability during the training process. The AE is also combined with other deep neural networks, e.g., Convolutional Neural Network (CNN) and attention-based Bidirectional Long Short-Term Memory (Bi-LSTM) [22] for encrypted traffic classification. This approach classifies encrypted traffic at two distinct levels using one dimensional CNN (1D-CNN), attention-based Bi-LSTM, and AE models. Their proposed approach classifies traffic types and applications based on a comprehensive set of session-level and packet-level with two types of features, i.e., spatial-temporal and statistical features. These features are derived from the relationships between packet content, the temporal relationships between packets within a session, and the statistical characteristics of a network session.

4.2.4 SOCIAL ANALYSIS

With the evolution of the Internet over the years, a multitude of online social media platforms have emerged, significantly transforming traditional social dynamics. People can now make friends no matter how far apart they are and share their interests, hobbies, and activities with each other. These different kinds of online interactions create large and complex social media networks, often called online social networks (OSNs). OSNs can also be categorized based on their content types, leading to distinctions like friendship networks, movie review networks, and music interaction networks, each derived from different platforms. Moreover, user-item networks in online shopping systems can be considered a form of social media network, as they also facilitate rich interactions among users on the Internet. Popular data sources for analyzing social media networks include Twitter, Facebook, Weibo, YouTube, and Instagram. GNNs have proven highly effective in tackling various challenges in online social networks (OSNs), particularly in malicious activity detection, emergency response, and anomaly identification.

Detecting malicious activities in OSNs involves analyzing user interactions to identify threats like fraud, cloned accounts, and cyberbullying. Sadhasivam et al. [39] proposed a GNN-based method to detect cloned profiles in real-time. Unlike conventional approaches that rely on activity monitoring, their model employed an Adaptive Random Subspace technique with k-Nearest Neighbors (k-NN) as the base classifier. By leveraging weighted graph theory, it assigned higher threshold values to suspected clones, improving trust evaluation and enabling effective mitigation of fraudulent accounts.

GNNs play a crucial role in crisis response by identifying and analyzing the spread of misinformation. Rumors—both false and true—can significantly impact social stability, making their early detection vital. To address this, a bi-directional graph attention network (Bi-GAT) model was introduced [40], which represents rumor propagation as a graph structure. The model applied graph attention networks (GATs) with a multi-head attention mechanism to extract key features from rumor nodes. After processing through two graph attention layers, the system classified misinformation with high accuracy, demonstrating GNNs' effectiveness in social media crisis detection.

To detect anomalous nodes in an OSN, the researchers can utilize the unsupervised and supervised learning approaches. The unsupervised anomaly detection is used to create a reconstructed graph, identifying nodes with higher reconstruction errors as anomalies. I. Ahmed et al. [41] proposed a dimensionality reduction method for anomaly detection. To effectively embed high-dimensional data, they proposed using a minimum spanning tree (MST), which is a graph-based algorithm, to approximate local neighborhood structures and generate structure-preserving distances among data points. By using a distance measure based on minimum spanning trees (MST) in AEs, they created a graph-regularized autoencoder that outperformed other methods on 20 standard anomaly detection datasets. Besides that, L. Zhao et al. [42] developed the method named GLAM, an end-to-end graph-level anomaly detection model utilizing GNNs. It addressed two key issues, i.e., the neglect of graph-level anomalies in graph databases and the challenge of selecting unsupervised models without labels.

In summary, representation learning has emerged as a powerful technique in cybersecurity, enabling more effective detection and classification of threats across various domains. By leveraging models such as autoencoders (AEs), contrastive representation learning (CRL), and graph neural networks (GNNs), cybersecurity systems can extract meaningful features from high-dimensional and complex data, improving detection accuracy and robustness. These techniques have been widely applied in malware detection, network traffic classification, and anomaly detection in social networks. The ability of representation learning to capture structural, spatial-temporal, and statistical relationships in data make it indispensable for identifying malicious behaviors, optimizing security operations, and mitigating emerging cyber threats. As cyberattacks become more sophisticated, and continued advancements in representation learning will play a crucial role in enhancing proactive threat detection and fortifying cybersecurity defenses.

4.3 SUMMARY

This chapter explores the application of representation learning techniques in the field of cybersecurity, emphasizing their ability to automatically extract meaningful features from complex and high-dimensional data. It introduces and analyzes three primary types of representation learning models: Autoencoder-based models, Contrastive Representation Learning models, and Graph Neural Network (GNN) models. Each of these approaches offers unique strengths in learning compact, informative embeddings that enhance the performance of downstream security tasks. The chapter demonstrates how these models are effectively applied to a range of cybersecurity domains, including attack detection, malware detection, network traffic analysis, and social network security. By leveraging the power of representation learning, these applications benefit from improved accuracy, adaptability, and robustness in identifying and responding to evolving cyber threats.

1. Geoffrey E Hinton and Ruslan R Salakhutdinov. Reducing the dimensionality of data with neural networks. *Science*, 313(5786):504–507, 2006.
2. Ian Goodfellow, Yoshua Bengio, and Aaron Courville. *Deep learning*. MIT press, 2016.

3. Yoshua Bengio, Aaron Courville, and Pascal Vincent. Representation learning: A review and new perspectives. *IEEE Transactions on Pattern Analysis and Machine Intelligence*, 35(8):1798–1828, 2013.
4. Pascal Vincent, Hugo Larochelle, Isabelle Lajoie, Yoshua Bengio, and Pierre-Antoine Manzagol. Stacked denoising autoencoders: Learning useful representations in a deep network with a local denoising criterion. *Journal of Machine Learning Research*, 11(Dec):3371–3408, 2010.
5. Andrew Ng et al. Sparse autoencoder. *CS294A Lecture notes*, 72(2011):1–19, 2011.
6. Raghavendra Chalapathy and Sanjay Chawla. Deep learning for anomaly detection: A survey. *arXiv preprint arXiv:1901.03407*, 2019.
7. Chong Zhou and Randy C Paffenroth. Anomaly detection with robust deep autoencoders. In *Proceedings of the 23rd ACM SIGKDD International Conference on Knowledge Discovery and Data Mining*, pages 665–674, 2017.
8. Miguel Nicolau, James McDermott, et al. Learning neural representations for network anomaly detection. *IEEE Transactions on Cybernetics*, 49(8):3074–3087, 2018.
9. Aaron van den Oord, Yazhe Li, and Oriol Vinyals. Representation learning with contrastive predictive coding. *arXiv preprint arXiv:1807.03748*, 2018.
10. Ting Chen, Simon Kornblith, Mohammad Norouzi, and Geoffrey Hinton. A simple framework for contrastive learning of visual representations. In *International Conference on Machine Learning*, pages 1597–1607. PMLR, 2020.
11. Izhak Golan and Ran El-Yaniv. Deep anomaly detection using geometric transformations. *Advances in Neural Information Processing Systems*, Vol. 31, 2018.
12. Vit Skvara, Jan Francaa, Matej Zorek, Tomas Pevny, and Vaclav Smidl. Comparison of anomaly detectors: Context matters. *IEEE Transactions on Neural Networks and Learning Systems*, 33:2494–2507, 2020.
13. Hao Zhou, Ke Yu, Xuan Zhang, Guanlin Wu, and Anis Yazidi. Contrastive autoencoder for anomaly detection in multivariate time series. *Information Sciences*, 610:266–280, 2022.
14. Petar Veličković, Guillem Cucurull, Arantxa Casanova, Adriana Romero, Pietro Lio, and Yoshua Bengio. Graph attention networks. *arXiv preprint arXiv:1710.10903*, 2017.
15. Yair Meidan, Michael Bohadana, Yael Mathov, Yisroel Mirsky, Asaf Shabtai, Dominik Breitenbacher, and Yuval Elovici. N-baiot—network-based detection of IoT botnet attacks using deep autoencoders. *IEEE Pervasive Computing*, 17(3):12–22, 2018.
16. J. Dromard, G. Roudière, and P. Owezarski. An Online and scalable unsupervised network anomaly detection method. *IEEE Transactions on Network and Service Management*, 14(1):34–47, March 2017.
17. O. Ibidunmoye, A. Rezaie, and E. Elmroth. Adaptive anomaly detection in performance metric streams. *IEEE Transactions on Network and Service Management*, 15(1):217–231, March 2018.
18. Ly Vu, Van Loi Cao, Quang Uy Nguyen, Diep N. Nguyen, Dinh Thai Hoang, and Eryk Dutkiewicz. Learning latent representation for iot anomaly detection. *IEEE Transactions on Cybernetics*, 52(5):3769–3782, 2022.
19. Li Ming, Han Dezhi, and Li Dun. A method combining improved mahalanobis distance and adversarial autoencoder to detect abnormal network traffic. In *Proceedings of the 27th International Database Engineered Applications Symposium*, IDEAS '23, page 161–169, New York, NY, USA, 2023. Association for Computing Machinery.
20. Luca Bergamin, Tommaso Carraro, Mirko Polato, and Fabio Aiolli. Novel applications for vae-based anomaly detection systems. *2022 International Joint Conference on Neural Networks (IJCNN)*, pages 1–8, 2022.

21. Touseef Iqbal and Shaima Qureshi. Reconstruction probability-based anomaly detection using variational auto-encoders. *International Journal of Computers and Applications*, 45:231–237, 2022.
22. Arul Jothi S, Harini S, Nivedha K., Selva Keerthana B G, Gokul Nithin Kumar R, and Jayasree B S. Data anomaly detection in wireless sensor networks using β-variational autoencoder. *2023 International Conference on Intelligent Systems for Communication, IoT and Security (ICISCoIS)*, pages 631–636, 2023.
23. Yixin Liu, Zhao Li, Shirui Pan, Chen Gong, Chuan Zhou, and George Karypis. Anomaly detection on attributed networks via contrastive self-supervised learning. *IEEE Transactions on Neural Networks and Learning Systems*, 33(6):2378–2392, 2021.
24. Rui Wang, Chongwei Liu, Xudong Mou, Kai Gao, Xiaohui Guo, Pin Liu, Tianyu Wo, and Xudong Liu. Deep contrastive one-class time series anomaly detection. In *Proceedings of the 2023 SIAM International Conference on Data Mining (SDM)*, pages 694–702. SIAM, 2023.
25. Huynh Cong Viet Ngu and Keon Myung Lee. Cl-tad: A contrastive-learning-based method for time series anomaly detection. *Applied Sciences*, 13(21), 2023.
26. Ömer Aslan Aslan and Refik Samet. A comprehensive review on malware detection approaches. *IEEE Accesses*, 8:6249–6271, 2020.
27. Ahmad Moawad, Ahmed Ismail Ebada, and Aya M Al-Zoghby. A survey on visualization-based malware detection. *Journal of Cybersecurity (2579-0072)*, 4(3), 2022.
28. Halit Bakır and Rezan Bakır. Droidencoder: Malware detection using auto-encoder based on feature extractors and machine learning algorithms. *Computers and Electrical Engineering*, 110:108804, 2023.
29. Sanjeev Kumar and Anil Kumar. Image-based malware detection based on convolution neural network with autoencoder in industrial internet of things using software defined networking honeypot. *Engineering Applications of Artificial Intelligence*, 133:108374, 2024.
30. Xiaofei Xing, Xiang Jin, Haroon Elahi, Hai Jiang, and Guojun Wang. A malware detection approach using autoencoder in deep learning. *IEEE Access*, 10:25696–25706, 2022.
31. Menaouer Brahami, Abdallah El Hadj Mohamed Islem, and Matta Nada. Android malware detection approach using stacked autoencoder and convolutional neural networks. *International Journal of Information Technology*, 19:1–22, 2023.
32. Jiayin Feng, Limin Shen, Zhen Chen, Yuying Wang, and Hui Li. A two-layer deep learning method for android malware detection using network traffic. *IEEE Access*, 8:125786–125796, 2020.
33. Yueming Wu, Shihan Dou, Deqing Zou, Wei Yang, Weizhong Qiang, and Hai Jin. Contrastive learning for robust android malware familial classification. *IEEE Transactions on Dependable and Secure Computing*, 2021.
34. Kailu Guo, Yang Xin, and Tianxiang Yu. Malware detection using contrastive learning based on multi-feature fusion. *2023 IEEE 22nd International Conference on Trust, Security and Privacy in Computing and Communications (TrustCom)*, pages 1681–1686, 2023.
35. Jonas Höchst, Lars Baumgärtner, Matthias Hollick, and Bernd Freisleben. Unsupervised traffic flow classification using a neural autoencoder. *2017 IEEE 42nd Conference on Local Computer Networks (LCN)*, pages 523–526, 2017.

36. Weijie Zhang, Lanping Zhang, Xixi Zhang, Yu Wang, Peng Liu, and Guan Gui. Intelligent unsupervised network traffic classification method using adversarial training and deep clustering for secure internet of things. *Future Internet*, 15:298, 2023.
37. Jun Cui, Longkun Bai, Guangxu Li, Zhigui Lin, and Penggao Zeng. Semi-2dcae: a semi-supervision 2d-cnn autoencoder model for feature representation and classification of encrypted traffic. *PeerJ Computer Science*, 9, 2023.
38. Mehdi Seydali, Farshad Khunjush, Behzad Akbari, and Javad Dogani. Cbs: A deep learning approach for encrypted traffic classification with mixed spatio-temporal and statistical features. *IEEE Access*, 11:141674–141702, 2023.
39. K. Dinakaran S. Sadhasivam, P. Valarmathie. Malicious activities prediction over online social networking using ensemble model. *Intelligent Automation & Soft Computing*, 36(1):461–479, 2023.
40. Chuanzheng Bai, Huidong Wang, Lu Wang, and Jinli Yao. Rumor detection based on bi-directional graph attention network. In *2021 China Automation Congress (CAC)*, pages 1728–1733, 2021.
41. Imtiaz Ahmed, Travis Galoppo, Xia Hu, and Yu Ding. Graph regularized autoencoder and its application in unsupervised anomaly detection. *IEEE Transactions on Pattern Analysis and Machine Intelligence*, 44(8):4110–4124, 2021.
42. Lingxiao Zhao, Saurabh Sawlani, Arvind Srinivasan, and Leman Akoglu. Graph anomaly detection with unsupervised GNNs. *arXiv preprint arXiv:2210.09535*, 2022.

5 Generative AI for Data Synthesis in Cybersecurity

Generative Artificial Intelligence (Generative AI), referring to generative deep neural networks have gained significant attention and shown great promise in cybersecurity in data generation. Machine Learning/Deep Learning (ML/DL) have proved to be effective method for analyzing network traffic data for cyber-security applications. To build a valuable ML/DL model, training datasets, e.g., network logs, user activity records, and incident reports, are very important. This type of data can provide valuable insights into patterns, anomalies, and risk factors, which are crucial for security analysis, threat detection, and incident response. However, working with real-world security data can present several challenges. The data may contain sensitive or confidential information, making it difficult to share or use for research and development purposes. Additionally, real-world security datasets may be imbalanced, limited in scope, or potentially biased, which can hinder the effectiveness of ML/DL models and algorithms. To address these challenges, researchers have proposed various extensions and modifications to generative deep neuron networks (or generative models) to generate data.

5.1 GENERATIVE AI FOR DATA SYNTHESIS

This section presents various generative models, including Variational AutoEncoder (VAE), Generative Adversarial Network (GAN), Conditional Generative Adversarial Network (CGAN), Adversarial AutoEncoder (AAE), and Table GAN (TGAN).

5.1.1 VARIATIONAL AUTOENCODER

Variational Autoencoder (VAE) [1] is a powerful variant of the AE architecture that is specifically designed for generative modeling. Unlike AE, which aims to learn a compact, deterministic representation of the input data, VAE takes a probabilistic approach to representation learning. The key idea of VAE is to model the data distribution as a latent variable model, where the observed data is generated from a set of unobserved, latent variables. During the training process, VAE learns to approximate the posterior distribution of the latent variables, given the observed data. This learned latent representation is then used to generate new data samples that closely resemble the original data distribution. By sampling from the learned latent distribution, VAE can produce data samples that capture the underlying characteristics of the training data.

A VAE architecture consists of two parts, i.e., encoder and decoder (Fig. 5.1). The difference between a VAE and an AE is that the bottleneck of the VAE is a Gaussian probability density ($q_\phi(\mathbf{z}|\mathbf{x})$). We can sample from this distribution to get noisy

DOI: 10.1201/9781003502531-5

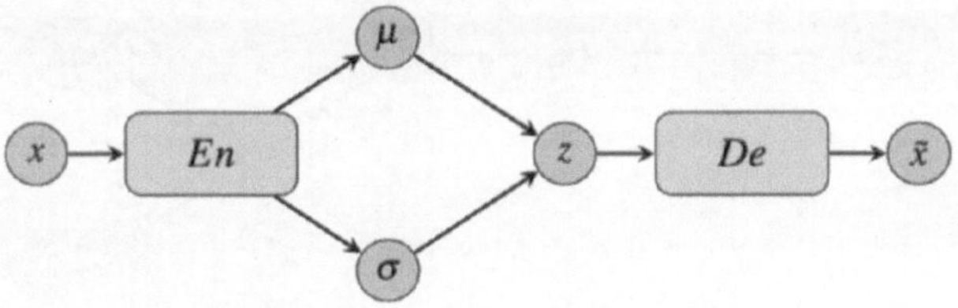

Figure 5.1 Architecture of Variational AutoEncdoer.

values of the representations $\mathbf{z}$. The decoder inputs a latent vector $\mathbf{z}$ and attempts to reconstruct the input. The decoder is denoted by $p_\theta(\mathbf{x}|\mathbf{z})$.

The loss function of a VAE, i.e., $\ell_{VAE}(x^i,\theta,\phi)$, for a data sample x^i includes two terms as follows:

$$\begin{aligned}\ell_{VAE}(x^i,\theta,\phi) = &-\mathbf{E}_{q_\phi(\mathbf{z}|x^i)}\left[log\, p_\theta(x^i|\mathbf{z})\right]\\ &+D_{KL}(q_\phi(\mathbf{z}|x^i)||p(\mathbf{z})).\end{aligned} \tag{5.1}$$

In Eq. 5.1, the first term is the expected negative log-likelihood of the i-th data sample. This term is also called the reconstruction error (RE) of VAE since it forces the decoder to learn to reconstruct the input data. The second term is the Kullback-Leibler (KL) divergence between the encoder's distribution $q_\phi(\mathbf{z}|\mathbf{x})$ and the expected distribution $p(\mathbf{z})$. This divergence measures how close q is to p [1]. In the VAE, $p(\mathbf{z})$ is specified as a standard Gaussian distribution with mean zero and standard deviation one, denoted as $\mathcal{N}(0,1)$. If the encoder outputs representations z that are different from those of a standard Gaussian distribution, it will receive a penalty in the loss. Since the gradient descent algorithm is not suitable to train a VAE with a random variable $\mathbf{z}$ sampled from $p(\mathbf{z})$, the loss function of the VAE is re-parameterized as a deterministic function as follows:

$$\begin{aligned}\ell_{VAE}(x^i,\theta,\phi) = &-\frac{1}{K}\sum_{k=1}^{K}\log p_\theta(x^i|z^{i,k})\\ &+D_{KL}(q_\phi(\mathbf{z}|x^i)||p(\mathbf{z})),\end{aligned} \tag{5.2}$$

where $z^{i,k} = g_\phi(\varepsilon^{i,k},x^i)$. g is a deterministic function. ε^k denotes $\mathcal{N}(0,1)$. K is the number of samples that is used to reparameterize $\mathbf{z}$ for the sample x^i.

To sample data from a training dataset using a VAE, we first need to train the VAE model on the original dataset. During this process, the encoder component of the VAE compresses the raw input data into a lower-dimensional latent space, while the decoder learns to reconstruct new data samples from this latent representation. Once the VAE is trained, we can sample a random vector from the learned latent distribution. This random vector is then fed into the decoder, which generates a new data sample at the output. The use of a VAE to augment training datasets not only enhances the diversity of the training data but also contributes to the effectiveness of various machine learning applications.

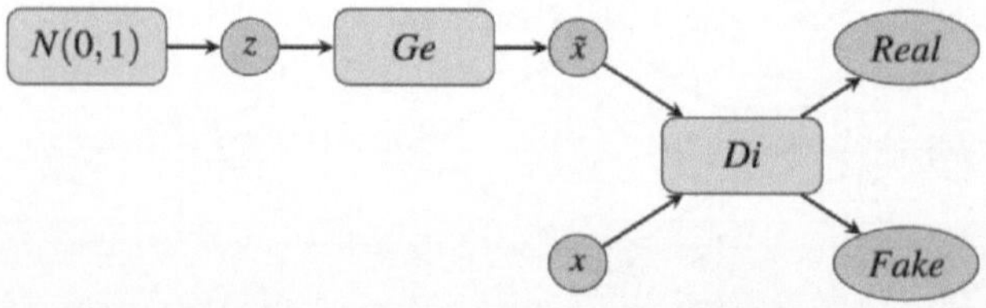

Figure 5.2 GAN architecture.

5.1.2 GENERATIVE ADVERSARIAL NETWORK

Generative Adversarial Networks (GANs) are a type of DL architecture introduced in 2014 by Ian Goodfellow et al. [2]. GAN consists of two main components, i.e., a generator and a discriminator, that are trained simultaneously in an adversarial manner. The generator aims to generate real data that fools the discriminator. The discriminator, on the other hand, is trained to distinguish between real data from the original dataset and synthetic data generated by the generator. This adversarial training process forces the generator to continually improve its outputs to make them more indistinguishable from the real data.

Fig. 5.2 illustrates the architecture of GAN where *Ge* and *Di* are the generator and discriminator, respectively. Based on the outcomes of *Di*, both *Di* and *Ge* fine-tune their respective model parameters in order to improve their performance. If *Di* makes the correct prediction and accurately identifies a generated sample as fake, then *Ge* will update its parameters in an attempt to generate even more realistic-looking fake samples that can better fool *Di* in the future. Conversely, if *Di* incorrectly classifies a generated sample as real, then *Di* will try to learn from this mistake in order to avoid similar errors going forward. The reward function for the discriminator *Di* is the number of correct predictions it makes, while the reward function for the generator *Ge* is the number of mistakes *Di* makes. This adversarial training dynamic continues iteratively until an equilibrium is established and *Di*'s classification accuracy is optimized. Through this iterative process, the generator *Ge* is incentivized to continually enhance the quality of its generated samples, while the discriminator *Di* becomes increasingly skilled at distinguishing real data from fakes. This competition between the generator and discriminator lies at the heart of the GAN training paradigm.

Specifically, the generator *Ge* inputs a noise sample $\mathbf{z}$ and outputs a generated sample $\tilde{\mathbf{x}}$. The discriminator *Di* is trained to maximize the difference between a generated sample $\tilde{\mathbf{x}}$ (outputs from the generator) and a real sample $\mathbf{x}$ (comes from the original data). The generator aims to fool the discriminator *Di* by minimizing the difference between $\tilde{\mathbf{x}}$ and $\mathbf{x}$.

$$L_{GAN} = E_{\mathbf{x}}[\log Di(\mathbf{x})] + E_{\mathbf{z}}[\log(1 - Di(Ge(\mathbf{z})))]. \tag{5.3}$$

The loss function used to train the GAN is presented in Eq. 5.3 in which *Di* is the discriminator used for predicting its input as a real or fake data sample. $Ge(\mathbf{z})$ is the output of *Ge* when given noise $\mathbf{z}$. $E_{\mathbf{x}}$ and $E_{\mathbf{z}}$ are the expected value (average value) of overall real and fake data samples, respectively. *Di* is trained to maximize this loss function, while *Ge* tries to minimize its second term.

After the training, the generator (*Ge*) of GAN can be used to generate synthesized data samples for attack datasets. Beyond just generating data, the GAN has found diverse applications in areas such as image-to-image translation, super-resolution, anomaly detection, semi-supervised learning, and even reinforcement learning. The flexibility and capability of the GAN framework have made it a significant advancement in deep learning. Researchers are actively exploring new GAN architectures and innovative loss functions. They are also developing improved training techniques. This ongoing research aims to expand the potential applications and performance of GANs. The goal is to push the boundaries of what these models can achieve, enhancing data generation and increasing realism in synthetic outputs across various fields, including image synthesis, video generation, and natural language processing.

One of the primary challenges encountered when training the GAN is the issue of instability and mode collapse [3]. It occurs when the generator model learns to produce a limited set of outputs that fail to capture the full diversity of the real data distribution. In other words, the generator starts generating similar or even identical samples, leading to a collapse in the modes of the data distribution. This situation can occur when the discriminator becomes too powerful, making it difficult for the generator to create diverse samples that can successfully trick the discriminator. As a result, the generator gets trapped in generating a narrow range of samples, rather than learning to capture the richness and complexity of the true data. Besides, the GAN highly sensitive to hyperparameter settings. The performance of the GAN is heavily dependent on factors like learning rates, batch sizes, network architectures, and other hyperparameters. Finding the optimal hyperparameter configurations can be a complex and time-consuming task, further compounding the difficulties in training a stable GAN.

5.1.3 CONDITIONAL GENERATIVE ADVERSARIAL NETWORK

A Conditional Generative Adversarial Network (CGAN) is an extension of the traditional GAN that incorporates additional information to guide the sample generation process. While the GAN generates data solely from random noise, the CGAN conditions the generation process on auxiliary input, such as class labels, text descriptions, or images. This conditioning enhances control over the generated output, making the CGAN more versatile and suitable for tasks requiring specific attributes in the generated data.

In a CGAN, the generator takes both a random noise vector and the conditional input to produce realistic samples that align with the given conditions. The discriminator also receives the conditional input along with either a real or generated sample, allowing it to assess authenticity while considering the provided context. By incorporating conditional information, the CGAN improves the quality and diversity of generated samples, making them highly effective in applications such as image synthesis, data augmentation, and style transfer.

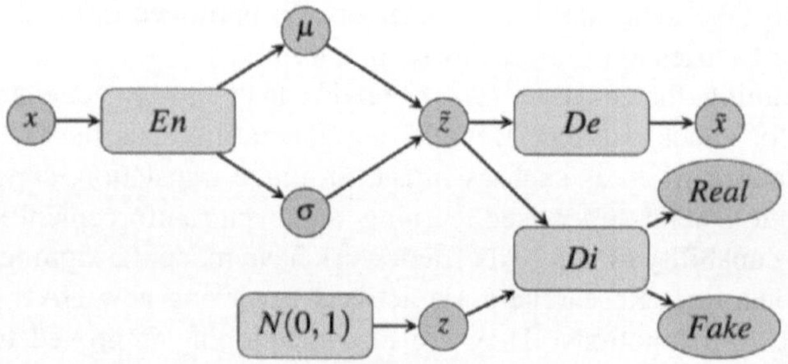

Figure 5.3 Architecture of Adversarial AutoEncoder.

To train a CGAN, the loss function of GAN is modified to integrate the conditional information, as shown in Eq. 5.4. This adjustment ensures that both the generator and discriminator effectively leverage the additional input during the adversarial training process.

$$L_{CGAN} = E_{\mathbf{x}}[\log D(\mathbf{x}|c)] + E_{\mathbf{z}}[\log(1 - D(G(\mathbf{z}|c)))], \tag{5.4}$$

where x is the input variable, c is the conditional variable, z is the noise vector, D and G is the discriminator and generator of the CGAN, respectively.

The CGAN has been widely applied in generating data with specific attributes, particularly in image, audio, and anomaly detection tasks. In image generation and manipulation, the CGAN enables the creation of realistic images based on text descriptions, as well as image-to-image translation tasks like sketch-to-photo conversion, colorization of grayscale images, and transforming aerial imagery into maps. In the audio and music domain, the CGAN can generate music or audio samples conditioned on metadata, lyrics, or other contextual cues. Additionally, the CGAN is valuable for anomaly detection and synthesis, as it can generate synthetic samples of rare or anomalous events, aiding in training more robust detection models.

5.1.4 ADVERSARIAL AUTOENCODER

Adversarial AutoEncoder (AAE) is a powerful variant of GAN that combines the advantages of AE and GAN. The AAE avoids using the KL divergence of the VAE to impose the prior by using adversarial learning. This allows the latent space, $p(\mathbf{z})$, to be learned from any distribution [4]. As illustrated in Fig. 5.3, training the AAE has two phases, i.e., reconstruction and regularization. In the reconstruction phase (RP), the latent sample $\tilde{\mathbf{z}}$ is drawn from the generator Ge. The sample $\tilde{\mathbf{z}}$ is then sent to the decoder (denoted by $p(\mathbf{x}|\tilde{\mathbf{z}})$) which generates $\tilde{\mathbf{x}}$ from $\tilde{\mathbf{z}}$. The recontruction error is computed by the difference between $\mathbf{x}$ and $\tilde{\mathbf{x}}$ as in Eq. 5.5.

$$L_{RP} = -\,\mathbb{E}_{\mathbf{x}}\left[\log p(\mathbf{x}|\tilde{\mathbf{z}})\right]. \tag{5.5}$$

In the regularization phase (RG), the discriminator Di receives $\tilde{\mathbf{z}}$ from the generator Ge and $\mathbf{z}$ is sampled from the true prior $p(\mathbf{z})$. The generator tries to generate the fake sample, i.e., $\tilde{\mathbf{z}}$, as similar as the real sample, i.e., $\mathbf{z}$, by minimizing the second term in Eq. 5.6. The discriminator then attempts to distinguish between $\tilde{\mathbf{z}}$ and $\mathbf{z}$ by maximizing this equation. The generator is also the encoder portion of the AE. Therefore, the training process in the regularization phase is the same as that of GAN.

$$L_{RG} = \mathbb{E}_{\mathbf{z}}\left[\log Di(\mathbf{z})\right] + \mathbb{E}_{\mathbf{x}}\left[\log(1 - Di(En(\mathbf{x})))\right]. \tag{5.6}$$

An extension of AAE is Supervised Adversarial AutoEncoder (SAAE) [4] where the label information is concatenated with the latent representation ($\mathbf{z}$) to form the input of Di. The class label information allows the SAAE to generate the data samples for a specific class. Another version of AAE is Denosing AAE (DAAE) [5] that attempts to match the intermediate conditional probability distribution $q(\tilde{\mathbf{z}}|\mathbf{x}_{\mathbf{noise}})$ with the prior $p(\mathbf{z})$ where $\mathbf{x}_{\mathbf{noise}}$ is the corrupted version of the input $\mathbf{x}$ by adding Gaussian noise vector to the original x 5.6. Because the DAAE reconstructs the original input data from the corrupted version of input data (input data with noise), its latent representation is often more robust than the representation in the AAE [6].

One key application of the AAE is in the realm of unsupervised representation learning. By training the AE to understand the main structure of the data and then adjusting this understanding to match a specific target distribution, the AAE can create valuable representations. Moreover, the adversarial process in training the AAE allows the AAE to generate high-quality samples that are indistinguishable from the original data distribution. The AAE has been used to generate diverse and realistic-looking samples, often with the ability to control specific attributes of the generated content through the use of conditional or disentangled representations.

5.1.5 CONDITIONAL TABLE GENERATIVE ADVERSARIAL NETWORK

Generating high-quality tabular data using GAN is a challenging task due to the unique characteristics of tabular data. Unlike images or text, tabular data consists of structured columns and rows, where each column represents a specific feature or attribute, and each row represents an individual data point. This structured nature of tabular data poses several challenges for the GAN. Conditional Generative Adversarial Network (CTGAN) [7] is an extension of GAN, designed specifically to address the challenges of modeling tabular data with a mix of continuous and discrete features. Like CGAN, CTGAN incorporates a conditional generator that learns to generate samples conditioned on the input data distribution. The CTGAN model is trained in an adversarial manner, where the generator and the discriminator network compete against each other. The generator tries to produce samples that are indistinguishable from the real data, while the discriminator aims to classify the generated samples as fake. The CTGAN has been successfully applied to a wide range of applications, including security-related tasks such as network traffic analysis, intrusion detection, and fraud detection. The generated synthetic data can be used to augment the original datasets to train machine learning models, and thus, improving model performance for these tasks.

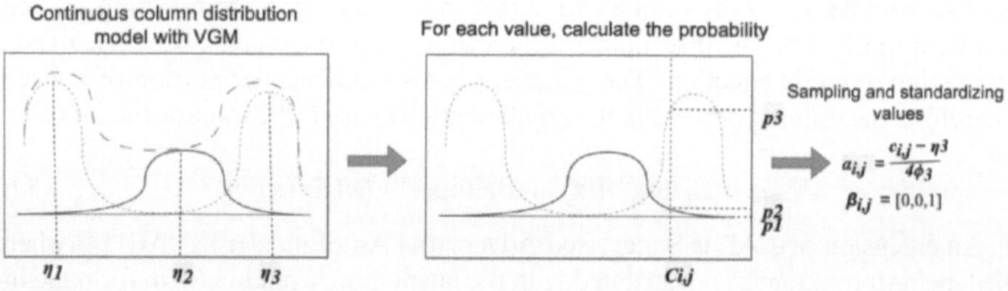

Figure 5.4 Normalization for continuous value vector.

To fully understand the CTGAN model, it is essential to first grasp the mode-specific normalization technique, which addresses the challenges posed by non-Gaussian and multi-modal distributions commonly found in tabular data. Unlike standard normalization techniques that assume a single-mode Gaussian distribution, mode-specific normalization identifies different distribution modes and applies tailored transformations to each. This approach significantly enhances the ability of GAN-based models to generate high-quality synthetic data by preserving complex feature distributions.

In contrast to image data, where pixel values generally follow a Gaussian-like distribution, continuous features in tabular data often exhibit multimodal distributions with multiple local maxima. To accurately model these characteristics, the CTGAN employs mode-specific normalization using a Variational Gaussian Mixture (VGM) model. Each continuous feature value is represented by a one-hot vector indicating its sampled mode and a scalar normalized according to that mode. This method enables CTGAN to capture the diverse statistical properties of tabular data, improving the realism and fidelity of the generated synthetic samples. Tabular data consists of both numerical and categorical attributes, requiring effective encoding methods to properly represent each type for CTGAN.

Numeric attributes

For numeric attributes in CTGAN, mode-specific normalization is employed to handle columns with complex distributions effectively. Each column is processed independently, with values represented using a combination of a one-hot vector indicating the selected mode and a scalar value normalized within that mode. This process consists of three main steps:

- Mode Estimation: For each numeric attribute C_i, a Variational Gaussian Mixture (VGM) model is used to estimate its modes and fit a Gaussian mixture distribution. As illustrated in Fig. 5.4, VGM finds three modes ($i = 3$) with the learned Gaussian mixture models as η_1, η_2, η_3:

$$P_{C_j}(C_{i,j}) = \sum_{k=1}^{3} \mu_k N(C_{i,j}, \eta_k, \phi_k), \tag{5.7}$$

where μ_k, ϕ_k are the weight and standard deviation of the mode $k-th$, respectively.

- Probability Computation: For each feature value in $c_{i,j}$ in C_i, we compute the probability of feature values coming from each model. For example, in Fig. 5.4, the probability densities are ρ_1, ρ_2, ρ_3 which are computed as follows:

$$\rho_k = \mu_k N(C_{i,j}, \eta_k, \phi_k). \tag{5.8}$$

- Mode Selection and Normalization: A Gaussian mode is sampled based on the computed probability densities, and the feature value is normalized accordingly. For instance, in Fig. 5.4, the third mode is selected, resulting in the representation of $C_{i,j}$ as a one-hot vector $\beta_{i,j} = [0,0,1]$

This mode-specific normalization technique ensures that continuous features with multi-modal distributions are effectively captured, improving the quality of synthetic data generation in CTGAN.

Categorical attributes

Traditionally, the generator in a GAN is initialized with a vector sampled from a Multivariate Normal Distribution (MVN). Through adversarial training with the discriminator, the generator learns to map this MVN to the distribution of real data. However, this approach does not address the imbalance problem in categorical columns. When training data is randomly sampled, smaller categories may be underrepresented, preventing the generator from learning their distribution effectively. Consequently, the generator ends up modeling a sampled data distribution that deviates from the actual data distribution. This issue resembles the "class imbalance" problem in classification models, but with an added complexity: instead of balancing a single categorical column, the model must preserve the overall distribution across multiple attributes.

CTGAN introduces a conditional generator to efficiently resample categorical or discrete attributes. Assuming that k^* is the value from the discrete column i^*, D_{i^*} needs to be fitted by the generated samples r. The generator can be interpreted as the conditional distribution of rows for a particular value at a particular column, i.e., $r \sim P_G(row|D_{i^*=k^*})$. This approach ensures that the generator learns to produce realistic data while preserving the relationships between categorical attributes.

In Fig. 5.5, a conditional vector presents the condition $D_{i^*} = k^*$. Note that, all discrete columns $D_1, D_2, \ldots, D_{N_d}$ are the one-hot vectors $d_1, d_2, \ldots, d_{N_d}$ that satisfy the one-hot vector $d_i = [d_{(k)}]$ with $k = 1, \ldots, |D_i|$. We set the mask vector of the one-hot vector d_i as $m_i = [m_i^{(k)}]$. Thus, the conditional vector can be represented as following:

$$m_i^{(k)} = \begin{cases} 1, & \text{if } i = i^* \text{ and } k = k^*. \\ 0, & \text{otherwise.} \end{cases} \tag{5.9}$$

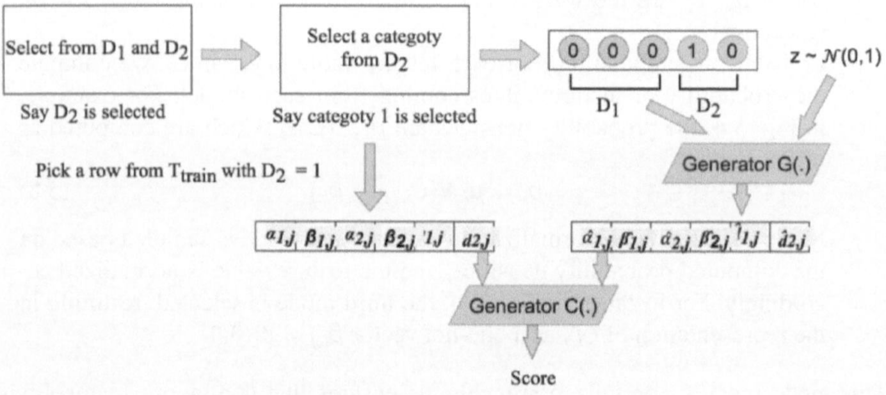

Figure 5.5 Representation of discrete columns.

Then, the conditional vector is defined as following:

$$cond = m_1 \oplus m_2 \oplus \ldots \oplus m_{N_d} \tag{5.10}$$

For example, in Fig. 5.5, the mask vectors of two columns $D_1 = 1,2,3$ and $D_2 = 1,2$ with the condition $D_2 = 1$ are $m_1 = [0,0,0]$ and $m_2 = [1,0]$, respectively, and thus, the conditional vector is $cond = [0,0,0,1,0]$.

Finally, we can represent a row in the tabular data is a combination of continuous and discrete features as following:

$$r_j = \alpha_{1,j} \oplus \beta_{1,j} \oplus \ldots \oplus \alpha_{N_c,j} \oplus \beta_{N_c,j} \oplus d_{1,j} \oplus \ldots \oplus d_{N_d,j}, \tag{5.11}$$

where $d_{i,j}$ is the one-hot representation of the discrete feature/column. This structured representation ensures that both numerical and categorical features are effectively encoded for model training.

After encoding both continuous and discrete attributes as described, the structure and training process of the CTGAN closely resemble those of CGAN. In the CTGAN, the generator takes as input a random noise vector along with a conditional vector that specifies the target category for discrete attributes. It then generates synthetic tabular data that aligns with the given conditions. Similarly, the discriminator receives both real and generated samples along with their corresponding conditional vectors, learning to distinguish between real and synthetic data while ensuring the generated samples adhere to the given conditions. By incorporating the mode-specific normalization technique for continuous attributes and conditional sampling for categorical features, the CTGAN effectively handles the complexities of tabular data. The adversarial training process follows the same min-max optimization as the CGAN, where the generator continuously improves to produce realistic data while the discriminator refines its ability to detect synthetic samples.

5.2 DATA GENERATION IN CYBERSECURITY

5.2.1 MALWARE DETECTION

Malware, or malicious software, is designed to infiltrate, damage, or exploit systems without user consent. It includes viruses, worms, Trojans, ransomware, and spyware, all of which threaten data security and system functionality. Detecting malware is essential to prevent data breaches, financial losses, and operational disruptions. Machine learning-based methods are proven to be one of the effective techniques for malware detection. However, machine learning-based malware detection faces challenges due to the difficulty of collecting diverse and representative malware samples. The continuous evolution of malware results in imbalanced or insufficient datasets, limiting model effectiveness. Generative techniques, e.g., GAN, VAE, help to address this issue by synthesizing realistic malware samples, enriching training data, and improving detection accuracy. By generating high-quality synthetic malware, these methods enhance the ability of machine learning and deep learning models to recognize malicious patterns, strengthening cybersecurity defenses.

In cybersecurity, generative models have emerged as a powerful tool for enhancing malware detection by synthesizing realistic malware samples [8]. One effective approach involves transforming malware binaries or hexadecimal data into 2D image representations, making structural patterns and behavioral features more distinguishable. This visual representation not only improves the ability to analyze and classify malware but also reveals intricate relationships within the data that are difficult to discern from raw binary formats. By leveraging image-based malware representations, GAN can generate synthetic malware samples that closely resemble real-world threats, thus enriching training datasets and improving the performance of detection systems. The study in [8] introduced a specialized GAN architecture called "MIGAN," designed for malware image synthesis. The MIGAN enhances dataset augmentation by producing high-quality synthetic malware samples, enabling classifiers to learn distinguishing characteristics of different malware families more effectively. As illustrated in Fig. 5.6, the MIGAN consists of a generator and a discriminator, similar to a Conditional GAN (CGAN), where label information is incorporated as a conditional vector. This conditional generation mechanism allows MIGAN to produce malware samples with specific attributes, making it highly valuable for improving machine learning-based malware detection and strengthening cybersecurity defenses.

Furthermore, representing malware as images enables the use of pre-trained computer vision models to enhance malware classification accuracy. By fine-tuning deep learning models originally trained for natural image recognition, we can leverage their advanced feature extraction capabilities and adapt them for malware detection [9]. This transfer learning approach allows malware classifiers to benefit from the robust pattern recognition and representation learning techniques developed for general image processing tasks. As a result, integrating pre-trained vision models with image-based malware representations significantly improves detection efficiency and accuracy, strengthening cybersecurity defenses against evolving threats.

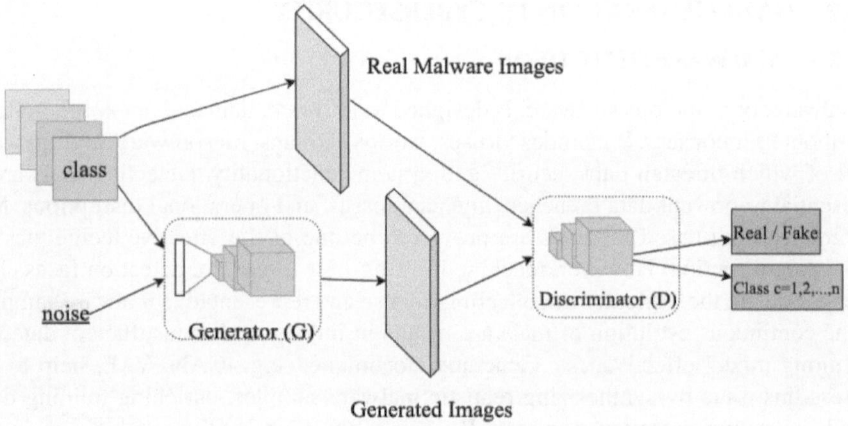

Figure 5.6 Synthetic Malware Image generation with MIGAN.

The GAN can generate malware samples in various forms, such as executable files, scripts, or network traffic patterns, allowing for a broader exploration of malware behaviors beyond image-based representations. One such approach is the Boundary Seeking GAN (BGAN), proposed by Moti et al. [10], which optimizes the generator to create malware samples positioned near the decision boundary of a malware classifier. By generating samples that closely resemble legitimate software while retaining malicious functionality, BGAN effectively fools detection models, making it a powerful tool for studying adversarial attacks against malware classifiers. Similarly, Peng et al. [11] developed a specialized GAN architecture that could generate adversarial samples tailored to evade their proposed malware detection system. GAN was trained to produce malware variants that maximized the classifier's prediction error, leading to an improvement in the overall evasion rate. This approach not only demonstrated the vulnerabilities in traditional malware detection systems but also highlighted the need for more robust and adaptive defense mechanisms against adversarial malware attacks.

5.2.2 ANOMALY AND INTRUSION DETECTION

An Intrusion Detection System (IDS) is a security mechanism designed to monitor network traffic and system activities for suspicious behavior, unauthorized access, or potential cyber threats, helping to detect and respond to security breaches in real time. The GAN has also been applied to generate synthetic data for improving intrusion detection systems (IDSs). This synthetic data is then used to enrich the training dataset, enabling more accurate identification of malicious activities. The work in [12] used the GAN model to generate adversarial samples that are used as validation sets, ensuring classifiers perform exceptionally well against even the most sophisticated adversarial attacks. This framework can effectively detect many types of intrusions that target Internet of Thing (IoT) environments, making it a strong solution for protecting these vulnerable connected devices. Besides that, Aloqaily et al. [13] proposed the GAN-based framework called as IDGAN that generates synthetic network traffic data, including both normal and intrusive patterns. The synthetic data can be used to train machine learning models for IDS, making them more effective at detecting a wider range of cyber attacks.

Different variants of GAN, such as Vanilla GAN, Wasserstein GAN, and CT-GAN, have been utilized to synthesize network traffic datasets [14]. These synthetic datasets enhance the training of intrusion detection systems (IDSs), improving their capability to identify cyber threats. Specifically, CTGAN has been employed to strengthen IDSs in detecting Distributed Denial of Service (DDoS) attacks in IoT networks [15]. In this approach, the generator produces synthetic network traffic data that closely mimics the statistical properties of legitimate traffic patterns, while the discriminator learns to differentiate between generated legitimate traffic and actual malicious traffic. By leveraging conditional generation and adversarial training, the CTGAN-based IDS effectively recognizes patterns in both normal and harmful IoT network traffic, thereby improving the detection accuracy of DDoS and DoS attacks and enhancing IoT security.

Beyond CTGAN, the Attention-GAN framework [16] has been introduced to improve IDSs by integrating attention mechanisms with GAN models. The attention component enhances the model's ability to focus on the most relevant features, making it particularly effective in detecting subtle and complex attack patterns. This targeted focus allows the IDS to better distinguish between benign and malicious traffic, ultimately increasing detection precision.

Another GAN-based model, AnoGAN [17], has been developed to detect anomalies by learning the normal data distribution and identifying deviations that indicate potential threats. Similarly, Ming et al. [18] proposed a method combining the Mahalanobis distance metric with an autoencoder (AE) to detect anomalous network traffic. This approach operates in an unsupervised manner, eliminating the need for labeled data. It employs an enhanced inverse Mahalanobis distance with a threshold to distinguish normal data from anomalies effectively.

A key innovation in Ming et al.'s work is the integration of AE with GAN, where the AE's output is fed into the GAN's discriminator for adversarial training. The loss function incorporates outputs from both the AE and the discriminator, enhancing the AE's feature extraction capabilities. This combined approach improves the detection of anomalous network traffic patterns, making it more effective in identifying cyber threats, even in the absence of labeled training data.

Using generative models to generate synthetic data can significantly enhance the accuracy of IDSs. By creating realistic network traffic samples, generative models help to address data scarcity and class imbalance issues, ensuring IDSs are trained on diverse attack scenarios. These models also enable IDSs to learn complex attack patterns, improving their ability to detect both known and emerging cyber threats. Additionally, adversarial training with generative models strengthens IDS robustness by exposing them to evasive attack samples, making them more resilient to sophisticated cyberattacks. Overall, incorporating generative models into IDSs leads to more effective and adaptive cybersecurity solutions.

5.2.3 NETWORK TRAFFIC CLASSIFICATION

Network traffic classification is an important task in network security and management, allowing for the identification of different types of network traffic such as web browsing, video streaming, file sharing, and more. Traditionally, network traffic classification has relied on approaches, e.g., deep packet inspection, which analyzes the contents of network packets to infer applications or protocols. However, these techniques can be computationally expensive and have difficulty keeping up with evolving network traffic patterns. Using generative models can produce synthetic network traffic samples that are statistically indistinguishable from real traffic, leading to improving accuracy of network traffic classification.

Data augmentation techniques are widely used to address the class imbalance problem in network traffic classification. One such approach is the GAN-based Traffic Augmentation (TA-GAN) framework [19], which integrates the generation of minority traffic samples with a feedback mechanism. This feedback mechanism helps guide the sample generation process while also assessing the quality of the synthesized traffic data, ensuring more effective augmentation. Another notable method is PacketCGAN [20], which employs a Conditional GAN (CGAN) to generate specific types of traffic. By leveraging CGAN's ability to conditionally generate samples based on class labels, PacketCGAN enhances the diversity and balance of network traffic datasets. Additionally, the FlowGAN [21] was introduced to tackle class imbalance in traffic classification by utilizing GAN's powerful data augmentation capabilities. The FlowGAN generates synthetic traffic data for underrepresented classes, improving model performance in recognizing minority traffic patterns. Furthermore, GAN-based data augmentation techniques have also been applied in encrypted traffic classification, as demonstrated in [22], further highlighting their versatility in enhancing network security applications.

Zhang et al. [23] developed a network called CAAE that combines a Convolutional Autoencoder (CAE) with adversarial training. As illustrated in Fig. 5.3, in the CAAE, an encoder maps input data x to a latent representation z, a generator reconstructs x' from z, and a discriminator that distinguishes real from fake samples. Unlike traditional GANs, where real and fake data are typically raw images, CAAE operates in the latent space, where the real latent vector z is obtained from the encoder, and the fake latent vector is sampled from a prior probability distribution. The discriminator then evaluates whether a given latent vector is real or fake based on its learned representation. This adversarial training process refines the generator, allowing it to produce more realistic latent representations, ultimately improving the quality of reconstructed data.

Generative models have proven to be effective in enhancing traffic classification by addressing class imbalance and improving dataset diversity. By generating high-quality synthetic traffic samples, these models help machine learning algorithms better recognize and classify minority traffic patterns. This leads to more robust and

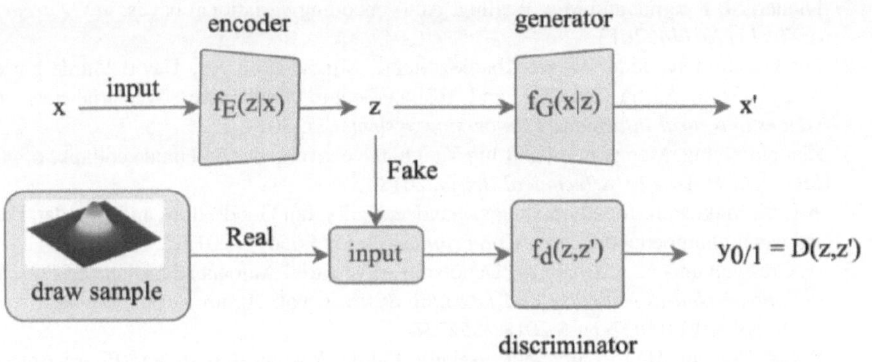

Figure 5.7 The framework of CAAE model for anomaly detection.

accurate traffic classification systems, ultimately strengthening network security and performance.

In conclusion, synthesizing data samples using generative models can significantly enhance security applications by providing high-quality, realistic data that can be used for building various security systems. These models can generate synthetic data that mimics the characteristics of real-world data, including benign and malicious data samples. This is particularly valuable in scenarios where acquiring labeled data is difficult, expensive, or time-consuming, such as in cybersecurity. By augmenting existing datasets with generated data samples, security applications can improve their ability to detect anomalies, identify intrusions, and respond to threats more effectively. Furthermore, synthetic data can help in simulating rare attack scenarios, allowing security systems to be better prepared for potential vulnerabilities. Overall, the use of generative models for data synthesis not only bolsters the robustness of security measures but also contributes to more comprehensive and adaptive defenses against evolving threats.

5.3 SUMMARY

This chapter focuses on the application of Generative AI techniques in cybersecurity, with an emphasis on synthetic data generation to enhance detection and classification systems. It introduces a variety of powerful generative models, including Variational Autoencoder (VAE), Generative Adversarial Network (GAN), Conditional GAN (CGAN), Adversarial Autoencoder (AAE), and Conditional Tabular GAN (CTGAN). These models are explored in the context of generating high-quality, realistic synthetic data to address challenges such as data scarcity and class imbalance. The chapter demonstrates how generative models are applied across key cybersecurity domains, including malware detection, anomaly/intrusion detection, and network traffic classification. By leveraging generative AI for data augmentation, these applications benefit from more robust and accurate machine learning models, ultimately improving the resilience of cybersecurity systems.

1. Diederik P Kingma and Max Welling. Auto-encoding variational bayes. *arXiv preprint arXiv:1312.6114*, 2013.
2. Ian Goodfellow, Jean Pouget-Abadie, Mehdi Mirza, Bing Xu, David Warde-Farley, Sherjil Ozair, Aaron Courville, and Yoshua Bengio. Generative adversarial nets. *Advances in Neural Information Processing Systems*, 27, 2014.
3. Zhaoyu Zhang, Mengyan Li, and Jun Yu. On the convergence and mode collapse of gan. *SIGGRAPH Asia 2018 Technical Briefs*, 2018.
4. Alireza Makhzani, Jonathon Shlens, Navdeep Jaitly, Ian Goodfellow, and Brendan Frey. Adversarial autoencoders. *arXiv preprint arXiv:1511.05644*, 2015.
5. A. Creswell and A. A. Bharath, "Denoising Adversarial Autoencoders," in *IEEE Transactions on Neural Networks and Learning Systems*, vol. 30, no. 4, pp. 968–984, April 2019, doi: 10.1109/TNNLS.2018.2852738.
6. Pascal Vincent, Hugo Larochelle, Isabelle Lajoie, Yoshua Bengio, and Pierre-Antoine Manzagol. Stacked denoising autoencoders: Learning useful representations in a deep network with a local denoising criterion. *Journal of Machine Learning Research*, 11(Dec):3371–3408, 2010.
7. Lei Xu, Maria Skoularidou, Alfredo Cuesta-Infante, and Kalyan Veeramachaneni. Modeling tabular data using conditional gan. *Advances in neural information processing systems*, 32, 2019.
8. Osho Sharma, Akashdeep Sharma, and Arvind Kalia. Migan: Gan for facilitating malware image synthesis with improved malware classification on novel dataset. *Expert Systems with Applications*, 241:122678, 2024.
9. Yusheng Dai, Hui Li, Yekui Qian, and Xidong Lu. A malware classification method based on memory dump grayscale image. *Digital Investigation*, 27:30–37, 2018.
10. Zahra Moti, Sattar Hashemi, Hadis Karimipour, Ali Dehghantanha, Amir Namavar Jahromi, Lida Abdi, and Fatemeh Alavi. Generative adversarial network to detect unseen internet of things malware. *Ad Hoc Networks*, 122:102591, 2021.
11. Xiaowei Peng, Hequn Xian, Qian Lu, and Xiuqing Lu. Semantics aware adversarial malware examples generation for black-box attacks. *Applied Soft Computing*, 109:107506, 2021.
12. Mohamed Amine Ferrag, Djallel Hamouda, Merouane Debbah, Leandros Maglaras, and Abderrahmane Lakas. Generative adversarial networks-driven cyber threat intelligence detection framework for securing internet of things. In *2023 19th International Conference on Distributed Computing in Smart Systems and the Internet of Things (DCOSS-IoT)*, pages 196–200, 2023.
13. Decentralized and resource-efficient self-calibration of visual sensor networks. *Ad Hoc Networks*, 88:112–128, 2019.
14. Xinxing Zhao, Kar Wai Fok, and Vrizlynn LL Thing. Enhancing network intrusion detection performance using generative adversarial networks. *arXiv preprint arXiv:2404.07464*, 2024.
15. Basim Ahmad Alabsi, Mohammed Anbar, and Shaza Dawood Ahmed Rihan. Conditional tabular generative adversarial based intrusion detection system for detecting ddos and dos attacks on the internet of things networks. *Sensors*, 23(12):5644, 2023.
16. Mohammed Abo Sen. Attention-gan for anomaly detection: A cutting-edge approach to cybersecurity threat management. *arXiv preprint arXiv:2402.15945*, 2024.
17. Thomas Schlegl, Philipp Seeböck, Sebastian M Waldstein, Ursula Schmidt-Erfurth, and Georg Langs. Unsupervised anomaly detection with generative adversarial networks to guide marker discovery. In *International Conference On Information Processing in Medical Imaging*, pages 146–157. Springer, 2017.

18. Li Ming, Han Dezhi, and Li Dun. A method combining improved mahalanobis distance and adversarial autoencoder to detect abnormal network traffic. In *Proceedings of the 27th International Database Engineered Applications Symposium*, IDEAS '23, page 161–169, New York, NY, USA, 2023. Association for Computing Machinery.
19. Yu Guo, Gang Xiong, Zhen Li, Junzheng Shi, Mingxin Cui, and Gaopeng Gou. Ta-gan: Gan based traffic augmentation for imbalanced network traffic classification. In *2021 International Joint Conference on Neural Networks (IJCNN)*, pages 1–8, 2021.
20. Pan Wang, Shuhang Li, Feng Ye, Zixuan Wang, and Moxuan Zhang. Packetcgan: Exploratory study of class imbalance for encrypted traffic classification using cgan. In *ICC 2020-2020 IEEE International Conference on Communications (ICC)*, pages 1–7. IEEE, 2020.
21. ZiXuan Wang, Pan Wang, Xiaokang Zhou, ShuHang Li, and MoXuan Zhang. Flowgan: unbalanced network encrypted traffic identification method based on gan. In *2019 IEEE Intl Conf on Parallel & Distributed Processing with Applications, Big Data & Cloud Computing, Sustainable Computing & Communications, Social Computing & Networking (ISPA/BDCloud/SocialCom/SustainCom)*, pages 975–983, 2019.
22. Mehdi Seydali, Farshad Khunjush, Behzad Akbari, and Javad Dogani. Cbs: A deep learning approach for encrypted traffic classification with mixed spatio-temporal and statistical features. *IEEE Access*, 11:141674–141702, 2023.
23. Weijie Zhang, Lanping Zhang, Xixi Zhang, Yu Wang, Peng Liu, and Guan Gui. Intelligent unsupervised network traffic classification method using adversarial training and deep clustering for secure internet of things. *Future Internet*, 15:298, 2023.

6 Generative AI for Sequence Learning in Cybersecurity

Generative models have received significant attention and demonstrated considerable promise across various domains, particularly in cybersecurity. In the field of sequence learning, which focuses on modeling and generating sequential data, these models have proven effective in addressing critical challenges in cybersecurity. Notable applications include phishing detection, spam filtering, and network data analysis, where generative models have shown their potential to enhance security measures and improve threat detection. This chapter presents and analyzes various generative models for sequence learning and their applications to important problems in cybersecurity.

6.1 AUTOREGRESSIVE MODELS FOR SEQUENCE LEARNING

Autoregressive models are a class of machine learning (ML) models that automatically predict the next component in a sequence by taking measurements from previous inputs in the sequence. Autoregression is a statistical technique that is often used in time-series analysis based on the assumption that the current value of a time series is a function of its past values. Autoregressive models use similar mathematical techniques to determine the probabilistic correlation between elements in a sequence. They then use the knowledge derived to predict the next element in an unknown sequence.

Autoregressive models are an important component of many advanced generative models for sequence learning. In this section, we will discuss some well-known generative models for sequence learning. Specifically, this chapter will investigate the variants of Recurrent Neural Networks, Encoder-Decoder Architecture, Attention mechanism, and Transformer models.

6.1.1 RECURRENT NEURAL NETWORK VARIANTS

A recurrent neural network (RNN) is a deep learning model that is trained to process and convert a sequential data input into a specific sequential output. Sequential data, such as words, sentences, or time-series, is a special data type where sequential components interrelate based on complex semantics and syntax rules. The fundamental principle of recurrent neural networks has been analyzed in Chapter 1. In this section, we will investigate some variants of RNN including, LSTM, Bi-LSTM and GRU.

Long Short Term Memory networks: Long Short Term Memory networks(LSTMs) are a special kind of RNNs that are specially designed to address the long-term dependency problem in RNN. LSTMs were introduced by Hochreiter

DOI: 10.1201/9781003502531-6

and Schmidhuber in 1997 [1]. After that, they were improved and widely applied to many problems, and they are proven to perform tremendously well on a large variety of problems.

LSTMs also have a chain structure similar to RNNs. However, the repeating module has four neural network layers as in Figure 6.1 instead of a simple layer in RNNs. Moreover, four networks in LSTMs interact in a very special way to learn long-term dependencies.

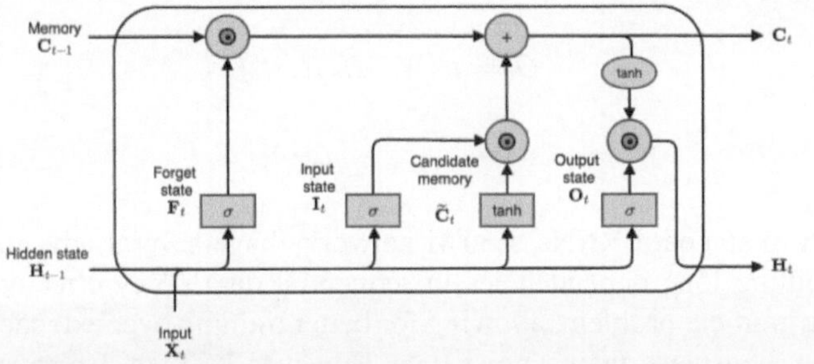

Figure 6.1 Repeating module in LSTMs

The core component in LSTMs is the cell state. The cell state attempts to convey the information from the current step to the next step. Other components in LSTMs are the gates. The gates are used to filter the information that is passed to the next step. The gates are implemented using a sigmoid layer and a pointwise multiplication operation. The sigmoid layer decides how much information is passed through. Each LSTM block has three gates to regulate information that passes through the cell state.

The flow of processing information in LSTMs includes three steps:

- First, the forget gate is used to eliminate the information from the cell state using a sigmoid layer. It receives the hidden state of the previous step H_{t-1} and the current input x_t and outputs a number in range [0, 1]. If the output of the sigmoid layer is 1, all the information is passed to the next step. Conversely, if the output is 0, all information is eliminated. Formally, the output of the sigmoid layer f_t is calculated as follows:

$$F_t = \sigma(W_f.[H_{t-1}, x_t]) \tag{6.1}$$

- Second, the input gate is used to store in the cell state. A sigmoid layer I_t in the input gate decides which values will be updated. Next, a vector of new candidate C'_t values is created through a tanh layer.

$$I_t = \sigma(W_i.[H_{t-1}, x_t]) \tag{6.2}$$

$$\tilde{C}_t = tanh(W_c.[H_{t-1}, x_t]) \tag{6.3}$$

- Third, the new cell state is updated from the old cell state. This is implemented by multiplying the old state C_{t-1} by F_t to forget the less relevant information from the previous step. Then we add it to the new updated information $\tilde{C}_t$ by scaling I_t with $\tilde{C}_t$.

$$C_t = F_t * C_{t-1} + I_t * C_t \tag{6.4}$$

- Last, the output is calculated based on the previous hidden state H_{t-1}, the input x_t, and the new hidden state is updated as follows:

$$O_t = \sigma(W_o.[H_{t-1}, x_t]) \tag{6.5}$$

$$H_t = O_t * tanh(C_t) \tag{6.6}$$

Compared to standard RNNs, LSTM networks have several advantages, particularly in handling long dependencies in sequential data. They effectively mitigate the vanishing gradient problem, allowing for better training over extended sequences and providing robustness against noisy data. However, LSTMs also have drawbacks, including their complexity and resource-intensive nature, which can lead to longer training times and higher computational demands. In addition, their ability to learn intricate patterns makes them prone to overfitting, especially with limited data. Furthermore, LSTMs can be difficult to interpret, complicating the understanding of their decision-making process, and they often require careful tuning of hyperparameters to achieve optimal performance.

Gated Recurrent Unit (GRU): (GRU) improves LSTM models by combining the forget and input gates into a unified gate. It also combines the cell state with the hidden state, thus the resulting model is simpler than the standard LSTM models while maintaining comparable performance. The structure of the GRUs block is presented in Figure 6.2, and the flow of processing information is as follows:

$$\begin{cases} Z_t = \sigma(W_o.[H_{t-1}, X_t]) \\ R_t = \sigma(W_r.[H_{t-1}, X_t]) \\ \tilde{H}_t = tanh(W.[R_t * H_{t-1}, X_t]) \\ O_t = \sigma(W_o.[H_{t-1}, X_t]) \\ H_t = (1 - Z_t) * H_{t-1} + Z_t * \tilde{H}_t \end{cases} \tag{6.7}$$

This simple design enables GRUs to efficiently process information over time. Their ability to selectively retain or discard information helps mitigate issues like the vanishing gradient problem, allowing for effective training on long sequences. Additionally, GRUs typically require fewer parameters than LSTMs, which can lead to faster training times and reduced computational requirements. However, like other recurrent architectures, they can still be prone to overfitting and may require careful hyperparameter tuning for optimal performance.

Bidirectional LSTM: BiLSTM [3] is an extension of traditional LSTMs to process the sequence information in both directions, backward (future to past) or forward (past to future). By processing information in both directions, the models are

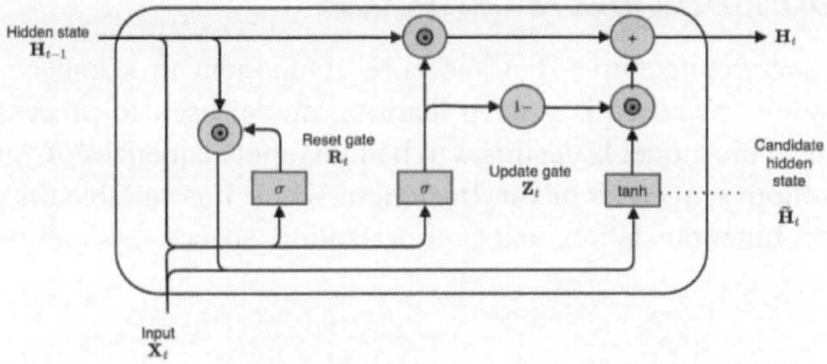

Figure 6.2 Repeating module in GRUs

able to better understand the relationship between sequences since they learn the relationship of both following and preceding elements in sequence.

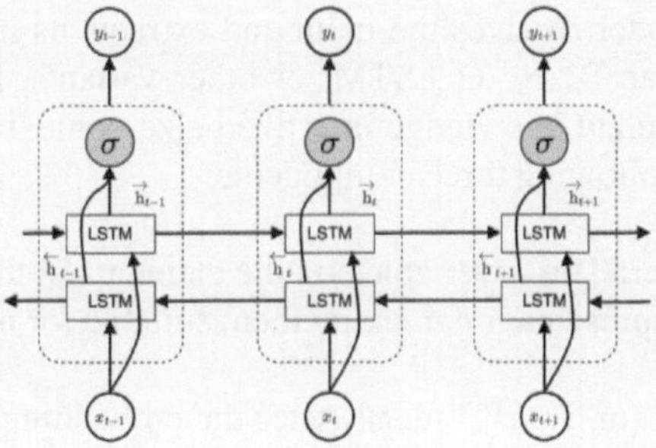

Figure 6.3 Bi-LSMT architecture

The architecture of BiLSTMs consists of two unidirectional LSTMs: one processes the sequence in forward and the other processes in backward directions. For a given input, both LSTM networks return a probability vector at the output, and the final output is the combination of the two probabilities. Figure 6.3 presents the architecture of a BiLSTM in detail.

The dual approach in bi-LSTMs allows them to capture contextual information from both directions, making them particularly effective for tasks where understanding the entire sequence is crucial. By combining the outputs from both directions, bi-LSTMs can better understand the context and variation of the data, leading to improved performance on various applications. However, this increased complexity comes with a higher computational cost and a greater number of parameters, which may necessitate more extensive training data and resources. Overall, bi-LSTMs are a powerful tool for modeling sequential data, offering enhanced context awareness and improved accuracy in many scenarios.

6.1.2 ENCODER-DECODER ARCHITECTURE

Encoder-decoder architecture [4] is the core component in sequence-to-sequence (Seq2Seq) models. Seq2Seq is a deep learning model used to process data in sequential format. This model is designed to handle input sequences of variable length and generate output sequences of varying length. Thus, it is suitable for various tasks in NLP like machine translation, text summarization, speech recognition, etc.

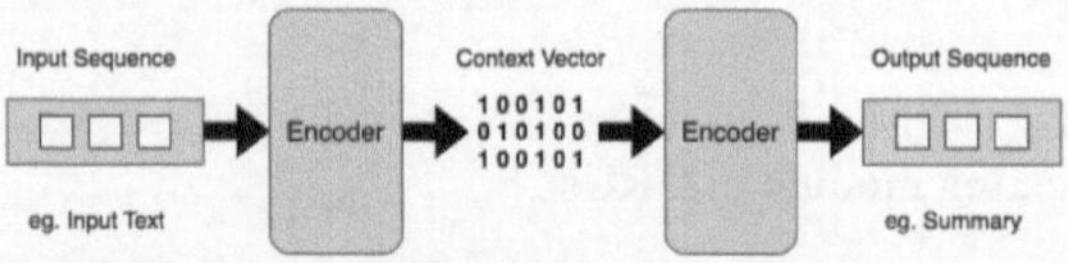

Figure 6.4 Encoder-Decoder architecture

An encoder-decoder architecture includes two subnetworks in Figure 6.4.

Encoder: The encoder receives the input and extracts its important information. The encoder is often an RNN, an LSTM, or other variants. It processes the input step by step to accumulate knowledge in a fixed-size context vector. Typically, the encoder includes the following three components:

- Embedding Layer: This layer converts the elements in the input into a continuous vector representation, making them suitable for neural network processing.
- Recurrent Layer (or Transformer): After the embedding layer is the recurrent layer, such as RNNs or LSTMs. This layer processes the embedded input sequence step by step. Modern encoder-decoder architecture may use parallelizable transformer-based encoders to improve performance.
- Context Vector: The encoder accumulates information from the input sequence into hidden states. The last hidden state (the state of the end of the input) is the context vector. This vector aggregates the essence of the input sequence and serves as the starting point for the decoder.

Decoder: The decoder receives the context vector as input. It then generates the output sequence over each step. At one step, the decoder gets the input from the previous output and the context vector. Like the encoder, the decoder are also an RNN, an LSTM, or other sequential modeling techniques to produce the output sequence. The decoder also consists of the following three layers:

- Embedding Layer: This layer aims to transform the input sequence (or individual elements) into continuous vector representations.
- Recurrent Layer (or Transformer): The next layer of the decoder is also the recurrent layer (or transformer) that processes the embedded input. This layer takes both the context vector and the previously generated elements as input to generate the output sequences step by step.

- Output Layer: In every step, the decoder outputs a probability distribution over all possible output values. The output element is then determined using this distribution. The output layer is often a softmax function that transforms the model output values into probabilities.

The encoder-decoder architecture is a powerful model for handling sequences of varying length. Thus, it has been applied to various problems. However, the standard encoder-decoder also has a limitation in capturing long sequence dependency. The reason is that it must learn to capture and compress all relevant information in the fixed context vector and effectively decode it into the output sequence. The attention techniques in the following subsection are proposed to address this limitation.

6.1.3 ATTENTION

The critical and apparent limitation of seq2seq models is that the context vector is fixed. This hinders the capability of remembering long sentences. Seq2seq models often forget the first parts of the sequence once they complete processing the whole input.

The attention mechanism was proposed in 2014 [5] to address the limitation of seq2seq models in memorizing long information in the input. Attention mechanism is different from the standard seq2seq model in the way it connects from the encoder to the context vector. Instead of connecting the context vector to only the last hidden state of the encoder, the attention mechanism creates shortcuts between the context vector and the entire input. Moreover, the weights of these shortcut connections are updated when training the model.

Because the context vectors are connected to the entire input sequence, the forgetting problem is solved. Moreover, the relationship between the input and the target is learned through the context vector. In other words, the more important input will likely have a greater weight value than the less relevant input. This is the reason why the mechanism is called attention (the output pays more attention to the more relevant input). Figure 6.5 shows the architecture of the attention scheme for a machine translation problem.

Formally, the attention mechanism is defined as follows. Let $x = [x_1, x_2, ...x_n]$ be the input sequence, $y = [y_1, y_2, ..., y_m]$ be the output sequence and $h = [h_1, h_2, ...h_n]$ be the hidden state of the seq2seq with attention mechanism, then the context vector for the output at position t is calculated as follows:

$$c_t = \sum_{i=1}^{n} \alpha_{t,i}.h_i \tag{6.8}$$

where $\alpha_{t,i}$ presents the alignment (the relevance) between the output at position t (y_t) and the input x_i. Specifically, it is calculated as follows:

$$\alpha_{i,t} = \frac{\exp(e_{t,i})}{\sum_{k=1}^{n} \exp(e_{t,k})} \tag{6.9}$$

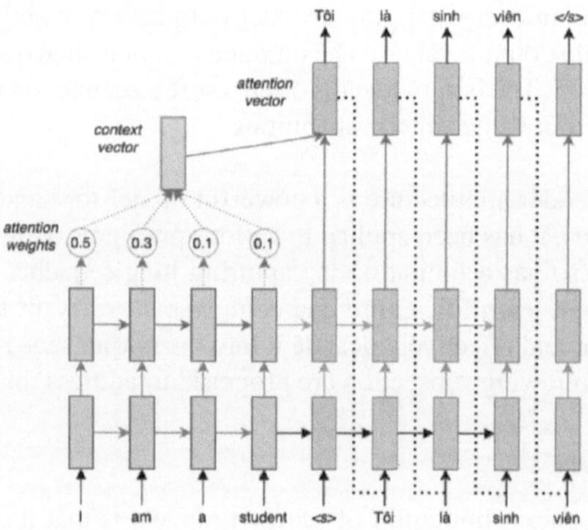

Figure 6.5 Attention architecture

Here, $e_{t,k} = score(s_t, h_k)$ is the score to measure the matching between the input at position k and the output at position t. This is often implemented by a feed-forward network with a single hidden layer and jointly trained with other parts of the model. For a network with tanh function, the score function is therefore defined as follows:

$$score(s_t, h_k) = W_c.tanh(W_a.[s_t, h_k]) \tag{6.10}$$

where W_c and W_a are two weight matrices of the alignment network.

Attention helps to solve the problem of the dependencies between the source and the target sequence. Thus, it leads to big improvements in machine translation [5]. Subsequently, it received paramount interest from the research community. There have been a large number of attention mechanisms developed for various applications. These attention mechanisms are often distinguished in the way the attention score is calculated. For example, Graves et al. proposed Content-based attention in [6] where the attention score is calculated based on the cosine similarity between the hidden state of the encoder and the decoder.

$$\text{score}(s_t, h_i) = \text{cosine}[s_t, h_i] \tag{6.11}$$

Luong et al. [7] proposed the Dot-Product attention, where the attention score is simply the dot product between the hidden state of the encoder and the decoder.

$$\text{score}(s_t, h_i) = s_t^\top h_i \tag{6.12}$$

Vaswani et al. [8] proposed Scaled Dot-Product, that is very similar to the Dot-Product attention except with a scaling factor corresponding to the dimension of the source hidden state.

$$\text{score}(s_t, h_i) = \frac{s_t^\top h_i}{\sqrt{n}} \tag{6.13}$$

where n is the dimension of the source hidden state.

6.1.4 TRANSFORMER

Transformer [8] aims to improve the performance of the attention mechanism based on RNN. Transformer is built based on three key concepts: Self-attention, Key, Value and Query, and Multi-Head attention.

Self-attention: Self-attention [9] is a type of attention mechanism that allows us to decide how important each part of an input sequence is. Self-attention makes it possible to find dependencies and connections in the input data. Specifically, self-attention calculates the weights between each element in the input sequence, allowing it to focus on the relevant input factors for a given task. This mechanism performs very well because it can learn the long-term dependencies and relationships in the input sequence, leading to performance improvement on many tasks.

The formal definition of self-attention is as follows. Given an input sequences, $\mathbf{x}_1, \ldots, \mathbf{x}_n$, where $\mathbf{x}_i \in \mathbb{R}^d$, the self-attention produces an output sequence $\mathbf{y}_1, \ldots, \mathbf{y}_n$ of the same length, where $\mathbf{y}_i$ is calculated based on the input at step i and the similarity of x_i with other values in the input sequence.

Self-attention is widely used in many problem domains especially in natural language processing (NLP). For example, self-attention significantly improves performance in various applications [10].

Key, Value, and Query: The second important concept in the transformer is Key, Value, and Query. The transformer considers the encoded representation of the input as a set of key-value pairs $(\mathbf{K}, \mathbf{V})$ and the output as the query vector. Figure 6.6 demonstrates how Key, Value and Query are calculated in an attention mechanism.

Key: The key vector $\mathbf{K}$ is the representation of each element in the sequence. It presents the context or the reference to support the prediction task. The key vector

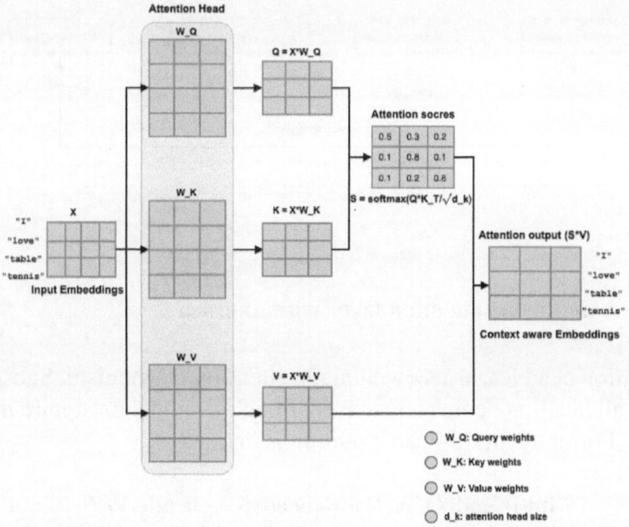

Figure 6.6 Key, Value, and Query in attention

K is calculated by multiplying the embedding vector X with a weights matrix W^k of the encoder as follows:

$$K = X \cdot W^k \tag{6.14}$$

Value: The value vector **V** presents the content that is being used to find the answer to the query. For example, in NLP problems, the value presents the meaning of each word in general, not specifically for this sentence. The value vector is also calculated by passing the embedding vector through another weight matrix W_v of the encoder, as in Equation 6.15.

$$V = X \cdot W^v \tag{6.15}$$

Query: The query vector presents the information that is expected to get at the output. The query vector **Q** is derived by passing the embedding of the previous output to a weight matrix W^q of the decoder, as in Equation 6.16.

$$Q = X \cdot W^q \tag{6.16}$$

The transformer employs scaled dot-product attention, producing an output that is a weighted sum of the values. The weight assigned to each value is determined by the dot product between the query and all the keys.

$$\text{Attention}(\mathbf{Q},\mathbf{K},\mathbf{V}) = \text{softmax}(\frac{\mathbf{QK}^\top}{\sqrt{n}})\mathbf{V} \tag{6.17}$$

where n is the size of the vector K .

Multi-Head Self-Attention: The multi-head mechanism executes the scaled dot-product attention several times concurrently, as illustrated in Figure 6.7.

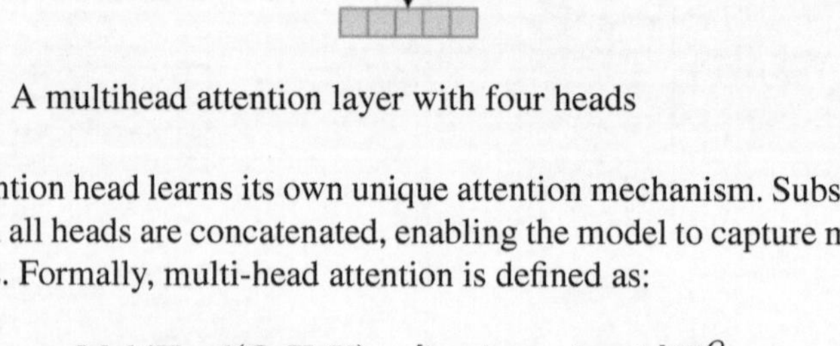

Figure 6.7 A multihead attention layer with four heads

Each attention head learns its own unique attention mechanism. Subsequently, the outputs from all heads are concatenated, enabling the model to capture more intricate relationships. Formally, multi-head attention is defined as:

$$\text{MultiHead}(\mathbf{Q},\mathbf{K},\mathbf{V}) = [\text{head}_1;\ldots;\text{head}_h]\mathbf{W}^O \tag{6.18}$$

where W^0 is the weights of the final matrix to project the concatenated vector into the desired output size and $\text{head}_i = \text{Attention}(\mathbf{QW}_i^Q, \mathbf{KW}_i^K, \mathbf{VW}_i^V)$ is the ith attention head.

It is noted that the learnable parameters of all multi-head attention layers consist of only three densely connected weights matrices (W_i^Q, W_i^K, W_i^V) and convolutions and recurrent mechanisms are not used in a multi-head attention layer.

Encoder: The transformer architecture introduced in [8] also includes an encoder and a decoder. The encoder creates representations that enable the model to pay attention to specific information within the input. In [8], the encoder is composed of a stack of six identical layers. Each layer contains two sub-components: a multi-head self-attention mechanism and a feed-forward network. Additionally, each sub-layer incorporates residual connections and layer normalization. All sub-layer outputs share the same dimensionality, $d_{\text{model}} = 512$.

Decoder: The decoder's purpose is to generate outputs based on the encoded representation. The decoder also consists of six identical layers, each containing two multi-head attention mechanisms and one fully connected feed-forward network. Each sub-layer incorporates layer normalization and residual connections. Additionally, the first multi-head attention sub-layer is masked to prevent it from attending to future positions in the sequence, ensuring that predictions for the current output are not influenced by subsequent tokens. This masking is necessary because future information cannot be accessed when generating the current output.

Overall, transformers are a revolutionary architecture in the field of deep learning, particularly for natural language processing tasks. Unlike RNNs and LSTMs, transformers do not process data sequentially, which significantly speeds up training and allows for better parallelization. Their layered structure, consisting of encoder and decoder components, enables them to perform well in various tasks, including translation, summarization, and sentiment analysis. Transformers have also paved the way for large-scale pre-trained models like BERT and GPT, which have set new benchmarks across multiple NLP tasks. Thus, transformers have transformed the landscape of machine learning, leading to impressive advancements in language understanding and generation. However, they do require substantial computational resources, particularly for training on large datasets, and may not perform as well on smaller datasets without fine-tuning.

6.2 CYBERATTACK DETECTION

This section presents the application of generative AI to cyberattack detection. Particularly, this section will highlight the application of auto-regressive models, other generative models on phishing detection, email spam detection, and network data analysis.

6.2.1 PHISHING DETECTION

Phishing attacks aim to steal confidential information from cyberspace users using various sophisticated techniques. Phishing attacks are considered one of the most dangerous cyber attacks. According to a report by Proofpoint [11], phishing attacks cost large organizations almost 15 million USD annually, or more than 1,500 USD per employee in 2021. In another report, the Anti-Phishing Working Group (APWG) shows that after the COVID-19 pandemic in March 2020, the number of phishing attacks dramatically increased everywhere [12].

There are various types of conducting phishing attacks including phishing through content injection, social engineering, online social networks, and mobile applications [13]. The most common form of phishing attacks is through emails that often send an email message to scare users to take some immediate actions. In addition, phishing attacks can also target other media types such as online social networks, forums, mobile apps, and messaging platforms [14]. Recently, phishing attacks have also targeted some emerging systems like blockchain platforms. These attacks may result in financial loss, the loss of Intellectual Property (IP), and also valuable confidential user information. More importantly, phishing attacks reduce the trust of the community and even influence national security [15]. Subsequently, phishing detection is more urgent nowadays than ever before.

There have been a number of techniques developed for phishing attack detection and prevention. The simplest solution is collecting information about attackers and adding them to a blacklist [16]. The blacklist can be gathered from some platforms such as PhishTank, SafeBrowsing, SmartScreen, and then added to the web browser [17]. The blacklist can also be added to some specialized security software, such as Intrusion Detection System (IDS) or Intrusion Prevention System (IPS) [18]. However, the shortcoming of the blacklist method is that it can not detect zero-day phishing attack, which is related to a newly designed phishing site. This is because the newly designed site will not be added to the blacklist for a while. Moreover, the effort to collect and manage blacklists is often very high since phishing websites are created quickly and they often stay for a very short time.

Due to these obstacles of the blacklists-based detection approach, recently, machine learning, specially generative AI, has been developed to detect phishing attacks. Generally, there are two approaches of applying generative models to phishing detection. The first approach is to use generative models such as large language models (LLMs), GAN, and VAE to generate synthetic phishing websites/emails to augment the training dataset. The second approach is to use autoagressive models and LLMs to learn the representation of phishing websites/emails.

In the first approach, generative models are used to create phishing data. This synthesized data is then used to enlarge the training datasets for phishing detection problems. There are several approaches to generate phishing data using generative models:

- Text generation: Generative models like LLMs can create realistic phishing emails or messages by mimicking common language patterns used in legitimate communications [19]. These models can be trained on large datasets of phishing and legitimate emails to learn the structure and style of deceptive messages.

- Email templates: Generative models can create email templates that resemble those of reputable organizations, including logos, formatting, and persuasive language designed to trick users into providing sensitive information [20].
- Website generation: Generative AI can generate fake websites that closely resemble legitimate sites, including similar layouts, graphics, and user interfaces [21]. This can involve using tools that automate the creation of web pages based on existing templates.
- Social engineering content: Generative AI can create entire phishing scenarios, including social engineering tactics that exploit psychological triggers [22]. This can include fake alerts, urgent requests, or offers that encourage victims to act quickly without thinking.

There have been many researchers who have paid attention to utilize generative models to augment phishing datasets during training detection models. For example, Anand et al. [23] addressed the imbalanced classes of the phishing URL detection problem using GANs. They trained a GAN to learn the string pattern statistics of URLs in the minority class and then used the GAN's generator to generate synthetic URLs. Both the generator and discriminator were LSTM networks. They conducted the experiments on the PhishTank and Common Crawl repositories, and the results demonstrate significant performance improvements after using the proposed solution. Robic-Butez and Win [24] proposed a novel method of detecting phishing websites using a GAN. They first extract features of phishing sites using internal structural, source code, and Metadata. They then train a GAN model where the discriminator attempts to distinguish between the phishing features and the generated features. After training, the discriminator is used to classify between phishing sites and legitimate sites. Sern et al. [25] propose PHISHGAN to generate a common type of phishing attacks: Homoglyph attacks. The authors convert the domain name into a grey image of shape $40x400x1$ using the Python Pillow package. After that, they train a GAN model where the generator attempts to generate a phishing domain image and the discriminator aims to identify the real and the fake images. After training, the generator is used to produce various sets of homoglyph images. The experiment results show that PHISHGAN is more efficient than state-of-the-art algorithms like Pix2Pix and CycleGAN in generating homoglyph images.

Recently, Al-Ahmadi [26] proposed PDGAN to detect phishing attacks using only the website's uniform resource locator (URL). The authors use a long short-term memory network (LSTM) as a generator to generate synthetic phishing URLs and a convolutional neural network (CNN) as a discriminator to determine whether the URLs are phishing or legitimate. The experiments conducted on the PhishTank and DomCop datasets show that PDGAN achieves a detection accuracy of 97.58% and a precision of 98.02% which are higher than the state-of-the-art models. Shirazi et al. [27] proposed two adversarial models, namely Adversarial Autoencoder (AAE) and Wasserstein Generative Adversarial Network (WGAN), for generating synthetic phishing data. The authors also investigate five hypotheses of the generated data, in which two interesting conclusions include: (i) synthesized samples are difficult to distinguish from actual data. (ii) augmenting the dataset with synthesized samples improves the accuracy of detection models with respect to the original samples. Moreover, adding some correctly labeled synthesized data in the training set helps models become more robust to exploratory attacks. Roy et al. [19] explore the potential of using four popular commercially available LLMs, i.e., GPT 3.5 Turbo, GPT 4, Claude, and Bard, to generate functional phishing attacks using a series of malicious prompts. They also discovered that these LLMs can generate both phishing websites and emails that can convincingly imitate well-known brands and elude detection mechanisms employed by anti-phishing systems. They then train a detection model using BERT and a transfer learning approach. Their model attains an average accuracy of 96% for phishing website prompts and 94% for phishing email prompts.

In the second approach, autoregressive models, e.g, variants of LSTM networks and transformers, are often used to directly learn from the text of a URL or a phishing site's content. By processing data in the sequence, these models can extract relevant features that characterize phishing attempts. They can also learn patterns related to common phishing phrases, URL structures, or specific formatting that distinguishes phishing from legitimate communications. For instance, Bahnsen et al. [28] proposed to apply LSTM for phishing detection. They use an LSTM network to directly learn a representation from the URL's character sequence instead of manually extracting the features. The experimental results on two well-known benchmarking PhishTank and Common Crawl show that the LSTM approach provided a higher accuracy rate and F1 score than the Random Forest classifier. Shivangi et al. [29] proposed a tool to analyze URLs and detect malicious ones using LSTM networks. These models extracted the features from the URLs automatically and used these features to classify the input URLs. The dataset used in the experiments was obtained through search engines, PhishTank, and the Twitter Streaming API. Their results show that the LSTM network achieved higher accuracy than the traditional neural networks, i.e. MLP. Moreover, they also deployed their model as an extension of the Google Chrome browser to provide the user with a safer browsing experience. Peng et al. [30] proposed a method based on the combination of Attention mechanism with Convolutional Neural Network and Long Short-Term Memory Network (Attention-Based CNN-LSTM) for detecting malicious Uniform Resource Locators (URLs). The attention mechanism allows us to identify the important role of different features in the input URLs. The malicious URLs were collected from PhishTank, while benign URLs were collected from popular websites. According to the results, the statistical features of the URL contributed most to detecting malicious URLs.

Recently, Large Language Models (LLMs) have been utilized for phishing detection due to their advanced capabilities in natural language processing tasks, as highlighted in [31]. Jamal and Wimmer [32] enhanced Transformer-based models for identifying phishing, spam, and ham emails by fine-tuning BERT family models specifically for this purpose. Their research demonstrated that these fine-tuned models effectively classify emails in both balanced and unbalanced datasets, improving detection accuracy for phishing and spam emails. This work can be considered as one of the first steps towards employing LLMs to improve phishing detection accuracy.

Amaz and Sarker [33] introduced an optimized, fine-tuned transformer-based DistilBERT model for phishing email detection. Through our experiments on the phishing email dataset, they showed that their fine-tuned model achieves higher accuracy than the pre-trained model. Moreover, they also demonstrate the explainable capability of the technique to explain how the model makes predictions in the context of text classification for phishing emails. Koide et al. [34] proposed a system that uses large language models (LLMs) to detect phishing emails called ChatSpamDetector. The ChatSpamDetector system first converts email data into a prompt suitable for LLM analysis. After that, the prompt is input to the LLMs to verify whether an email is phishing or not. They conducted an evaluation using a large phishing email dataset and compared their system to several LLMs and baseline systems. The experimental results show that their system using GPT-4 achieves an accuracy of 99.70%. This is evidence for the ability of LLMs as a potentially powerful tool in the fight against email-based phishing threats.

Overall, generative AI has significantly enhanced phishing detection by learning the underlying characteristics of phishing data and augmenting training datasets. Techniques such as Generative Adversarial Networks (GANs) and Variational Autoencoders (VAEs) can create convincing phishing email templates that mimic legitimate communications, providing valuable training data for detection systems. These models, such as LLMs, also learn complex patterns and features characteristic of phishing attempts, improving the ability to identify deceptive tactics. Additionally, generative AI can model the sequential nature of phishing emails, allowing for better contextual understanding and adaptation to evolving phishing strategies. By generating synthetic user profiles and diverse phishing examples, generative AI has strengthened the effectiveness and robustness of phishing detection systems.

6.2.2 EMAIL SPAM DETECTION

Email spam, also often referred to as junk email, consists of unsolicited messages sent in bulk, typically for advertising purposes or to promote scams. Email spam is a persistent problem that affects individuals, businesses, and organizations worldwide. According to research by Dekker published in Eftsure [35], in 2023, spam accounts for 14.5 million messages globally per day, and this makes up 45% of all emails.

This pervasive issue affects individuals and organizations, leading to significant economic and security concerns [36]. Spam not only makes it difficult for users to find legitimate messages, but it also poses serious risks by often containing phishing attempts, malware, or other malicious content. The financial impact of spam can be substantial, incurring costs related to lost productivity, network bandwidth, and potential data breaches. Moreover, as spammers become increasingly sophisticated, the threat to cybersecurity intensifies, necessitating robust filtering and detection measures to protect users and maintain the integrity of email communications.

The widespread use of email spam has led to an increased need for effective spam detection mechanisms to filter out unwanted and potentially harmful messages from users. Email spam detection involves the application of various techniques, ranging from rule-based filters to sophisticated machine learning algorithms, to identify and classify spam emails accurately.

Rule-based filtering is one of the traditional and widely used techniques for email spam detection. It involves the application of predefined rules or heuristics to identify spam emails based on specific characteristics or patterns. The popular rule-based techniques include keyword matching [37], blacklist filtering [38], and email structure analysis [39]. Rule-based filtering is relatively simple to implement and can provide effective spam detection in some cases. However, it has limitations, such as being less adaptable to evolving spamming techniques and potentially generating false positives or false negatives.

To address the limitations of rule-based techniques, more advanced techniques including machine learning [40] and deep learning [41] have been employed to improve spam detection accuracy. Machine learning algorithms can identify patterns in the spam email and adapt their detection criteria based on new data. Thus, they improve the detection accuracy over time. However, traditional machine learning algorithms often require experts to handcraft relevant features from the email data, which can be time-consuming and may not capture the full complexity of spam email patterns. Moreover, machine learning models may struggle to effectively process unstructured data like email text, images, or attachments, which limits their ability to leverage the rich information present in emails. Conversely, deep learning models can effectively process unstructured data like email text, images, and attachments, enabling them to leverage the rich information present in emails to achieve better accuracy. Additionally, deep learning models can automatically extract relevant features from raw data, reducing the need for manual feature engineering. This end-to-end learning approach allows them to handle complex and diverse spam email patterns without relying on handcrafted rules or heuristics [42].

Recently, generative models have been employed for spam email detection. Most of the previous research on applying the generative models for spam email detection employs autoagressive models to learn representation of spam email. For example, Wanda et al. [43] proposed the GRU model for effective spam detection in emails. The proposed GRUSpam model automatically extracts high-level features from spam messages to enhance the accuracy of spam detection. The experiments show the promise results in addressing the evolving landscape of spam and improving detection efficiency. Specifically, GRUSpam outperforms traditional classifiers like decision trees, K-Nearest Neighbor, and Support Vector Machines. Zavrak and Yilmaz [44] proposed a novel technique for email spam detection that combines convolutional neural networks, gated recurrent units, and attention mechanisms. The attention mechanism allows the model to focus on the necessary parts of the email text. Moreover, the combination of CNN and GRN helps to extract both temporal and spatial features from the emails. Based on the cross-dataset evaluation results, the proposed technique advances the results of the present attention-based techniques by utilizing temporal convolutions. Ghanem and Erbay [45] proposed a novel deep learning model called Contextualized Bi-directional Long Short Term Memory neural network (CBLSTM) to address the spam texts problem on social networks. They used a bidirectional long short-term neural network with embedding from language models to handle the "out of vocabulary" problem in spam email. The experimental results on three benchmark datasets show that the proposed method achieves high accuracy and outperforms the existing state-of-the-art methods to detect spam on social networks.

More recently, LLMs have also been utilized for spam email detection. Due to the ability to understand and analyze natural language, LLMs enhance the effectiveness and accuracy of spam detection systems, making them more robust to evolving spam strategies. For instance, Labonne and Moran [46] investigated the effectiveness of large language models (LLMs) in email spam detection. They compare prominent models from three distinct families: BERT-like, Sentence Transformers, and Seq2Seq in four public datasets. The experimental results reveal that LLMs often surpass the performance of the popular baseline techniques, particularly in few-shot scenarios. The authors also introduce Spam-T5, a model that has been specifically adapted and fine-tuned for the purpose of detecting email spam, and this model surpasses baseline models and other LLMs in the majority of scenarios, particularly when there is a limited number of training samples available. Wu et al. [47] evaluated the performance of ChatGPT for spam email detection in English and Chinese email datasets. The authors employ ChatGPT for spam email detection using in-context learning, which requires a prompt instruction and a few demonstrations. The experiments show that the performance of ChatGPT is not better than deep supervised learning methods in the large English dataset. However, ChatGPT presents superior performance on the low-resourced Chinese dataset, even outperforming BERT. Roumeliotis et al. [48] address the persistent challenges of spam emails and phishing attacks by introducing a cutting-edge approach to email filtering. They combine the capabilities of advanced language models, particularly the state-of-the-art GPT-4 Large Language Model (LLM), along with BERT and RoBERTa Natural Language Processing (NLP) models. Through an extensive literature review, experimentation, and evaluation, they demonstrate the effectiveness of the proposed approach in accurately identifying spam and phishing emails while minimizing false positives. They also show the potential of fine-tuning LLMs for specialized tasks like spam classification, offering enhanced protection against evolving spam.

Generally, the research on the application of generative AI techniques, such as Large Language Models (LLMs), generative adversarial networks (GANs) or variational autoencoders (VAEs) and Autoregressive models for spam email detection has received limited attention from the research community. Despite the advancements in generative AI, the majority of existing literature on spam email detection has predominantly focused on discriminative models and traditional machine learning algorithms. The lack of research papers exploring the application of generative AI to spam email detection presents an opportunity for future research. Future studies could investigate how generative AI models can be used to generate synthetic spam emails for training robust detection models. Research could also explore the use of generative models to generate countermeasures to enhance the resilience of spam detection systems. Furthermore, the application of LLMs also opens a potential direction to improve the accuracy and efficiency of spam email detection systems. Overall, there is significant potential for future research to explore the application of generative AI in spam email detection. Generative AI could offer new insights and approaches to tackle this persistent problem.

6.3 NETWORK DATA ANALYSIS

This section discusses the application of generative AI in network data analysis. Specifically, we will discuss the applications of generative models in network traffic analysis, software logs analysis, and system call analysis.

6.3.1 NETWORK TRAFFIC ANALYSIS

Network traffic analysis [49] is the process of inspecting and examining data packets flowing through a computer network. It involves monitoring and analyzing network traffic to gain insights into network behavior, identify potential security threats, and optimize network performance. Network traffic analysis allows the examination of network traffic patterns, protocols, and packet payloads to detect anomalies, such as malicious activities unauthorized access attempts. Moreover, it also aids in network troubleshooting by identifying latency issues and performance bottlenecks. Overall, network traffic analysis is a fundamental practice for maintaining a secure and efficient network infrastructure.

There are several approaches to network traffic analysis that are commonly used in practice, such as signature-based analysis, statistical analysis and machine learning-based analysis. Signature-based techniques [50] for network traffic analysis identify specific patterns or signatures associated with malicious activities. These techniques rely on predefined rules or signatures that represent known vulnerabilities, exploits, and attack types. When network packets are analyzed, any match against these signatures triggers an alert, allowing security systems to respond swiftly to potential threats. This method is particularly effective for detecting established malware, network intrusions, and other forms of attacks that have been previously cataloged. However, signature-based detection struggles with zero-day attacks and sophisticated, evolving malware that may not have identifiable signatures.

Statistical analysis for network traffic analysis involves examining data patterns and trends within network communications to identify anomalies and potential security threats [51]. Using statistical methods allows for the detecting deviations that may indicate malicious activity, such as unusual spikes in traffic, abnormal packet sizes, or irregular communication patterns between devices. Techniques such as time-series analysis, clustering, and hypothesis testing are employed to evaluate the frequency and distribution of network events. This approach provides valuable insights into network performance and security, helping to uncover hidden threats that signature-based methods might miss. Furthermore, statistical analysis can facilitate proactive monitoring, enabling organizations to respond promptly to emerging vulnerabilities and optimize their network infrastructure for better security and efficiency. However, statistical analysis often fails to detect threats if they do not significantly deviate from established patterns. Additionally, the reliance on baseline behavior means that any changes in legitimate traffic, such as those due to network upgrades or seasonal variations, can distort the analysis. Furthermore, statistical methods often lack adaptability to new and sophisticated attacks, making them less effective against evolving threats.

Recently, machine learning has emerged as a powerful tool for network traffic analysis [52]. Machine learning offers advanced capabilities to detect and respond to security threats in real-time. ML models allow learning from historical data and identify patterns to detect unknown or emerging threats. Additionally, ML-based models can learn intricate relationships in network traffic data and make accurate predictions, leading to more reliable detection outcomes. Moreover, ML models can adapt to changes in network infrastructure and new applications without requiring manual updates, ensuring their effectiveness in dynamic network environments.

Recently, generative models utilizing deep neural networks have been applied to improve network traffic analysis. For instance, models like GANs and VAEs are employed to synthesize data for network analysis tasks. Ring et al. [53] introduced a methodology for generating realistic flow-based network traffic using Generative Adversarial Networks (GANs). They proposed three preprocessing techniques to convert flow-based data into continuous values and developed a domain knowledge-based evaluation method to assess the quality of generated traffic. Their experiments, conducted on the CIDDS-001 dataset, revealed that two of the three preprocessing approaches produced high-quality synthetic data.

Xu et al. [26] presented STAN (Synthetic network Traffic generation with Autoregressive Neural models), which generate realistic synthetic network traffic datasets for downstream applications. The model captures both temporal dependencies and attribute relationships at any given time, handling both continuous and discrete variables. To evaluate the synthetic data's effectiveness, the authors trained two anomaly detection models using self-supervised learning, showing only a minor drop in accuracy when models were trained exclusively on synthetic data. Shahid et al. [55] proposed combining an autoencoder with a GAN to generate sequences of packet sizes corresponding to bidirectional flows. The autoencoder learns latent representations of real packet size sequences, while the GAN generates latent vectors that are decoded into realistic sequences. Experiments using bidirectional flows from a Google Home Mini demonstrated that this approach produces packet size sequences closely resembling real bidirectional flows.

With the development and the success of LLMs in various application domains, researchers have also paid more attention to applying LLMs for network traffic analysis. For instance, Kholgh et al. [56] addressed challenges posed by the scarcity of realistic datasets for network traffic analysis by proposing a novel framework to generate reliable synthetic data, namely PAC-GPT. The core components of this framework include two modules: a Flow Generator, which is responsible for capturing and regenerating patterns in a series of network packets, and Packet Generator, which can generate individual network packets given the network flow. The authors also propose a packet generator based on LLM chaining. The experimental results conclude that transformers are a suitable approach for synthetic packet generation with minimal fine-tuning performed. Qu et al. [57] addressed the token length limitation problem of LLMs in generating network traffic. They introduce TrafficGPT, a deep learning model that can tackle complex challenges related to long flow classification and generation tasks. This model uses generative pre-training with the linear attention mechanism, which allows for a substantially increased capacity of up to 12,032 tokens from the previous limit of only 512 tokens. Moreover, TrafficGPT shows superior performance in classification tasks, reaching state-of-the-art levels. Kan et al. [58] presented an open-source mobile network-specialized LLM called Mobile-LLaMA, which is an instruction-finetuned variant of the LLaMA 2 13B model. They built Mobile-LLaMA by instruction fine-tuning LLaMA 2 13B with their own network analysis data collected from publicly available, real-world 5G network datasets, and expanded its capabilities through a self-instruct framework utilizing OpenAI's pre-trained models (PMs). Mobile-LLaMA has three main functions: packet analysis, IP routing analysis, and performance analysis enabling it to provide network analysis and contribute to the automation and artificial intelligence (AI) required for 5G network management and data analysis. Their evaluation demonstrates Mobile-LLaMA's proficiency in network analysis code generation, achieving a score of 247 out of 300, surpassing GPT-3.5's score of 209.

Overall, Large Language Models (LLMs) provide a promising tool for enhancing accuracy in network traffic analysis. LLMs, with their advanced natural language processing and machine learning capabilities, offer unique advantages in understanding and analyzing network traffic data. It is expected that there will be more research of applying LLMs to network traffic analysis in the coming time.

6.3.2 SOFTWARE LOGS ANALYSIS

Software logs analysis is the process of examining and extracting valuable insights from the logs generated by software applications, systems, or services [59]. Software logs are records of events, activities, errors, and other relevant information that occur during the execution of software. Log analysis involves collecting and parsing logs to identify patterns, anomalies, and trends, which can provide valuable information about system behavior, performance, and potential issues. By analyzing software logs, one can gain visibility into the health, security, and operational aspects of their software systems. Log analysis can also help in troubleshooting and diagnosing software problems, detecting and mitigating security incidents, optimizing system performance, and monitoring compliance with regulatory requirements.

There are several techniques used in software log analysis to extract valuable insights and information about software and systems. Traditional logs analysis techniques include pattern matching, log parsing, and statistical analysis. Pattern matching involves searching for specific patterns or regular expressions within log data. This technique helps identify predefined events or errors that can provide important information about system behavior or potential issues [60]. Log parsing breaks down log entries into structured data for easier analysis. It facilitates subsequent analysis and enables querying and filtering based on specific criteria [61]. Statistical analysis techniques are used to derive meaningful insights from log data. These techniques can identify trends, patterns, and anomalies within the log entries, aiding in troubleshooting, performance optimization, and anomaly detection [62].

Advanced log analysis techniques such as machine learning and deep learning are also widely used for logs analysis. By training models on historical log data, machine learning can learn patterns of normal behavior and flag any deviations as potential anomalies or security threats. Thus, machine learning techniques can enhance the accuracy and efficiency of log analysis, especially in detecting complex or previously unseen issues [63]. Deep learning even shows better results in software log analysis due to its ability to automatically learn complex patterns and representations from raw log data [64]. Specifically, Deep learning models are excellent at feature extraction, automatically learning relevant characteristics from raw log data without requiring extensive manual preprocessing. Additionally, the ability to handle unstructured data makes deep learning particularly suitable for log analysis, as logs often consist of varied formats and structures.

Recently, Generative models, including Variational Autoencoders (VAEs) and Generative Adversarial Networks (GANs) have improved software log analysis in various ways. First, these models can be used to generate synthetic log data. Generative models can be trained on existing log data to generate synthetic log entries that closely resemble real log data. This synthetic log can be useful in some scenarios such as augmenting datasets, creating realistic test scenarios, or simulating logs for training anomaly detection models [65, 66]. Second, generative models can be used for anomaly detection in logs. These models first learn the normal patterns and structures present in log data. By comparing new log entries against the learned model, deviations from the learned distribution can be flagged as potential anomalies [67, 68]. The benefit of using generative models is that they can capture complex relationships and identify subtle deviations that may be missed by traditional rule-based, statistical methods or traditional machine learning. Last but not least, generative models can be used for log parsing. These models can be applied to automatically extract structured information from unstructured log data, thus helping to reduce the manual effort required for log parsing and enabling more efficient and accurate analysis of log data [69].

Besides GANs and VAEs, LLMs are also used for log analysis. For example, LLMs can assist in log parsing by understanding the natural language content of log messages. They can identify and extract relevant fields and attributes from log entries, such as timestamps, log levels, error codes, or message content. LLMs can leverage their contextual understanding to interpret the log messages and accurately extract structured information [70]. LLMs can be used for log summarization. LLMs can generate summaries or abstracts of log entries, capturing the key information and reducing the need to analyze each log entry individually. By training on a large corpus of logs, LLMs can generate concise summaries that provide a high-level overview of the log content, aiding in efficient log analysis and troubleshooting [71].

Overall, LLMs provide a powerful tool for log analysis by combining language understanding and contextual knowledge. However, it's important to note that LLMs require substantial computational resources for training and inference and may have limitations in handling large-scale log datasets.

6.3.3 SYSTEM CALLS ANALYSIS

System call analysis is the process of examining the system calls made by a software application or program during its execution [72]. System calls are the interface between user applications and the operating system, allowing programs to request services from the underlying operating system kernel. By analyzing system calls, we can gain insights into the behavior and interactions of software with the operating system. System call analysis can reveal important information such as file operations, network communication, process management, and resource utilization. It can help identify security vulnerabilities, detect malicious activities, and monitor the performance of software applications.

Popular system call analysis techniques include profiling, behavior analysis, dependency analysis, and machine learning-based analysis. Profiling aims to collect information about the system calls made by a set of programs [73]. This can be done by coding to log system call invocations or by using tools like to trace system calls in real-time. Profiling helps in understanding the frequency and timing of system calls, identifying bottlenecks, and optimizing system performance. Behavior analysis focuses on understanding the patterns and sequences of system calls made by programs [74]. Behavior analysis helps to identify normal or expected behavior and detect anomalies or suspicious activities. Behavioral analysis is commonly used in intrusion detection systems, malware analysis, and system monitoring. Dependency analysis aims to identify the dependencies and relationships between system calls of programs [75]. It examines the interactions between system calls and the resources they access, such as files, network sockets, or shared memory. Dependency analysis helps to identify potential conflicts or synchronization issues resulting from programs. Machine learning algorithms can be used to classify system call behavior or identify malicious activities. These algorithms can learn to recognize patterns and make predictions about the nature of system call activities [76].

Recently, generative models, especially LSTMs, have been used to analyze system calls. For example, Xiao et al. [27] applied LSTM to analyze system calls to detect Android malwares. They treat one system call sequence as a sentence in the language and construct a classifier based on the LSTM model. Specifically, two LSTM models are trained using the system call sequences from malware and benign applications, respectively. After that, two similarity scores are computed and used to determine whether the application under analysis is malicious or trusted by the greater score. The experiments show that the proposed approach achieves a high recall of 96.6% with a low false positive rate of 9.3%. Xie et al. [78] propose a sensitivity-based LSTM model to design a System-call Behavioral Language (SBL) system for malware detection. The proposed model consists of two parts: the behavior language learning to extract features from system call sequences, and the sensitivity-based attention to pay more attention to important features. The model achieves 78% specificity on the unknown attack datasets compared to only 66% of the classic language models. Soni et al. [79] compared and LSTM sequence-to-sequence (Seq-Seq) model and one-class support vector machines (OCSVM) for anomalous system call sequence detection. The evaluated results show that the LSTM Seq-Seq model with a sequence length of five provides a higher detection accuracy of 97.2%.

Other generative models, such as GANs, are also used to facilitate system call analysis. Alsheraifi et al. [80] proposed to use a GAN model to simulate system calls of malicious Android processes. They tested tabular-based and pictorial-based models in multiple trials to determine the better one for classification. However, the testing quantity was insufficient for definitive evidence of the quality of proposed models. For VAEs and LLMs, there is very limited research employing these models in system calls analysis. In fact, we could not find any related research paper in Google Scholar that applies VAEs and LLMs for system calls analysis. This opens an opportunity for further future search on applying generative models, particularly VAE and LLMs to system calls analysis.

6.4 SUMMARY

In this chapter, we investigated the applications of generative models in general and autoregressive models in particular to cybersecurity. We began by examining four important architectures for sequence learning, i.e., variants of the recurrent neural network, encoder-decoder architecture, Attention, and Transformer. The chapter then highlighted the prominent application of autoregressive models to two important problems in cybersecurity, including cyberattack detection and network data analysis.

1. Sepp Hochreiter and Jürgen Schmidhuber. Long short-term memory. *Neural Computation*, 9(8):1735–1780, 1997.
2. Kyunghyun Cho, Bart Van Merrienboer, Caglar Gulcehre, Dzmitry Bahdanau, Fethi Bougares, Holger Schwenk, and Yoshua Bengio. Learning phrase representations using rnn encoder-decoder for statistical machine translation. arxiv 2014. *arXiv preprint arXiv:1406.1078*, 2020.
3. Zhiheng Huang, Wei Xu, and Kai Yu. Bidirectional lstm-crf models for sequence tagging. *arXiv preprint arXiv:1508.01991*, 2015.
4. Kyunghyun Cho, Bart Van Merriënboer, Dzmitry Bahdanau, and Yoshua Bengio. On the properties of neural machine translation: Encoder-decoder approaches. *arXiv preprint arXiv:1409.1259*, 2014.
5. Dzmitry Bahdanau. Neural machine translation by jointly learning to align and translate. *arXiv preprint arXiv:1409.0473*, 2014.
6. Alex Graves. Neural turing machines. *arXiv preprint arXiv:1410.5401*, 2014.
7. Minh-Thang Luong. Effective approaches to attention-based neural machine translation. *arXiv preprint arXiv:1508.04025*, 2015.
8. A Vaswani. Attention is all you need. *Advances in Neural Information Processing Systems*, 2017.
9. Jianpeng Cheng. Long short-term memory-networks for machine reading. *arXiv preprint arXiv:1601.06733*, 2016.
10. Peter Shaw, Jakob Uszkoreit, and Ashish Vaswani. Self-attention with relative position representations. *arXiv preprint arXiv:1803.02155*, 2018.
11. Proofpoint. The 2021 ponemon cost of phishing study, 2021.
12. APWG. Manning r (2020) phishing reports, 2020.

13. Kang Leng Chiew, Kelvin Sheng Chek Yong, and Choon Lin Tan. A survey of phishing attacks: Their types, vectors and technical approaches. *Expert Systems with Applications*, 106:1–20, 2018.
14. Abdul Basit, Maham Zafar, Xuan Liu, Abdul Rehman Javed, Zunera Jalil, and Kashif Kifayat. A comprehensive survey of AI-enabled phishing attacks detection techniques. *Telecommunication Systems*, 76:139–154, 2021.
15. Zulfikar Ramzan and Candid Wüest. Phishing attacks: Analyzing trends in 2006. In *CEAS*. Citeseer, 2007.
16. Pawan Prakash, Manish Kumar, Ramana Rao Kompella, and Minaxi Gupta. Phishnet: predictive blacklisting to detect phishing attacks. In *2010 Proceedings IEEE INFOCOM*, pages 1–5. IEEE, 2010.
17. Carlo Marcelo Revoredo da Silva, Eduardo Luzeiro Feitosa, and Vinicius Cardoso Garcia. Heuristic-based strategy for phishing prediction: A survey of url-based approach. *Computers & Security*, 88:101613, 2020.
18. R Gowtham and Ilango Krishnamurthi. A comprehensive and efficacious architecture for detecting phishing webpages. *Computers & Security*, 40:23–37, 2014.
19. Sayak Saha Roy, Poojitha Thota, Krishna Vamsi Naragam, and Shirin Nilizadeh. From chatbots to phishbots?: Phishing scam generation in commercial large language models. In *2024 IEEE Symposium on Security and Privacy (SP)*, pages 221–221. IEEE Computer Society, 2024.
20. Lepe Khanum. Phishing emails analysis and design automatic text generation tool for writing and sending phishing mail. Master's thesis, 2020.
21. Mengxiang Wan, Hanbing Yao, and Xin Yan. Generation of malicious webpage samples based on gan. In *2020 IEEE 19th International Conference on Trust, Security and Privacy in Computing and Communications (TrustCom)*, pages 864–869. IEEE, 2020.
22. Sean Gallagher, Ben Gelman, Salma Taoufiq, Tamás Vörös, Younghoo Lee, Adarsh Kyadige, and Sean Bergeron. Phishing and social engineering in the age of llms. In *Large Language Models in Cybersecurity: Threats, Exposure and Mitigation*, pages 81–86. Springer Nature Switzerland Cham, 2024.
23. Ankesh Anand, Kshitij Gorde, Joel Ruben Antony Moniz, Noseong Park, Tanmoy Chakraborty, and Bei-Tseng Chu. Phishing url detection with oversampling based on text generative adversarial networks. In *2018 IEEE International Conference on Big Data (Big Data)*, pages 1168–1177. IEEE, 2018.
24. Pierrick Robic-Butez and Thu Yein Win. Detection of phishing websites using generative adversarial network. In *2019 IEEE International Conference on Big Data (Big Data)*, pages 3216–3221, 2019.
25. Lee Joon Sern, Yam Gui Peng David, and Chan Jin Hao. Phishgan: Data augmentation and identification of homoglyph attacks. In *2020 International Conference on Communications, Computing, Cybersecurity, and Informatics (CCCI)*, pages 1–6, 2020.
26. Saad Al-Ahmadi, Afrah Alotaibi, and Omar Alsaleh. Pdgan: Phishing detection with generative adversarial networks. *IEEE Access*, 10:42459–42468, 2022.
27. Hossein Shirazi, Shashika R. Muramudalige, Indrakshi Ray, Anura P. Jayasumana, and Haonan Wang. Adversarial autoencoder data synthesis for enhancing machine learning-based phishing detection algorithms. *IEEE Transactions on Services Computing*, 16(4):2411–2422, 2023.
28. Alejandro Correa Bahnsen, Eduardo Contreras Bohorquez, Sergio Villegas, Javier Vargas, and Fabio A. González. Classifying phishing urls using recurrent neural networks. In *2017 APWG Symposium on Electronic Crime Research (eCrime)*, pages 1–8, 2017.

29. S. Shivangi, Pratyush Debnath, K. Sajeevan, and D. Annapurna. Chrome extension for malicious urls detection in social media applications using artificial neural networks and long short term memory networks. In *2018 International Conference on Advances in Computing, Communications and Informatics (ICACCI)*, pages 1993–1997, 2018.
30. Yongfang Peng, Shengwei Tian, Long Yu, Yalong Lv, and Ruijin Wang. Malicious url recognition and detection using attention-based cnn-lstm. *KSII Transactions on Internet and Information Systems (TIIS)*, 13(11):5580–5593, 2019.
31. Dinil Mon Divakaran and Sai Teja Peddinti. Llms for cyber security: New opportunities. *arXiv preprint arXiv:2404.11338*, 2024.
32. Suhaima Jamal and Hayden Wimmer. An improved transformer-based model for detecting phishing, spam, and ham: A large language model approach. *arXiv preprint arXiv:2311.04913*, 2023.
33. Mohammad Amaz Uddin and Iqbal H Sarker. An explainable transformer-based model for phishing email detection: A large language model approach. *arXiv e-prints*, pages arXiv–2402, 2024.
34. Takashi Koide, Naoki Fukushi, Hiroki Nakano, and Daiki Chiba. Chatspamdetector: Leveraging large language models for effective phishing email detection. *arXiv preprint arXiv:2402.18093*, 2024.
35. Niek Dekker. Spam statistics: a deep dive into unwanted emails. *Eftsure Detection and Classification of Spam Email: A Machine*, Vol. 259, 2023.
36. Mostafa Raad, Norizan Mohd Yeassen, Gazi Mahabubul Alam, Bilal Bahaa Zaidan, and Aws Alaa Zaidan. Impact of spam advertisement through e-mail: A study to assess the influence of the anti-spam on the e-mail marketing. *African Journal of Business Management*, 4(11):2362, 2010.
37. Cihan Varol and Hezha M Tareq Abdulhadi. Comparision of string matching algorithms on spam email detection. In *2018 International Congress on Big Data, Deep Learning and Fighting Cyber Terrorism (IBIGDELFT)*, pages 6–11. IEEE, 2018.
38. Nikhil Agrawal and Shailendra Singh. Origin (dynamic blacklisting) based spammer detection and spam mail filtering approach. In *2016 Third International Conference on Digital Information Processing, Data Mining, and Wireless Communications (DIPDMWC)*, pages 99–104. IEEE, 2016.
39. Luíz Henrique Gomes, Fernando DO Castro, Virgílio AF Almeida, Jussara M Almeida, Rodrigo B Almeida, and Luis MA Bettencourt. Improving spam detection based on structural similarity. *SRUTI*, 5:12–12, 2005.
40. Asif Karim, Sami Azam, Bharanidharan Shanmugam, Krishnan Kannoorpatti, and Mamoun Alazab. A comprehensive survey for intelligent spam email detection. *IEEE Access*, 7:168261–168295, 2019.
41. Qussai Yaseen et al. Spam email detection using deep learning techniques. *Procedia Computer Science*, 184:853–858, 2021.
42. Sriram Srinivasan, Vinayakumar Ravi, Mamoun Alazab, Simran Ketha, Ala' M Al-Zoubi, and Soman Kotti Padannayil. Spam emails detection based on distributed word embedding with deep learning. *Machine Intelligence and Big Data Analytics for Cybersecurity Applications*, pages 161–189, 2021.
43. Putra Wanda. Gruspam: robust e-mail spam detection using gated recurrent unit (gru) algorithm. *International Journal of Information Technology*, 15(8):4315–4322, 2023.
44. Sultan Zavrak and Seyhmus Yilmaz. Email spam detection using hierarchical attention hybrid deep learning method. *Expert Systems with Applications*, 233:120977, 2023.

45. Razan Ghanem and Hasan Erbay. Spam detection on social networks using deep contextualized word representation. *Multimedia Tools and Applications*, 82(3):3697–3712, 2023.
46. Maxime Labonne and Sean Moran. Spam-t5: Benchmarking large language models for few-shot email spam detection. *arXiv preprint arXiv:2304.01238*, 2023.
47. Yuwei Wu, Shijing Si, Yugui Zhang, Jiawen Gu, and Jedrek Wosik. Evaluating the performance of chatgpt for spam email detection. *arXiv preprint arXiv:2402.15537*, 2024.
48. Konstantinos I Roumeliotis, Nikolaos D Tselikas, and Dimitrios K Nasiopoulos. Next-generation spam filtering: Comparative fine-tuning of llms, nlps, and cnn models for email spam classification. *Electronics*, 13(11):2034, 2024.
49. Yuantian Miao, Zichan Ruan, Lei Pan, Jun Zhang, and Yang Xiang. Comprehensive analysis of network traffic data. *Concurrency and Computation: Practice and Experience*, 30(5):e4181, 2018.
50. Farzaneh Izak Shiri, Bharanidharan Shanmugam, and Norbik Bashah Idris. A parallel technique for improving the performance of signature-based network intrusion detection system. In *2011 IEEE 3rd International Conference on Communication Software and Networks*, pages 692–696. IEEE, 2011.
51. Zhengbing Hu, Roman Odarchenko, Sergiy Gnatyuk, Maksym Zaliskyi, Anastasia Chaplits, Sergiy Bondar, and Vadim Borovik. Statistical techniques for detecting cyberattacks on computer networks based on an analysis of abnormal traffic behavior. *International Journal of Computer Network and Information Security*, 9(6):1, 2020.
52. Sanjeev Patel, Akash Gupta, Suman Kumari, Manjeet Singh, Vishnu Sharma, et al. Network traffic classification analysis using machine learning algorithms. In *2018 International Conference on Advances in Computing, Communication Control and Networking (ICACCCN)*, pages 1182–1187. IEEE, 2018.
53. Markus Ring, Daniel Schlör, Dieter Landes, and Andreas Hotho. Flow-based network traffic generation using generative adversarial networks. *Computers & Security*, 82:156–172, 2019.
54. Shengzhe Xu, Manish Marwah, Martin Arlitt, and Naren Ramakrishnan. Stan: Synthetic network traffic generation with generative neural models. In *Deployable Machine Learning for Security Defense: Second International Workshop, MLHat 2021, Virtual Event, August 15, 2021, Proceedings 2*, pages 3–29. Springer, 2021.
55. Mustafizur R. Shahid, Gregory Blanc, Houda Jmila, Zonghua Zhang, and Hervé Debar. Generative deep learning for internet of things network traffic generation. In *2020 IEEE 25th Pacific Rim International Symposium on Dependable Computing (PRDC)*, pages 70–79, 2020.
56. Danial Khosh Kholgh and Panos Kostakos. Pac-gpt: A novel approach to generating synthetic network traffic with gpt-3. *IEEE Access*, 11:114936–114951, 2023.
57. Jian Qu, Xiaobo Ma, and Jianfeng Li. Trafficgpt: Breaking the token barrier for efficient long traffic analysis and generation, 2024.
58. Khen Bo Kan, Hyunsu Mun, Guohong Cao, and Youngseok Lee. Mobile-llama: Instruction fine-tuning open-source llm for network analysis in 5g networks. *IEEE Network*, 2024.
59. Shilin He, Pinjia He, Zhuangbin Chen, Tianyi Yang, Yuxin Su, and Michael R Lyu. A survey on automated log analysis for reliability engineering. *ACM Computing Surveys (CSUR)*, 54(6):1–37, 2021.
60. Anne Bouillard, Marc-Olivier Buob, Maxime Raynal, and Achille Salaün. Log analysis via space-time pattern matching. In *2018 14th International Conference on Network and Service Management (CNSM)*, pages 303–307. IEEE, 2018.

61. Tianzhu Zhang, Han Qiu, Gabriele Castellano, Myriana Rifai, Chung Shue Chen, and Fabio Pianese. System log parsing: A survey. *IEEE Transactions on Knowledge and Data Engineering*, 2023.
62. Nimrod Busany and Shahar Maoz. Behavioral log analysis with statistical guarantees. In *Proceedings of the 38th International Conference on Software Engineering*, pages 877–887, 2016.
63. Qimin Cao, Yinrong Qiao, and Zhong Lyu. Machine learning to detect anomalies in web log analysis. In *2017 3rd IEEE International Conference on Computer and Communications (ICCC)*, pages 519–523. IEEE, 2017.
64. Rakesh Bahadur Yadav, P Santosh Kumar, and Sunita Vikrant Dhavale. A survey on log anomaly detection using deep learning. In *2020 8th International Conference on Reliability, Infocom Technologies and Optimization (Trends and Future Directions)(ICRITO)*, pages 1215–1220. IEEE, 2020.
65. Christian Toemmel. Catch me if you gan: Using artificial intelligence for fake log generation. *arXiv preprint arXiv:2112.12006*, 2021.
66. He Mingshu, Jin Lei, Wang Xiaojuan, and Li Yuan. Web log classification framework with data augmentation based on gans. *The Journal of China Universities of Posts and Telecommunications*, 27(5):34, 2020.
67. Shiyi Kong, Jun Ai, Minyan Lu, and Yiang Gong. Grand: Gan-based software runtime anomaly detection method using trace information. *Neural Networks*, 169:365–377, 2024.
68. Amey Wadekar, Tanishq Gupta, Rohit Vijan, and Faruk Kazi. Hybrid cae-vae for unsupervised anomaly detection in log file systems. In *2019 10th International Conference on Computing, Communication and Networking Technologies (ICCCNT)*, pages 1–7. IEEE, 2019.
69. Wanli Yuan, Shi Ying, Xiaoyu Duan, Hailong Cheng, Yishi Zhao, and Jianga Shang. Pve: A log parsing method based on vae using embedding vectors. *Information Processing & Management*, 60(5):103476, 2023.
70. Zhihan Jiang, Jinyang Liu, Zhuangbin Chen, Yichen Li, Junjie Huang, Yintong Huo, Pinjia He, Jiazhen Gu, and Michael R Lyu. Llmparser: A llm-based log parsing framework. *arXiv preprint arXiv:2310.01796*, 2023.
71. Priyanka Mudgal and Rita Wouhaybi. An assessment of chatgpt on log data. In *International Conference on AI-generated Content*, pages 148–169. Springer, 2023.
72. F Tchakounté and P Dayang. System calls analysis of malwares on android. *International Journal of Science and Technology*, 2(9):669–674, 2013.
73. Mikhail Dmitriev. Profiling java applications using code hotswapping and dynamic call graph revelation. In *Proceedings of the 4th International Workshop on Software and Performance*, pages 139–150, 2004.
74. Yongkai Fan, Jing Lei, Cong Peng, Jinghan Wang, Jiaxu Liu, Guanqun Zhao, and Jianrong Bai. Software malicious behavior analysis model based on system call and function interface. In *2019 IEEE 9th Annual International Conference on CYBER Technology in Automation, Control, and Intelligent Systems (CYBER)*, pages 59–64. IEEE, 2019.
75. Amitabh Srivastava, Jay Thiagarajan, and Craig Schertz. Efficient integration testing using dependency analysis. *Microsoft Research, TechReport MSR-TR-2005-94*, 2005.
76. Yasir Malik, Carlos Renato Salim Campos, and Fehmi Jaafar. Detecting android security vulnerabilities using machine learning and system calls analysis. In *2019 IEEE 19th International Conference on Software Quality, Reliability and Security Companion (QRS-C)*, pages 109–113. IEEE, 2019.

77. Xi Xiao, Shaofeng Zhang, Francesco Mercaldo, Guangwu Hu, and Arun Kumar Sangaiah. Android malware detection based on system call sequences and lstm. *Multimedia Tools and Applications*, 78:3979–3999, 2019.
78. Wenqi Xie, Shengwei Xu, Shihong Zou, and Jinwen Xi. A system-call behavior language system for malware detection using a sensitivity-based lstm model. In *Proceedings of the 3rd International Conference on Computer Science and Software Engineering*, pages 112–118, 2020.
79. Jayesh Soni, Suresh K Peddoju, Nagarajan Prabakar, and Himanshu Upadhyay. Comparative analysis of lstm, one-class svm, and pca to monitor real-time malware threats using system call sequences and virtual machine introspection. In *International Conference on Communication, Computing and Electronics Systems: Proceedings of ICCCES 2020*, pages 113–127. Springer, 2021.
80. Hamad H Alsheraifi, Hussain M Sajwani, Saeed M Aljaberi, Abdelrahman A Alblooshi, Ali H Alhashmi, Saoud A Sharif, and Ernesto Damiani. Using generative adversarial networks to simulate system calls of malicious android processes. In *2022 IEEE International Conference on Big Data (Big Data)*, pages 6516–6521. IEEE, 2022.

7 Generative AI for Adversarial Learning in Cybersecurity

Deep learning has evolved into a powerful and efficient framework capable of tackling a wide range of complex learning challenges that were previously difficult to address with traditional machine learning techniques. Over the past few years, deep learning has advanced significantly, achieving human-level performance or even surpassing it in several tasks. However, deep learning systems are susceptible to security threats posed by crafted adversarial examples, which, although imperceptible to humans, can cause models to misclassify outputs. This chapter will delve into the application of generative AI in adversarial learning within cybersecurity and discuss strategies for defending against such attacks.

7.1 ADVERSARIAL MACHINE LEARNING

Recently, deep learning has been used extensively for diverse applications thanks to its convincing performance. More importantly, these methods are the core components in a number of safety-critical applications such as autonomous vehicles, health-care applications and intelligent defense systems. However, emerging adversarial learning attacks pose a serious security threat to deep learning systems. The existence of the adversarial phenomenon in machine learning is one of the main obstacles hindering the development of these models. Subsequently, investigating and addressing the challenges posed by adversarial phenomena is of great importance and it has received paramount interest from the research community.

In this section, we will investigate the basic concepts of adversarial machine learning. The applications of generative AI to adversarial learning and the defense strategies will be discussed in the next two sections.

7.1.1 OBJECTIVE OF ADVERSARIAL LEARNING

Adversarial machine learning involves the study of attacks targeted at machine learning models and the development of strategies to defend against these attacks. Specifically, adversarial machine learning studies the techniques that are used to exploit vulnerabilities of a given machine learning model [1] and the methods to mitigate these attacks. The adversarial phenomenon of machine learning is the phenomenon when a machine learning model makes inconsistent or unexpected predictions for some specific inputs. In other words, the adversarial phenomenon is the scenario when the model could give totally different predictions from human prediction.

DOI: 10.1201/9781003502531-7

The objectives of adversarial machine learning include several key areas aimed at enhancing the security and robustness of machine learning models. The following are the primary objectives of adversarial learning:

- **Understanding Vulnerabilities**: Adversarial learning aims to analyze and comprehend the inherent weaknesses in machine learning models that make them susceptible to adversarial attacks. This involves studying the factors that contribute to these vulnerabilities, such as model architecture, training data, and decision boundaries.
- **Developing Attack Techniques**: Adversarial learning attempts to create and evaluate various adversarial attack methods, including white-box, black-box, and grey-box attacks. This helps in illustrating the potential risks and understanding how adversaries might exploit model weaknesses.
- **Designing Robust Defenses**: Another objective of adversarial learning is to develop effective strategies and techniques to enhance the resilience of the machine learning models against adversarial inputs. This includes methods like adversarial training, input preprocessing, ensemble approaches, and regularization techniques.
- **Evaluating Security**: Adversarial learning also rigorously assesses the security of machine learning models by testing them against a range of adversarial attacks. This evaluation helps determine the effectiveness of existing defenses and identify areas for improvement.

While conventional machine learning attempts to improve the consistency between machine learning models and humans, adversarial learning focuses on exploring the adversarial phenomenon to exploit machine learning models. The objective of the adversary attack is to cause the model to predict incorrectly. In essence, an adversarial attack involves providing an input x^* to a machine learning system with the intention of causing it to make an incorrect prediction. The objectives of adversarial attacks can be broadly categorized based on their impact on the system's output as follows:

1. **Confidence reduction**: The attackers attempt to reduce the confidence score of the target model when predicting the legitimate samples. For example, the attackers try to influence the model to make it return a low score for a malignant input image. Subsequently, the cancer patients could be ignored without treatment, and this may result in serious consequences for them.
2. **Mis-classification**: Attackers aim to alter the model's output for a given input example so that it is classified as any class other than its original class. For instance, a legitimate input might be misclassified as any class other than the legitimate class.
3. **Targeted mis-classification**: The attackers attempt to create data samples that force the model to classify them to a specific target class. For example, the attackers contaminate the training data to make the model to predict any input sample to be a legitimate class.

4. **Source/Target mis-classification**: Attackers attempt to manipulate the model into predicting a specific input as a particular target class. For example, they might try to make the model classify a malignant input as belonging to the normal class.

Generally, three different paradigms have been developed to exploit the adversarial phenomenon in different stages of the machine learning life-cycle. These paradigms include training-time adversarial attack, testing-time adversarial attack and exploratory attack [2]. Although the objectives of these paradigms are the same, their developments are almost independent and different. In the following subsection, we will further investigate them in detail.

7.1.2 ADVERSARIAL ATTACKS CLASSIFICATION

There are usually two methods to categorize adversarial attacks. The first method is based on the stages of the life-cycle of a machine learning system in which attackers launch attacks. The second method is based on the information that attackers can use to conduct attacks.

a) Classification based on the stages of the machine learning life-cycle

In the first method of categorizing, there are three main types of adversarial attacks including poisoning attack at training phase, evasion attack at the testing phase and exploratory attack at the deploying and running phase [3].

Evasion attack: Evasion attack is one of the most common types of attacks in the adversarial setting [4]. This attack consists of carefully crafting the input samples at test time to make them misclassified. The attackers do not have access to the training data. Specifically, an adversary may subtly alter the features of an input—such as modifying pixel values in an image or slightly changing the text in a message—while maintaining its original intent, aiming to exploit vulnerabilities in the model's decision-making process. These perturbations are often imperceptible to human observers but can significantly confuse machine learning algorithms, leading to incorrect predictions and potentially allowing harmful activities to go undetected. Figure 7.1 presents an example of evasion attacks in which the image of cat is misclassified into the fox when the input is modified by adding with a noise.

As machine learning continues to be integrated into security systems, understanding and mitigating evasion attacks is crucial for maintaining the integrity and

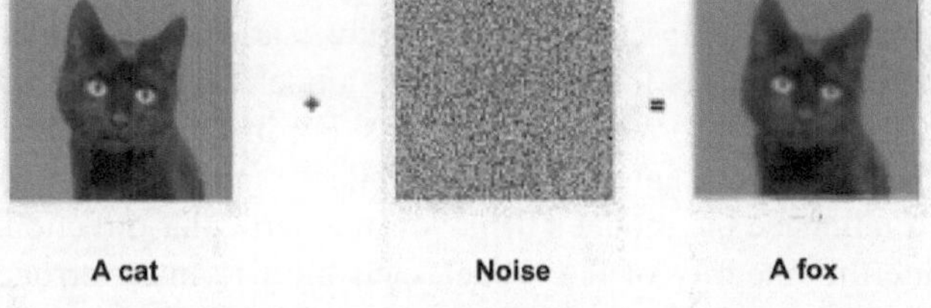

Figure 7.1 A demonstration evasion attack

effectiveness of these models. One of the approaches to mitigate evasion attacks is building robust machine learning models during the training and validation steps. This often involves exposing the model to a diverse range of adversarial examples to improve its resilience. Another approach is implementing preprocessing techniques, such as normalization or feature scaling, that can help reduce the impact of small perturbations in input data. Additionally, techniques like image denoising can mitigate the effects of adversarial manipulations in visual data.

Poisoning attack: Poisoning attacks, also known as contamination attacks, are a prevalent type of threat against machine learning systems. These attacks occur during the training phase of the machine learning lifecycle. An attacker attempts to compromise the training process by either injecting specially crafted samples into the training data or altering the labels of existing data samples [6]. There are generally two main strategies for poisoning training data, as illustrated in Figure 7.2.

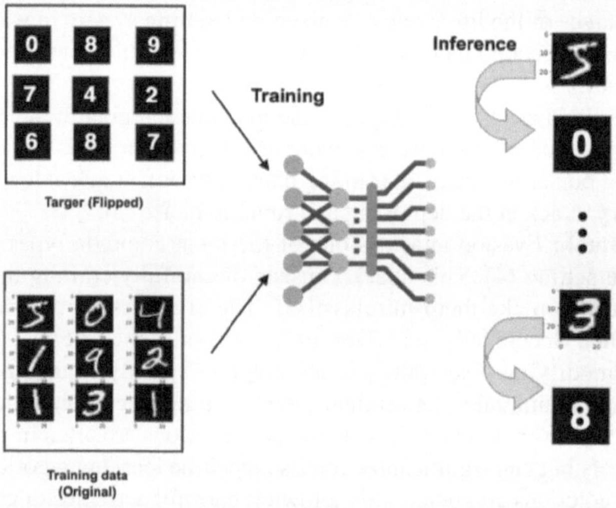

Figure 7.2 Poisoning attack by label-flipping

- *Data injection*: In this strategy, the attacker lacks direct access to the existing training data, but has the capability to add new data to the training set. Consequently, they can compromise the target model by inserting adversarial samples into the dataset. Specifically, adversaries strategically craft and insert poisoned samples that can skew the model's learning, leading to biased predictions or incorrect classifications. For instance, an attacker might add mislabeled data points or data that reflects specific characteristics intended to influence the model's behavior in a particular direction. This can undermine the integrity of the model, causing it to make erroneous decisions in real-world applications. The impact of data injection can be severe, especially in critical areas such as finance, healthcare, and cybersecurity, where compromised model performance can lead to significant operational risks. To combat this threat, developers must implement robust data validation techniques, anomaly detection systems, and continuous monitoring of model inputs to ensure the integrity and reliability of their machine learning systems.

- *Data modification*: In this strategy, the attacker has full access to the training data. He poisons the training data by directly altering its labels or some feature values. After that, the poisoned data is used for training the target model. More precisely, adversaries may change labels, introduce subtle perturbations, or replace legitimate data with malicious content, all aimed at influencing the model's output in a desired direction. For example, an attacker might modify a few features of training samples to cause the model to misclassify similar, unaltered inputs during inference. This can lead to reduced accuracy, biased predictions, or even the introduction of vulnerabilities that can be exploited in real-world scenarios. The consequences of data modification can be particularly severe in sensitive domains like finance and healthcare, where reliance on accurate model predictions is critical. To mitigate the risk of such attacks, researchers and practitioners should implement rigorous data integrity checks, employ robust training algorithms that can withstand adversarial influences, and continually monitor model performance for unexpected behavior.

Poisoning attacks are particularly challenging in federated learning [7] due to the decentralized nature of the training process, where multiple clients collaboratively update a global model without sharing their raw data. In these attacks, malicious participants can submit harmful model updates that introduce biases or inaccuracies, effectively compromising the integrity of the global model. For instance, an attacker might manipulate their local data or model parameters to skew the training process, leading to degraded performance or biased predictions. Figure 7.3 shows an example of a data poisoning attack on a federated learning system where the attack contaminates the local data of Client 1 and then sends the poisoned model to the server for aggregation. This leads to the global model being poisoned and this poisoned model is then broadcast to all other clients.

Since federated learning relies on aggregating diverse updates from various clients, detecting the poisoning attacks can be particularly difficult [8]. To mitigate the risks associated with poisoning attacks, federated learning systems can employ robust aggregation techniques to identify and down-weight suspicious updates, implement anomaly detection algorithms to flag outlier behaviors, and continuously monitor the global model's performance to ensure its reliability. These strategies aim to enhance the resilience of federated learning against potential vulnerabilities introduced by malicious participants.

Exploratory attack: Exploratory attacks aim to acquire as much information as possible about the machine learning model and the patterns within its training data [9]. These attacks often involve systematically testing the model's behavior by providing various inputs to identify weaknesses or biases in its predictions. For

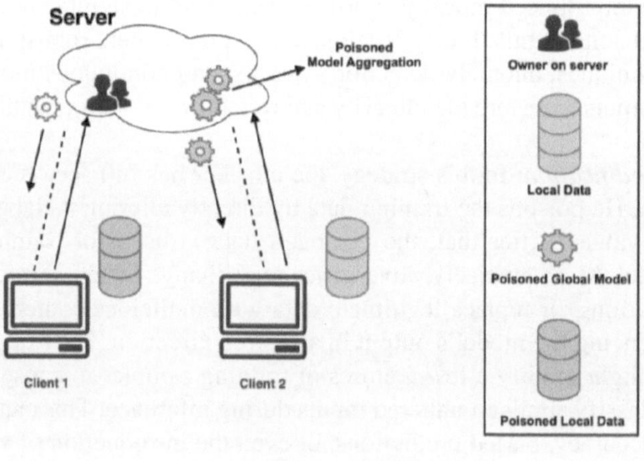

Figure 7.3 Data poisoning attack to federated learning

instance, attackers craft the adversarial examples, input them to the machine learning system and analyze the output of the system to understand its behavior. An attacker might use techniques such as input perturbation, where they subtly modify input features to observe changes in the model's output, thus gaining insights into its decision boundaries and potential blind spots. Figure 7.4 demonstrates an example of an exploratory attack where the attacker sends different input questions to a machine learning service for spam filters. After getting the knowledge about the behavior of the service, the attacker can craft the spam message to bypass the filtering system.

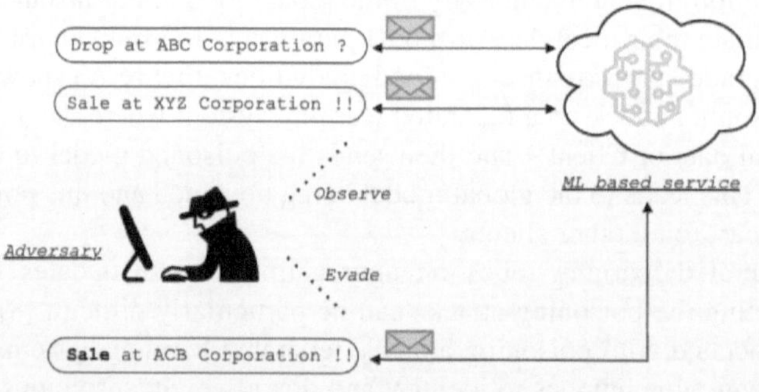

Figure 7.4 Exploratory attacks on a machine learning based spam filtering system.

The goal of exploratory attacks is to gather information that can later be used for more targeted attacks, such as adversarial attacks or data poisoning. By understanding how the model responds to different types of input, adversaries can identify specific vulnerabilities that can be exploited to degrade performance or manipulate outcomes. There are four common types of exploratory attacks as follows:

- *Model Inversion Attack*: Model Inversion is a type of attack that aims at inverting a given pre-trained model to recover a private training dataset, such as images, texts, and graphs. Different from the membership inference attack or the property inference attack that only reveal partial information about the training data, MI enables an adversary to fully reconstruct private training samples, which has raised growing concern.
- *Inference Attack*: Inference attacks are a type of adversarial attack that seeks to extract knowledge about pre-trained machine learning models from partial information. By analyzing large datasets, attackers can illicitly gain insights into the training dataset or its contents without direct access. Inference attacks can be categorized into two main types: membership inference attacks (MIAs) and property inference attacks (PIAs). Membership inference attacks aim to determine if a specific record is part of the target model's training set. For example, attackers might use MIAs to identify common customers between themselves and their competitors to develop targeted business strategies. Property inference attacks primarily use publicly available information about an instance to infer unknown attributes, such as sexual orientation, political inclination, or gender, often in online social networks. Over time, the concept of PIAs has expanded to infer broader properties of the training set, like gender or age ratios, becoming more prevalent.

There are several countermeasures to mitigate the exploratory attacks on machine learning models [9]. One effective strategy is to implement input validation and sanitization techniques that ensure only legitimate and expected data formats are processed by the model. Another is to employ adversarial training that helps the model learn to recognize and withstand perturbations, making it less susceptible to exploratory probing. Regular audits and monitoring of model behavior can also be beneficial, as they allow for the detection of unusual patterns or input distributions that may indicate exploratory attacks. Furthermore, incorporating ensemble methods can enhance model robustness by aggregating predictions from multiple models, making it more challenging for attackers to discern vulnerabilities. Finally, establishing strict access controls and employing differential privacy techniques can limit the information available to potential attackers, thereby reducing the likelihood of successful exploratory attacks and safeguarding the integrity of the machine learning system.

b) Classification based on the information used by attackers

In the second method of categorizing, the adversarial attacks can be broadly classified into either White-Box or Black-Box attacks.

White-box attacks: The first attack in this category is White-box. White-box attack [10] assumes that the attacker has total knowledge about the model used in machine learning systems (e.g., type of neural networks along with number of layers, number of neurons in each layer, etc.). The attacker also knows the algorithm used in training (e.g., gradient-descent optimization), and he/she can access the training data. Additionally, the attacker has access to the optimized parameters of the trained model, providing them with comprehensive insights. This allows attackers to create highly effective adversarial examples tailored to exploit specific vulnerabilities in the model. For example, they can compute the gradients of the loss function with respect to the input data, enabling them to make precise adjustments that result in misclassification while ensuring these changes remain imperceptible to human observers.

Generally, there are six main techniques used to conduct white-box attacks:

- The Fast Gradient Sign Method (FGSM) generates adversarial examples by leveraging the gradient of the loss function with respect to the input data. It modifies the input by perturbing it in the direction of the gradient to maximize the loss, resulting in a manipulated input that is likely to be misclassified.
- Projected Gradient Descent (PGD) [12] extends the FGSM method by applying iterative small perturbations to the input. After each iteration, the perturbed input is projected back onto a predefined norm ball to ensure it remains within a specified distance from the original input. This iterative process typically produces stronger adversarial examples.
- Carlini & Wagner (C&W) Attack [13]: This advanced attack formulates the creation of adversarial examples as an optimization problem. Its objective is to minimize input distortion while ensuring the model produces incorrect classifications. The C&W attack is highly effective, even against models equipped with robust defense mechanisms.
- DeepFool [14]: DeepFool is a simple and accurate method to fool deep neural networks. This attack iteratively computes the smallest perturbation necessary to change the classification of an input. It iteratively approximates the decision boundary of the model, making it a precise and efficient approach for generating adversarial examples.
- Jacobian-based Saliency Map Attack (JSMA) [15]: JSMA uses the Jacobian matrix of the model's output to identify which features to perturb for maximum impact on classification. By selectively modifying certain features, it creates adversarial examples that are often less noticeable to human observers.
- Transfer Attacks [16]: Transfer learning is not only applied to train a machine learning model that can leverage the knowledge from one domain to another domain. This technique can also be employed by attacks. An adversary can train a surrogate model and generate adversarial examples that may transfer effectively to the target model.

To defend against white-box attacks, we can employ various strategies such as adversarial training, which involves training the model on a mixture of clean and adversarial examples to enhance its robustness [17]. Additionally, implementing input preprocessing techniques and using ensemble methods can help mitigate the risks posed by such attacks. Continuous monitoring and updating of models are also crucial to ensure they remain resilient against evolving attack strategies.

Black-box attacks. The second attack is a Black-box. Black-box attack [18] is different from the white-box attack in the way that it assumes no knowledge about the model is available to attackers. Instead of exploiting detailed knowledge of the model, attackers rely on querying the model with various inputs to observe its outputs, thereby inferring information about its decision boundaries and vulnerabilities. Black-box attacks can be particularly concerning because they can be executed without insider knowledge, making them applicable in real-world scenarios where adversaries may only interact with the model through its API.

Black box attacks can be further categorized into two classes. The first class is the non-adaptive black-box attack [19]. In a non-adaptive black-box attack, the attacker lacks knowledge about the model itself but has access to the target model's training data. The attacker can train a new model using a different algorithm on this dataset to approximate the behavior of the target model. Subsequently, they craft adversarial examples using their trained model and apply these crafted inputs to the target model to induce mispredictions.

The second class is the adaptive black-box attack [20]. In an adaptive black-box attack, the attacker lacks knowledge about the model's architecture, the distribution of the training data, and the training process. However, they can interact with the target model by sending queries and observing the responses. The attacker queries the target model, labels the samples based on the returned values, and then trains a surrogate model using this labeled data. This surrogate model is used to generate adversarial samples that cause the target model to mispredict malicious data.

There are various techniques used to conduct black-box attacks. These techniques often rely on limited information about the model, primarily using its outputs to infer vulnerabilities. Following are some common techniques used in black-box attacks:

- Query-based Attacks [21]: In these attacks, adversaries query the model with multiple inputs to collect output predictions. By analyzing these outputs, they can infer the model's decision boundaries. Two main approaches to carry out query-based attacks include Random Sampling and Gradient Estimation. Generating random inputs and observing the model's predictions to identify patterns. Gradient Estimation uses the outputs of the model to approximate gradients, allowing for the generation of adversarial examples.
- Zeroth Order Optimization (ZOO) [22]: ZOO techniques estimate the gradients of the model using only the output scores. By perturbing the input and measuring the change in output, attackers can perform optimization to create adversarial examples without direct access to the model's internals.
- Boundary Attack [23]: This approach starts with a misclassified input and iteratively moves it towards the decision boundary of the model in a manner that maintains its adversarial nature. By querying the model's output, the attacker can navigate the input space to create effective adversarial examples.
- Sign-Optimized Perturbations: In this technique, the attacker adds small, carefully chosen perturbations to the input based on the sign of the gradient estimated through query outputs. This method efficiently creates adversarial examples while minimizing the number of queries needed.

- Adversarial Patch [24]: This method involves creating a physical or digital patch that, when placed on an input (like an image), causes the model to misclassify it. The patch can be optimized in a black-box setting by iteratively testing its effectiveness through model queries.
- Model Distillation [25]: Knowledge distillation, or model distillation, involves transferring knowledge from a larger model to a smaller one. This technique is commonly used to simplify deep neural network models, but is also exploited by attackers for black-box attacks. Specifically, adversaries train a simplified model that mimics the behavior of the target model by learning from its outputs. Using this distilled model, they can craft adversarial examples that are likely to impact the original model's predictions.

Mitigating black-box attacks on machine learning models requires a combination of strategies aimed at enhancing model robustness and security. Popular countermeasures include adversarial training, defensive distillation, and input preprocessing.

Grey-box attacks. The third type of attack is the grey-box attack [26]. Grey-box attacks represent a hybrid approach, combining elements of both white-box and black-box attacks. In this scenario, the attacker possesses partial knowledge about the model's architecture, parameters, or training data. Unlike white-box attacks, where the adversary has complete access, or black-box attacks, where they have no internal knowledge, grey-box attacks leverage this intermediate level of information to create effective adversarial examples.

In grey-box scenarios, attackers might possess knowledge about the model's input-output behavior or its general structure but lack complete details. This allows them to employ techniques that leverage both the observed outputs from the model and the known characteristics of the model type. For instance, an attacker might use a known architecture (like a convolutional neural network) to apply targeted perturbations to the input, informed by their understanding of how similar models behave.

Grey-box attacks can be particularly dangerous because they combine the advantages of both white-box and black-box strategies, allowing for more nuanced and targeted adversarial examples. Defending against such attacks requires robust strategies, including adversarial training, model regularization, and continuous monitoring of model performance to detect unusual patterns that may indicate an ongoing attack.

7.2 GENERATIVE AI FOR ADVERSARIAL ATTACKS

This section presents an overview of applying generative AI to create adversarial attacks. The section first analyses the utilisation of generative models for evasion attacks. After that, their usage in poisoning attacks is highlighted. Finally, the section discusses some researches on the application of generative models to exploratory attacks.

7.2.1 EVASION ATTACKS

Evasion attacks are perhaps the most prevalent type of attack on machine learning systems. These attacks typically involve carefully altering input examples to deceive the model into making incorrect predictions and evading detection. Adversarial examples are crafted from input samples with the intent of causing a machine learning model to produce outputs that contradict human expectations. For instance, an input originally belonging to class A might be subtly modified in a way that remains imperceptible to humans, yet the model incorrectly classifies it as class B [4].

Attackers employ several methods to modify input samples and create adversarial examples. One common approach involves using information from the gradient of the loss function. The attacker treats the crafting process as an optimization problem, starting with a benign example. They then take steps in the direction opposite to the gradient of the loss function with respect to the input, targeting a specific class [5].

While crafting input examples is a common method for creating adversarial samples, it may not always be feasible due to the unavailability of real data samples. In such cases, attackers can conduct evasion attacks by utilizing generative models to produce artificial examples that deceive target systems. The approach involves training a generative model on the original benign data to learn its distribution. Then, the generator of the trained model is used to create adversarial examples by manipulating the latent space. This process involves two steps: first, a benign data point is projected into the latent space, and then, the attacker searches nearby in the latent space to find another data point that, when translated back to the original space, is misclassified by the target model. This method enables attackers to generate entirely new adversarial examples without relying on perturbations.

Using generative models for crafting adversarial examples is popular in the image process domain. For example, Xiao et al. proposed AdvGAN [27] to generate adversarial examples. The authors train a conditional adversarial network to directly produce adversarial examples and apply these examples to both black-box attacks and black-box attacks. Their experimental results show the effectiveness of the generative examples against some popular defenses on the MNIST dataset. Song et al. [28] proposed to use Auxiliary Classifier Generative Adversarial Network (AC-GAN) to synthesize unrestricted adversarial examples entirely from scratch. Specifically, they train an AC-GAN model to estimate the class-conditional distribution over data samples. Then, the trained AC-GAN model is used to generate adversarial samples that are mis-predicted by the target system. Xiao et al. [29] propose to use a pre-trained model based on GAN models to generate adversarial patches to face recognition models. They show that the adversarial examples optimized on a manifold support to build a surrogate model that responds similarly to the target models. Dolatabadi et al. proposed AdvFlow [30] that uses normalizing flows to model the density function of adversarial examples around a given target input. They then apply the generated adversarial example to a novel black-box attack on image classifiers. The experimental results show the competitive performance of AdvFlow against defended models.

Although modifying input examples is a common approach for generating adversarial samples, it is not always practical due to the lack of access to real data samples. In such situations, attackers can leverage generative models to create artificial examples that deceive target systems. This technique involves training a generative model on benign data to capture its distribution and then using the model's generator to produce adversarial examples by manipulating the latent space. The process consists of two steps: first, a benign data point is mapped into the latent space, and then the attacker searches within this space to identify another point that, when transformed back into the original space, is misclassified by the target model. This method enables attackers to generate entirely new adversarial examples without relying on perturbations.

Hu et al. [33] introduced MalGAN, a technique for crafting adversarial examples aimed at malware detection systems. These systems operate as black boxes for attackers, concealing their internal processes. As a result, attackers utilize black-box attacks to deduce the features leveraged by the malware detection algorithm. The fundamental idea behind MalGAN is to trick the malware detection system into incorrectly classifying benign programs as malware. It utilizes a dataset of programs represented as binary vectors based on API features. MalGAN incorporates a black-box detection model (oracle) within its discriminator and generator framework, where the generator creates adversarial examples and the discriminator simulates the oracle. These adversarial examples can effectively evade the black-box detector, showcasing their transferability across different classifiers. Nevertheless, retraining the detector using these adversarial examples strengthens its resistance to such attacks.

Lin et al.[34] developed the IDSGAN model to generate adversarial attacks that can deceive and circumvent intrusion detection systems (IDS). This model treats the IDS as a black box, employing adversarial techniques to mislead it. IDS systems typically utilize classifiers like DNN or SVM on cybersecurity datasets such as NSL-KDD. IDSGAN comprises a generator and a discriminator, with the discriminator mimicking the behavior of the black-box IDS. The generator produces adversarial examples by making minimal perturbations only to non-functional features of the attack data. Experimental results show that IDSGAN effectively generates adversarial examples that evade detection by the black-box IDS, resulting in lower detection rates and higher evasion rates. Similarly, Chauhan et al.[35] introduced an attack model based on GAN to create adversarial polymorphic attacks. This model aims to assess the vulnerability and resilience of IDS systems. The polymorphic attacks continuously modify their features to produce various iterations, with the goal of identifying the most effective attack that can bypass the IDS while maintaining its functional characteristics.

Recent progress in large language models (LLMs) has significantly increased the prevalence of adversarial attacks leveraging these models. For instance, ChatGPT has been shown to facilitate advanced attack scenarios, including social engineering schemes [36], phishing campaigns [37], and the creation of malicious software [38]. Karanjai [37] highlights that LLMs can generate highly convincing phishing emails capable of bypassing spam filters and deceiving recipients, though the effectiveness of such emails depends on the specific LLM and its training data. Additionally, ChatGPT has demonstrated the ability to produce sophisticated and obfuscated malware, incorporating evasive coding techniques. Beckerich [38] presents a proof-of-concept showing that ChatGPT can be utilized to distribute malicious software that evades detection and establishes communication with a server to execute commands targeting victim systems. These findings underscore the significant cybersecurity risks posed by the misuse of LLM services.

Generative models offer a significant advantage in creating adversarial examples, particularly in domains where valid perturbations are challenging to identify. For instance, in malware analysis, obtaining malware samples for perturbation can be difficult. In such cases, researchers can use generative models like GANs [39] or auto-aggressive models [40] to produce adversarial examples that appear benign to the target classifier. Furthermore, adding noise to benign samples may alter their meaning, making tasks like generating adversarial text more complex without changing the intended meaning of a sentence [41]. Importantly, generative models can create adversarial examples without relying on perturbation, allowing them to bypass certain defenses designed to detect and eliminate perturbations. As a result, these systems struggle to identify adversarial examples generated by such models.

It should be noted that generative models are not only used to carry out adversarial attacks but they themselves can be the victim of these attacks. For example, adversarial attacks have been used to launch attacks on generative models in natural language processing [42] and image compression [43].Wallace et al. [42] demonstrate the existence of universal adversarial triggers for deep generative models in NLP tasks. These triggers are token sequences that, when appended to any input, consistently cause the target model to produce specific predictions. The authors also show that the appended words cause the GPT-2 model to output offensive text. Interestingly many of the appended sentences can bypass all models tested. Kos et al. propose three classes of attacks against VAE and VAE-GAN when these models are trained on three image classification datasets, i.e., MNIST, SVHN and CelebA. The first attack adds a classifier to the encoder of the target generative model, which can then be used to indirectly manipulate the latent representation. The second attack directly uses the VAE loss function to generate a target reconstruction image from an adversarial example. The third directly optimizes against differences in source and target latent representations. They argue that an attacker could potentially trick the decoder to return an entirely different image.

7.2.2 POISONING ATTACKS

Poisoning attacks differ from evasion attacks in that they modify inputs during the training phase rather than the testing phase. The objective of poisoning attacks is to compromise the target model. A common type of poisoning attack is the backdoor attack, which involves adding a "trigger" (such as a small pattern or a single value) to data points. This trigger is added to numerous training samples, which are then labeled as a specific class. As the model trains, it learns to associate the trigger with the target class, allowing the attacker to manipulate the model's predictions by including the trigger in new inputs.

The triggered samples can be generated by generative models. Among generative models, GAN and its variants are very widely used to create triggered samples in poisoning attacks. For example, Turner et al. [44] propose label-consistent backdoor attacks that create adversarial samples similar to their original samples using generative models. The triggered samples are then used to force the model to learn the trigger pattern. This approach has the advantage that the triggered examples are less likely to be filtered out by defenders. Munozgonzalez et al. [45] propose to use Generative Adversarial Net to craft systematic poisoning attacks against machine learning classifiers. They attempt to generate adversarial training examples that look like genuine data but can degrade the classifier's accuracy when used for training. Using GAN allows them to identify the regions of the underlying data distribution that are more vulnerable to data poisoning. The experimental results show the effectiveness of their method against machine learning models including deep networks. Keshav2021 et al. [46] leverage Generative Adversarial Text to Image Synthesis to create triggers against machine learning classifiers. Their method has three components, i.e., a generator, a discriminator, and a target classifier. They perform an extensive evaluation to prove the efficiency of their attack to compromise machine learning models, including deep networks.

Another generative model that is often used for poisoning attack is Variational AutoEncoder (VAE). Specifically, the adversarial samples are usually generated by sampling from the latent space of a VAE. For example, Ujjwal and [47] proposed a novel framework to generate adversarial examples where the latent space representation of a disentangled variational autoencoder is utilized to tamper with the inherent structure of the image while maintaining the perceptual quality intact and to act as legitimate data samples. The latent space poisoning exploits the inclination of classifiers to model the independent and identically distribution of training dataset and tricks it by producing out-of-distribution samples. The authors train a disentangled variational autoencoder, and then we add noise perturbations using a class-conditioned distribution function to the latent space under the constraint that it is misclassified to the target label. Kumar et al. [48] developed a generative adversarial model based on a Variational Autoencoder to mimic the attack on Ensemble Clustering. In particular, the generative model will simulate behaviors of both the clean basic partition and the perturbed key basic partition algorithm, and thus can launch effective attacks with less attention. The authors have conducted extensive experiments on eleven clustering benchmarks and have demonstrated that their proposed approach is effective in attacking the Ensemble Clustering algorithm under both transductive and inductive settings.

In addition, generative models can be used to speed up the generation of poisoned samples. Yang et al. [49] propose a generative method to accelerate the generation rate of the poisoned data. Specifically, they use a GAN model to generate poisoned data by updating a reward function of the loss, and the target model (discriminator) receives the poisoned data to calculate the loss w.r.t. the normal data. The experimental results show that the generative method speeds up the generating rate up to 239 times compared with the gradient method.

Generative models are also used to launch poisoning attacks to federated learning systems. Zhang et al. [50] used GANs to attack federated learning networks. An attacker first attempts to train a GAN to mimic prototypical samples of the benign participants. Then these generated samples are fully controlled by the attacker to generate the poisoning updates to the server. In the evaluation, they show that the attacker can successfully generate samples of other benign participants. Later, Zhang et al. [51] proposed a poison data generation method, named Data˙Gen, based on the generative adversarial networks (GANs). Data˙Gen relies on the iteratively updated global model parameters to regenerate samples of interested victims. They then propose a novel generative poisoning attack model, named PoisonGAN, against the federated learning framework. This model uses Data˙Gen to reduce the attack assumptions and make attacks feasible in practice. The experimental results demonstrate that these two attack models are effective in federated learning. Recently, Chen et al. [52] introduced an innovative end-to-end poisoning framework, P-GAN. Specifically, the malicious participant initially employs semi-supervised learning to train a surrogate target model. Subsequently, this participant employs a GAN-based method to produce adversarial perturbations to degrade the surrogate target model's performance. Finally, the generator is obtained and tailored for VFL poisoning.

The poisoning attacks are also conducted to generative models. Wallace et al. [53] applies poisoning attacks to NLP translation models. They insert a small number of pseudo triggers into the model, making this model difficult to identify which training samples result in the error. Their experimental results show the potential vulnerability of generative models to poisoning. Wan et al. [54] optimize their inputs and outputs using a bag-of-words approximation to the LLMs. This allows them to contribute poison examples to training datasets for LLMs allowing them to manipulate model predictions whenever a desired trigger phrase appears in the input. They evaluate their method on open-source instruction-tuned LMs. By using as few as 100 poison examples, they can cause arbitrary phrases to have consistent negative polarity or induce degenerate outputs across hundreds of held-out tasks. Moreover, they also show that larger LMs are increasingly vulnerable to poisoning and that defenses based on data filtering or reducing model capacity provide only moderate protections while reducing test accuracy. Shu et al. [55] proposed AutoPoison, an automated data poisoning pipeline. This approach naturally and coherently incorporates versatile attack goals into poisoned data with the help of an oracle LLM. They show two example attacks: content injection and over-refusal attacks, each aiming to induce a specific exploitable behavior. They quantify and benchmark the strength and the stealthiness of their data poisoning scheme. The results show that AutoPoison allows an adversary to change a model's behavior by poisoning only a small fraction of data while maintaining a high level of stealthiness in the poisoned examples. This research raises awareness of the importance of data quality for responsible deployments of LLMs.

Finally, generative models can also be used to defend agaist poisining attacks. Zhao et al. [56] proposed a poisoning defense mechanism to detect and mitigate poisoning attacks in federated learning. This mechanism uses generative adversarial networks to generate auditing data in the training procedure and removes adversaries by auditing their model accuracy. The experiments conducted on two well-known datasets, MNIST, and Fashion-MNIST show that the proposed defense method can detect and mitigate the poisoning attack. Alonso et al. [57] explored the application of autoencoders as a means to detect anomalous or fraudulent updates to Differentially Private Federated Learning (DP-FL). They assess the effectiveness in identifying anomalies by leveraging the reconstruction errors generated by autoencoders. Specifically, they proposed an approach to ensure data privacy by adding noise to the model updates before aggregation, thus preventing any individual contributor's data from being compromised.

Generally, generative models, GAN, VAEs, and LLMs have been used for poisoning attacks. On the one hand, these model can be used to carry out poisoning attacks to various machine learning systems. One the other hand, these models can help to mitigate and prevent poisoning attacks. Moreover, these generative models can themselves be the victim of poisoning attacks.

7.2.3 EXPLORATORY ATTACKS

Exploratory attacks differ from other adversarial techniques as they do not involve contaminating the training data or creating adversarial samples to evade machine learning systems. Instead, their focus is on extracting information about the target model. Specifically, exploratory attacks aim to uncover details such as the training data, model parameters, or hyperparameters [58]. These attacks compromise the privacy of the training data and can steal sensitive and valuable information about machine learning systems.

The first common exploratory attack is a model inversion attack, where the attackers attempt to extract sensitive information from machine learning models. In this attack, adversaries leverage access to a model's outputs to reconstruct input data or infer private attributes associated with the training data. By systematically querying the model and analyzing its predictions, attackers can infer details about individual training samples, potentially compromising the confidentiality of sensitive information, such as personal data or proprietary datasets. For instance, in image classification tasks, an attacker might generate images that maximize the model's output for a specific class, effectively reconstructing features of the original training images. The implications of model inversion attacks are severe, particularly in applications involving sensitive data, such as healthcare or finance, highlighting the need for robust privacy-preserving techniques, such as differential privacy, to safeguard against such vulnerabilities.

Some researchers have applied generative models to model inversion attacks and achieve state-of-the-art performance. For example, Zhang et al. [59] present a novel attack method, termed the generative model-inversion attack, which can invert deep neural networks with high success rates. Rather than reconstructing private training data from scratch, they leverage partial public information, which can be very generic, to learn a distributional prior via generative adversarial networks (GANs) and use it to guide the inversion process. The extensive experiments demonstrate that the proposed attack improves identification accuracy over the existing work by about 75% for reconstructing face images from a state-of-the-art face recognition classifier. We also show that differential privacy, in its canonical form, is of little avail to defend against our attacks. Chen et al. [60] present a novel inversion-specific GAN that can better distill knowledge useful for performing attacks on private models from public data. Particularly, they train the discriminator to differentiate not only the real and fake samples but also the soft-labels provided by the target model. Moreover, unlike previous work that directly searches for a single data point to represent a target class, they propose to model a private data distribution for each target class. The experiments show that the combination of these techniques can significantly boost the success rate of the state-of-the-art MI attacks by 150%, and generalize better to a variety of datasets and models.

Recently, Struppek et al. [61] proposed Plug & Play Attacks to address the time- and resource-consuming, inflexible, and susceptible to distributional shifts between datasets or the previous methods. Their methods relax the dependency between the target model and image prior, and enable the use of a single GAN to attack a wide range of targets, requiring only minor adjustments to the attack. Moreover, we show that powerful MIAs are possible even with publicly available pre-trained GANs and under strong distributional shifts, for which previous approaches fail to produce meaningful results. Our extensive evaluation confirms the improved robustness and flexibility of Plug & Play Attacks and their ability to create high-quality images revealing sensitive class characteristics. Nguyen et al. [62] analyzed the optimization objective of state-of-the-art model inversion algorithms, arguing that the objective is sub-optimal for achieving model inversion. Based on this, they proposed an improved optimization objective that boosts attack performance significantly. Their proposed solutions are simple and improve all state-of-the-art model inversion attack accuracy significantly. Specifically in the standard CelebA benchmark, their solutions improve accuracy by 11.8% and achieve, for the first time, over 90% attack accuracy. Their findings demonstrate that there is a clear risk of leaking sensitive information from deep learning models.

Not only being applied to carry out model inversion attacks, generative models have been used to defend against model inversion attacks. For example, Gong et al. [63] introduced a novel GAN-based defense approach against model inversion attacks. Unlike previous works that perturb the prediction vector of the model, they manipulate the training procedure of the victim model by incorporating carefully-designed GAN-based fake samples. They also adjust the loss of the inverted samples to inject misleading features into the protected label of the victim model. Additionally, they adopt the concept of continual learning to improve the utility of the model. Extensive experiments were conducted on the CelebA, VGG-Face, and VGG-Face2 datasets demonstrate that our proposed method outperforms existing defenses against state-of-the-art model inversion attacks.

The second prevalent type of exploratory attack is the membership inference attack [64], where the attacker seeks to determine if a specific individual's data was included in the training dataset. These attacks exploit the fact that machine learning models often make different predictions for data points that were part of the training dataset versus those that were not. This issue is particularly pronounced with overfitted models. However, recent studies have shown that this problem also affects models that are not overfitted [65].

Membership inference attacks have been carried out to generative models to exploit their weakness. For example, Hayes et al. [66] present the first membership inference attacks against generative models. They leverage Generative Adversarial Networks (GANs) to detect overfitting and recognize inputs that were part of training datasets, using the discriminator's capacity to learn statistical differences in distributions. They present attacks based on both white-box and black-box access to the target model, against several state-of-the-art generative models, over datasets of complex representations of faces (LFW), objects (CIFAR-10), and medical images (Diabetic Retinopathy). Hilprecht et al. [67] present two information leakage attacks that outperform previous work on membership inference against generative models. The first attack allows membership inference without assumptions on the type of the generative model. The second attack specifically targets Variational Autoencoders, achieving high membership inference accuracy. Their attacks are evaluated on two generative model architectures, Generative Adversarial Networks (GANs) and Variational Autoencoders (VAEs), trained on standard image datasets. The results show that the two attacks yield success rates superior to previous work on most data sets while at the same time having only very mild assumptions. Sun et al. [68] proposed a series of attribute-based Membership inference attacks that consider both black-box and white-box settings. The attacks are performed on the generator, and the inferences are derived by overfitting the non-trivial attributes. Additionally, they put forward a novel perspective on model generalization by evaluating the overfitting status of each individual attribute. A series of empirical evaluations in both settings demonstrates that the attacks remain stable and successful when using non-IID candidate samples. Further experiments illustrate that each attribute exhibits a distinct overfitting status.

Recently, membership inference attacks were also carried to LLMs. Galli et al. [69] introduced an efficient methodology to conduct the Membership Inference Attacks (MIA). They generate *noisy neighbors* for a target sample by adding stochastic noise in the embedding space, requiring operating the target model in inference mode only. Their findings demonstrate that this approach closely matches the effectiveness of employing shadow models, showing its usability in practical privacy auditing scenarios for LLMs. Duan et al. [70] performed a large-scale evaluation of MIAs over a suite of language models (LMs) ranging from 160M to 12B parameters. They find that MIAs barely outperform random guessing for most settings across varying LLM sizes and domains. Based on further analysis, they can identify specific settings where LLMs have been shown to be vulnerable to membership inference and show that the apparent success in such settings can be attributed to a distribution shift, such as when members and non-members are drawn from the seemingly identical domain but with different temporal ranges. Wen et al [71] present the first membership inference attack tailored to Large Language Models (LLMs) in In-Context Learning (ICL) by relying solely on generated texts without their associated probabilities. They propose four attack strategies tailored to various constrained scenarios and conduct extensive experiments on four popular large language models. Empirical results show that their attacks can accurately determine membership status in most cases, e.g., 95% accuracy advantage against LLaMA, indicating that the associated risks are much higher than those shown by existing probability-based attacks. Additionally, they propose a hybrid attack that synthesizes the strengths of the aforementioned strategies, achieving an accuracy advantage of over 95% in most cases.

Generative models are also employed for exploratory attacks. Shi et al. [72] utilize a generative adversarial network (GAN) for this purpose. Initially, the adversary conducts an exploratory (inference) attack by querying the API of an online machine learning system, specifically a classifier, with input data samples and collecting the returned labels. The GAN is then trained on this gathered data to generate synthetic training data. This generated data is used to train a classifier, which subsequently selects training and test samples based on the confidence scores it produces. Ultimately, this selected data is used to execute poisoning attacks (intended to disrupt the training process) and evasion attacks (aimed at misleading the classifier). This research demonstrates the effectiveness of GANs in enhancing exploratory attacks with minimal resource consumption (via a limited number of API calls). Overall, there is a scarcity of research on the use of generative models for exploratory attacks, highlighting a potential area for future investigation.

7.3 DEFENSE STRATEGIES

Previous research has shown that many modern machine learning algorithms can be surprisingly vulnerable to adversarial attacks. However, developing an effective defense strategy against these attacks is equally challenging. This difficulty arises from several factors:

- Constructing a theoretical model for adversarial attacks is challenging due to the complexity of crafting adversarial samples, which involves solving non-linear and non-convex optimization problems inherent to most machine learning models. The lack of appropriate theoretical tools to analyze and characterize solutions to these optimization challenges makes it difficult to provide a solid theoretical basis for evaluating the effectiveness of specific defenses in mitigating certain adversarial examples.
- The need for robustness in machine learning models: Machine learning models must typically deliver accurate outputs for all possible inputs. Enhancing the model to be robust against adversarial examples can alter its fundamental objective, potentially resulting in inaccurate predictions for valid inputs.

Most existing defense strategies struggle to address all types of adversarial attacks effectively. A defense mechanism that mitigates one type of attack may inadvertently expose vulnerabilities that attackers familiar with its design can exploit. Furthermore, implementing these defenses often introduces performance overheads, potentially degrading the model's prediction accuracy. In this section, we explore recent advancements in defense mechanisms against adversarial attacks, along with their associated implementation challenges.

A significant body of research has focused on developing practical defenses against adversarial examples. Generally, these defense mechanisms can be categorized based on their implementation methods.

7.3.1 ADVERSARIAL TRAINING

Adversarial training [73] is a technique designed to enhance model robustness by incorporating adversarial examples into the training dataset. This approach involves generating numerous adversarial samples and using them to strengthen the model's resilience. These adversarial examples can be created either by introducing noise to original samples or by perturbing their content.

The adversarial training is one of the first defense strategies used for training robust deep neural networks. For example, Shaham et al. [74] generate perturbed examples by adding the original data a small amount with respect to the sign of the gradient of the loss function. Specifically, the adversarial sample x' is created using the following equation:

$$x' = x + \varepsilon . \frac{\partial J_x(\theta)}{\partial \theta} \tag{7.1}$$

where $\frac{\partial J_x(\theta)}{\partial \theta}$ is the gradient of the loss function at the input x. After that, the crafted samples are used during the training process through an iterative minimization-maximization procedure, where the network parameters are updated with respect to worst-case data, rather than to the original training data. The authors reported that their approach increases the robustness of the network to existing adversarial examples, while making it harder to generate new ones.

In another work, Lyu [75] proposed a unified framework to build a robust machine learning models against adversarial attacks. In this paper, the small perturbation is calculated as follows:

$$\varepsilon = \sigma . sign(\frac{\partial J_x(\theta)}{\partial \theta}) . (\frac{|\partial J_x(\theta)|}{||\partial J_x(\theta)||_{p*}})^{\frac{p}{1-p}} \tag{7.2}$$

where σ is a small constant, p is a hyper-parameter and $p*$ is where the dual of p. After that, the regular optimization problem is converted to the new optimization as follows:

$$min_{\theta} J(x) + \sigma . || \frac{\partial J_x(\theta)}{\partial \theta} ||_{p*} \tag{7.3}$$

That is, instead of minimizing $J(x)$, the authors minimize $J(x+\varepsilon)$. The authors proven that the proposed framework offers a unified view to deal with adversarial examples.

Recently, Tramer et al. introduced Ensemble Adversarial Training (EAT), which involves augmenting training data with adversarial examples generated from multiple target models rather than a single one [76]. The benefit of EAT lies in its ability to mitigate the sharp curvatures resulting from single-step attacks. However, designing the interactions among different target models can be challenging, potentially leading to similar predictions and shared adversarial subspaces, which may compromise its performance. For a more in-depth review of adversarial training, readers are directed to [73].

Adversarial training is effective in defending against adversarial examples, particularly those designed for the original model. However, this method lacks robustness against black-box attacks [77], where adversarial samples are generated from a locally trained substitute model. Additionally, adversarial training is susceptible to two-step attacks, where random perturbations are first applied to an instance, followed by the application of any classical attack method, such as black-box or white-box attacks, using the modified sample.

7.3.2 DEFENSIVE DISTILLATION

Knowledge distillation, introduced by Hinton et al. [78], is a method for transferring knowledge from a larger model to a smaller, more deployable one suitable for real-world limitations. This technique is mainly used with neural network models that feature complex architectures with numerous layers and parameters. As deep learning has advanced over the past decade, achieving success in areas like speech recognition, image recognition, and natural language processing, knowledge distillation methods have garnered considerable interest for practical applications.

Deploying large deep neural network models presents considerable challenges, especially for edge devices with restricted memory and computational power. To tackle this problem, an initial model compression method was proposed to transfer knowledge from a larger model to train a smaller one, with the goal of preserving performance while minimizing loss [79]. Hinton and his colleagues formalized this process of creating a compact model from a larger one as the "Knowledge Distillation" framework [78]. Recently, this approach has also been used as a defense strategy against adversarial example generation methods.

Assume we have a neural network F that has been trained on a dataset $D = (X,Y)$, where X represents the input samples and Y is the corresponding label. The final softmax layer of F generates a probability distribution over Y. Next, we aim to train a second neural network F' using the input X to achieve comparable performance. Instead of relying on the target class labels Y from the training dataset, we utilize the outputs from network F as the labels for training F'. Consequently, the new labels reflect a distribution over all possible values in Y rather than a single value from the original labels.

Papernot et al. [80] were the first authors to use defensive distillation to defend against adversarial attacks. The benefit of using soft-targets as training labels is that additional knowledge found in probability vectors is used to train the distilled model. Moreover, training a network with the relative information about classes prevents models from over-fitting to the original data, and contributes to better generalization. Subsequently, the distilled models possess the ability to combat the adversarial example attacks.

Recently, Gao et al. [81] used a defensive distillation to build a robust autoencoder for an end-to-end communication system. Distillation is introduced to modify the structure of the neural network so that it assists to enhance the robustness of the system model against perturbation. Imam et al. [82] proposed Self-Ensembling Vision Transformer (ViT) with defensive Distillation and Adversarial training (SEDA). SEDA utilizes efficient CNN blocks to learn spatial features with various levels of abstraction from feature representations extracted from intermediate ViT blocks, which are largely unaffected by adversarial perturbations. Furthermore, SEDA leverages adversarial training in combination with defensive distillation for improved robustness against adversaries. Distillation using soft probabilities introduces uncertainty and variation into the output probabilities, making it more difficult for adversarial and privacy attacks.

7.3.3 BLOCKING THE TRANSFERABILITY

Blocking transferability is a defensive strategy aimed at protecting machine learning models from black-box attacks. Many established defense mechanisms often fail due to the strong transferability characteristic of neural networks, where adversarial examples created for one classifier can lead to similar mistakes in another classifier. This transferability remains effective even among classifiers with different architectures or trained on distinct datasets. Thus, an essential method for defending against black-box attacks is to disrupt the transferability of adversarial examples. The main idea is to inhibit the transferability of the machine learning model, making it more difficult for adversaries to generate effective adversarial examples.

For instance, Hosseini et al. [83] mitigated the transferability of models by employing a three-step NULL labeling technique. The concept involves adding a new sample with a NULL label to the original dataset. Subsequently, the authors train the model to classify adversarial examples as NULL, effectively rejecting them. More specifically, the NULL labeling method comprises three steps:

1. Training the model on a clean dataset to establish the decision boundaries for samples within each class.
2. Calculating the probability of belonging to the NULL for the adversarial example generated based on the amount of perturbation that example has.
3. Retraining the original classifier using both clean samples and different perturbed inputs of the samples. The label for the training data is decided using the NULL probabilities obtained in step 2.

The experimental results indicate that the NULL labeling method can effectively reject adversarial examples while maintaining high accuracy on the clean data.

Recently, Agarwal et al. [84] introduced a method to safeguard machine learning models against adversarial examples by utilizing non-deep learning techniques. This approach employs the Discrete Wavelet Transform and Discrete Sine Transform to extract image features, followed by classification using a Support Vector Machine. The results indicate that this method effectively mitigates the impact of adversarial perturbation attacks. Meanwhile, McCarthy et al. [85] proposed a novel defense strategy against adversarial attacks on network traffic classifiers by leveraging hierarchical learning. This strategy aims to reduce the vulnerability of the attack surface within the parameter space constraints, thereby enhancing resilience to crafted adversarial attacks. The experimental findings demonstrate that the proposed model maintains classification accuracy comparable to the original model when not under attack, while effectively withstanding adversarial attacks.

7.3.4 OTHER TECHNIQUES

Beside three above techniques for defending against adversarial attacks. There are also a number of other techniques used for the same purpose. These techniques include: Gradient hiding, Feature squeezing, Defense-GAN and

Gradient hiding: Gradient hiding is an inherent defense strategy against gradient-based attacks. This method seeks to obscure details about the model's gradient from potential attackers. For instance, using non-differentiable techniques like Decision Trees or Random Forests in the model makes it more challenging for gradient-based attacks to succeed [76]. Recently, Qiu et al. [86] addressed the challenges posed by advanced gradient-based attack techniques (e.g., BPDA and EOT). Specifically, the authors make two steps towards mitigating those advanced gradient-based attacks with two major contributions. First, they perform an in-depth analysis about the root causes of those attacks, and propose four properties that can break the fundamental assumptions of those attacks. Second, they identify a set of operations that can meet those properties. based on these operations, they design two preprocessing functions that can invalidate these powerful attacks. Extensive evaluations indicate that their solutions can effectively mitigate all existing standard and advanced attack techniques.

Feature squeezing: Feature squeezing is an additional method for enhancing model robustness. This technique focuses on simplifying the representation of data within models [87]. As a result, model outputs become less sensitive, allowing for the removal of adversarial perturbations. For input images, two primary approaches can be used to decrease data complexity. The first involves lowering the color depth of each pixel, which means encoding colors with fewer values. The second approach uses a smoothing filter to consolidate multiple inputs into a single value.

Recently, Rosenberg et al. [88] presented a novel defense method, referred to as sequence squeezing, aimed at making Long Short Term Memory classifiers more robust against adversarial attacks. This method differs from existing defense methods, which were designed only for non-sequence based models. Using sequence squeezing, they were able to decrease the effectiveness of such adversarial attacks from 99.9% to 15%, outperforming all of the baseline defense methods. Zheng et al. [89] proposed a new method to effectively detect adversarial examples presented to a person re-identification (ReID) network. The proposed method utilizes parts-based feature squeezing to detect the adversarial examples. They apply two types of squeezing to segmented body parts to better detect adversarial examples. The experimental results show that the proposed method can effectively detect the adversarial examples, and has the potential to avoid significant decreases in person ReID performance caused by adversarial examples.

Generally, the feature squeezing technique is a potential approach to mitigate the influence of adversarial attacks. These techniques help to effectively prevent adversarial attacks However, the side effect is that they sometimes result in the worse performance of the model on the original data.

Defense-GAN: Defense-GAN [90] is the approach to combat both white-box and black-box attacks. The idea is to train a GAN model in which the generator of the GAN model approximates the input samples. During the inference time, every test sample is first transformed by the generator of GAN, and the output of GAN is used as the input to the classifier instead of the original input. More specifically, a GAN is trained to model the distribution of unperturbed samples. Then, at inference time, for a given input x, a new sample $z*$ that minimizes Equation 7.4 is first found. Finally, $z*$ is input to the model instead of x.

$$z* = min_z||G(z) - x||^2 \tag{7.4}$$

Figure 7.5 provides an overview of the Defense-GAN mechanism. Although Defense-GAN has demonstrated considerable effectiveness against adversarial attacks, its success relies heavily on the expressiveness and generative abilities of the GAN. Furthermore, training a GAN can be challenging, and improper execution may lead to a significant decline in the performance of Defense-GAN.

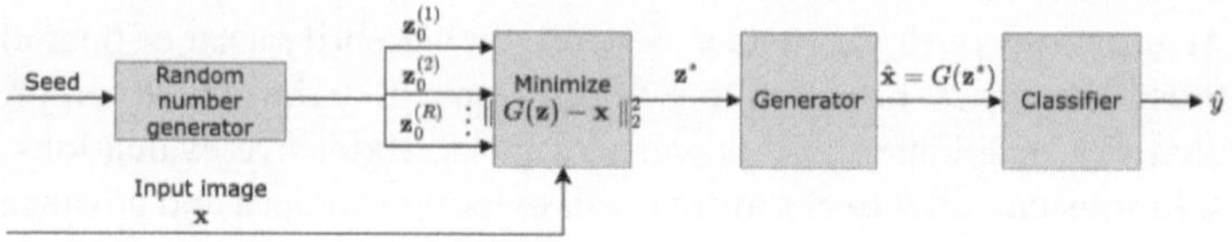

Figure 7.5 Architecture of Defense-GAN

MagNet: Meng et al. [91] proposed MagNet as a defense mechanism against adversarial examples. This approach treats the classifier as a black box, analyzing only the output of its final layer without accessing internal layers or modifying the classifier itself. MagNet incorporates detectors to differentiate between normal and adversarial inputs by measuring the distance between the test sample and the data manifold, rejecting inputs that exceed a predefined threshold. Additionally, it employs a reformer, based on an autoencoder, to adjust adversarial examples into legitimate ones. An overview of this technique is illustrated in Figure 7.6.

Although MagNet demonstrated strong effectiveness against various black-box attacks, its performance deteriorated significantly under white-box attacks, where attackers are presumed to have knowledge of MagNet's parameters. To counter this vulnerability, the authors proposed enhancing the defense by employing multiple types of autoencoders and randomly selecting one for each instance. This strategy increases unpredictability, making it more challenging for adversaries to determine which autoencoder was used.

Currently, existing defense mechanisms have limitations as they can only offer robustness against specific attacks under particular conditions. Developing a machine

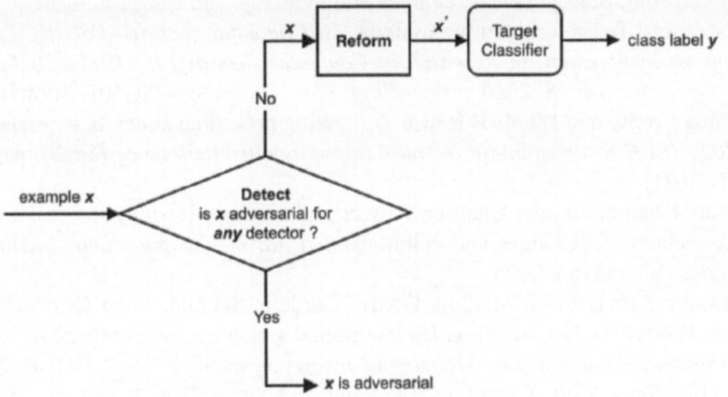

Figure 7.6 Architecture of MagNet

learning model that is universally robust against all types of adversarial examples remains an unresolved research challenge.

7.4 SUMMARY

In this chapter, we examined the application of generative models in adversarial learning. Initially, we covered the objectives of adversarial learning and categorized adversarial attacks. We then delved into using generative models to execute three primary types of adversarial attacks: evasion attacks, poisoning attacks, and exploratory attacks. Finally, the chapter concluded by presenting various techniques and strategies for defending against these adversarial attacks.

1. Anirban Chakraborty, Manaar Alam, Vishal Dey, Anupam Chattopadhyay, and Debdeep Mukhopadhyay. Adversarial attacks and defences: A survey. *arXiv preprint arXiv:1810.00069*, 2018.
2. Yevgeniy Vorobeychik and Murat Kantarcioglu. *Adversarial Machine Learning*. Springer Nature, 2022.
3. Zewen Li, Fan Liu, Wenjie Yang, Shouheng Peng, and Jun Zhou. A survey of convolutional neural networks: Analysis, applications, and prospects. *IEEE Transactions on Neural Networks and Learning Systems*, 33(12):6999–7019, 2022.
4. Kevin Eykholt, Ivan Evtimov, Earlence Fernandes, Bo Li, Amir Rahmati, Chaowei Xiao, Atul Prakash, Tadayoshi Kohno, and Dawn Song. Robust physical-world attacks on deep learning visual classification. In *2018 IEEE/CVF Conference on Computer Vision and Pattern Recognition*, pages 1625–1634, 2018.
5. Ian J. Goodfellow, Jean Pouget-Abadie, Mehdi Mirza, Bing Xu, David Warde-Farley, Sherjil Ozair, Aaron Courville, and Yoshua Bengio. Generative adversarial networks, 2014.
6. Zhiyi Tian, Lei Cui, Jie Liang, and Shui Yu. A comprehensive survey on poisoning attacks and countermeasures in machine learning. *ACM Computing Surveys*, 55(8):1–35, 2022.

7. Vale Tolpegin, Stacey Truex, Mehmet Emre Gursoy, and Ling Liu. Data poisoning attacks against federated learning systems. In *Computer security–ESORICs 2020: 25th European Symposium on Research in Computer Security, ESORICs 2020, guildford, UK, September 14–18, 2020, proceedings, part i 25*, pages 480–501. Springer, 2020.
8. Aashma Uprety and Danda B Rawat. Mitigating poisoning attack in federated learning. In *2021 IEEE Symposium Series on Computational Intelligence (SSCI)*, pages 01–07. IEEE, 2021.
9. Muskan Khan and Laiba Ghafoor. Adversarial machine learning in the context of network security: Challenges and solutions. *Journal of Computational Intelligence and Robotics*, 4(1):51–63, 2024.
10. Alexandre Sablayrolles, Matthijs Douze, Cordelia Schmid, Yann Ollivier, and Hervé Jégou. White-box vs black-box: Bayes optimal strategies for membership inference. In *International Conference on Machine Learning*, pages 5558–5567. PMLR, 2019.
11. Yujie Liu, Shuai Mao, Xiang Mei, Tao Yang, and Xuran Zhao. Sensitivity of adversarial perturbation in fast gradient sign method. In *2019 IEEE symposium Series on Computational Intelligence (SSCI)*, pages 433–436. IEEE, 2019.
12. Yingpeng Deng and Lina J Karam. Universal adversarial attack via enhanced projected gradient descent. In *2020 IEEE International Conference on Image Processing (ICIP)*, pages 1241–1245. IEEE, 2020.
13. Medha Pujari, Bhanu Prakash Cherukuri, Ahmad Y Javaid, and Weiqing Sun. An approach to improve the robustness of machine learning based intrusion detection system models against the carlini-wagner attack. In *2022 IEEE International Conference on Cyber Security and Resilience (CSR)*, pages 62–67. IEEE, 2022.
14. Seyed-Mohsen Moosavi-Dezfooli, Alhussein Fawzi, and Pascal Frossard. Deepfool: a simple and accurate method to fool deep neural networks. In *Proceedings of the IEEE Conference on Computer Vision and Pattern Recognition*, pages 2574–2582, 2016.
15. Rey Wiyatno and Anqi Xu. Maximal jacobian-based saliency map attack. *arXiv preprint arXiv:1808.07945*, 2018.
16. Ambra Demontis, Marco Melis, Maura Pintor, Matthew Jagielski, Battista Biggio, Alina Oprea, Cristina Nita-Rotaru, and Fabio Roli. Why do adversarial attacks transfer? explaining transferability of evasion and poisoning attacks. In *28th USENIX Security Symposium (USENIX security 19)*, pages 321–338, 2019.
17. Riya Singhal, Meet Soni, Shruti Bhatt, Manav Khorasiya, and Devesh C Jinwala. Enhancing robustness of malware detection model against white box adversarial attacks. In *International Conference on Distributed Computing and Intelligent Technology*, pages 181–196. Springer, 2023.
18. Nicolas Papernot, Patrick McDaniel, Ian Goodfellow, Somesh Jha, Z Berkay Celik, and Ananthram Swami. Practical black-box attacks against machine learning. In *Proceedings of the 2017 ACM on Asia Conference on Computer and Communications Cecurity*, pages 506–519, 2017.
19. Hadi Zanddizari, Behnam Zeinali, and J Morris Chang. Generating black-box adversarial examples in sparse domain. *IEEE Transactions on Emerging Topics in Computational Intelligence*, 6(4):795–804, 2021.
20. Zhuang Liu, Hanzi Mao, Chao-Yuan Wu, Christoph Feichtenhofer, Trevor Darrell, and Saining Xie. A convnet for the 2020s. In *Proceedings of the IEEE/CVF Conference on Computer Vision and Pattern Recognition*, pages 11976–11986, 2022.
21. Zeyu Qin, Yanbo Fan, Hongyuan Zha, and Baoyuan Wu. Random noise defense against query-based black-box attacks. *Advances in Neural Information Processing Systems*, 34:7650–7663, 2021.

22. Pin-Yu Chen, Huan Zhang, Yash Sharma, Jinfeng Yi, and Cho-Jui Hsieh. Zoo: Zeroth order optimization based black-box attacks to deep neural networks without training substitute models. In *Proceedings of the 10th ACM Workshop on Artificial Intelligence and Security*, pages 15–26, 2017.
23. Huichen Li, Xiaojun Xu, Xiaolu Zhang, Shuang Yang, and Bo Li. Qeba: Query-efficient boundary-based blackbox attack. In *Proceedings of the IEEE/CVF Conference on Computer Vision and Pattern Recognition*, pages 1221–1230, 2020.
24. Tom B Brown, Dandelion Mané, Aurko Roy, Martín Abadi, and Justin Gilmer. Adversarial patch. *arXiv preprint arXiv:1712.09665*, 2017.
25. Jianping Gou, Baosheng Yu, Stephen J Maybank, and Dacheng Tao. Knowledge distillation: A survey. *International Journal of Computer Vision*, 129(6):1789–1819, 2021.
26. Shengzhe Xu, Manish Marwah, Martin Arlitt, and Naren Ramakrishnan. Stan: Synthetic network traffic generation with generative neural models. In *Deployable Machine Learning for Security Defense: Second International Workshop, MLHat 2021, Virtual Event, August 15, 2021, Proceedings 2*, pages 3–29. Springer, 2021.
27. Xi Xiao, Shaofeng Zhang, Francesco Mercaldo, Guangwu Hu, and Arun Kumar Sangaiah. Android malware detection based on system call sequences and lstm. *Multimedia Tools and Applications*, 78:3979–3999, 2019.
28. Yang Song, Rui Shu, Nate Kushman, and Stefano Ermon. Constructing unrestricted adversarial examples with generative models. *Advances in Neural Information Processing Systems*, Vol. 31, 2018.
29. Z. Xiao, X. Gao, C. Fu, Y. Dong, W. Gao, X. Zhang, J. Zhou, and J. Zhu. Improving transferability of adversarial patches on face recognition with generative models. In *2021 IEEE/CVF Conference on Computer Vision and Pattern Recognition (CVPR)*, pages 11840–11849, Los Alamitos, CA, USA, jun 2021. IEEE Computer Society.
30. Hadi M. Dolatabadi, Sarah Erfani, and Christopher Leckie. Advflow: Inconspicuous black-box adversarial attacks using normalizing flows. *Advances in Neural Information Processing Systems*, Vol. 33, pp. 15871-15884, 2020.
31. Isaac Corley, Jonathan Lwowski, and Justin Hoffman. Domaingan: Generating adversarial examples to attack domain generation algorithm classifiers. *arXiv preprint arXiv:1911.06285*, 2019.
32. Charles A. Kamhoua Conrad S. Tucker Dule Shu, Nandi O. Leslie. Generative adversarial attacks against intrusion detection systems using active learning. In *Proceedings of the 2nd ACM Workshop on Wireless Security and Machine Learning*, pages 1–6, 2018.
33. Weiwei Hu and Ying Tan. Generating adversarial malware examples for black-box attacks based on gan. *International Conference on Data Mining and Big Data*, pp. 409-423, 2022.
34. Zilong Lin, Yong Shi, and Zhi Xue. *IDSGAN: Generative Adversarial Networks for Attack Generation Against Intrusion Detection*, page 79–91. Springer International Publishing, 2022.
35. Ravi Chauhan, Ulya Sabeel, Alireza Izaddoost, and Shahram Shah Heydari. Polymorphic adversarial cyberattacks using wgan. *Journal of Cybersecurity and Privacy*, 1(4):767–792, 2021.
36. M. Asfour and J. C. Murillo. Harnessing large language models to simulate realistic human responses to social engineering attacks: A case study. *International Journal of Cybersecurity Intelligence & Cybercrime*, 6(2):21–49, 2023.
37. Rabimba Karanjai. Targeted phishing campaigns using large scale language models. *arXiv preprint arXiv:2301.00665*, 2023.

38. Mika Beckerich, Laura Plein, and Sergio Coronado. Ratgpt: Turning online llms into proxies for malware attacks. *arXiv preprint arXiv: 2308.09183*, 2023.
39. Weiwei Hu and Ying Tan. Generating adversarial malware examples for black-box attacks based on gan. *arXiv preprint arXiv: 1702.05983* , 2017.
40. Weiwei Hu and Ying Tan. Black-box attacks against RNN based malware detection algorithms. *arXiv preprint arXiv: 1705.08131*, 2017.
41. Zhengli Zhao, Dheeru Dua, and Sameer Singh. Generating natural adversarial examples. *arXiv preprint arXiv: 1710.11342*, 2018.
42. Eric Wallace, Shi Feng, Nikhil Kandpal, Matt Gardner, and Sameer Singh. Universal adversarial triggers for attacking and analyzing nlp. *arXiv preprint arXiv: 1908.07125*, 2019.
43. Jernej Kos, Ian Fischer, and Dawn Song. Adversarial examples for generative models. *arXiv preprint arXiv: 1702.06832*, 2017.
44. Alexander Turner, Dimitris Tsipras, and Aleksander Madry. Label-consistent backdoor attacks. *arXiv preprint arXiv: 1912.02771*, 2019.
45. Luis Munoz-Gonzalez, Bjarne Pfitzner, Matteo Russo, Javier Carnerero-Cano, and Emil C. Lupu. Poisoning attacks with generative adversarial nets. *arXiv preprint arXiv: 1906.07773*, 2019.
46. Keshav Kasichainula, Hadi Mansourifar, and Weidong Shi. Poisoning attacks via generative adversarial text to image synthesis. In *2021 51st Annual IEEE/IFIP International Conference on Dependable Systems and Networks Workshops (DSN-W)*, pages 158–165, 2021.
47. Ujjwal Upadhyay and Prerana Mukherjee. Generating out of distribution adversarial attack using latent space poisoning. *IEEE Signal Processing Letters*, 28:523–527, 2021.
48. Chetan Kumar, Deepak Kumar, and Ming Shao. Generative adversarial attack on ensemble clustering. In *Proceedings of the IEEE/CVF Winter Conference on Applications of Computer Vision*, pages 2848–2857, 2022.
49. Chaofei Yang, Qing Wu, Hai Li, and Yiran Chen. Generative poisoning attack method against neural networks, 2017.
50. Han Zhang, Ian Goodfellow, Dimitris Metaxas, and Augustus Odena. Self-attention generative adversarial networks. In *International Conference on Machine Learning*, pages 7354–7363. PMLR, 2019.
51. Jiale Zhang, Bing Chen, Xiang Cheng, Huynh Thi Thanh Binh, and Shui Yu. Poisongan: Generative poisoning attacks against federated learning in edge computing systems. *IEEE Internet of Things Journal*, 8(5):3310–3322, 2021.
52. Xiaolin Chen, Daoguang Zan, Wei Li, Bei Guan, and Yongji Wang. A gan-based data poisoning framework against anomaly detection in vertical federated learning. *arXiv preprint arXiv:2401.08984*, 2024.
53. Eric Wallace, Tony Z. Zhao, Shi Feng, and Sameer Singh. Concealed data poisoning attacks on NLP models, 2021.
54. Alexander Wan, Eric Wallace, Sheng Shen, and Dan Klein. Poisoning language models during instruction tuning. In *International Conference on Machine Learning*, pages 35413–35425. PMLR, 2023.
55. Manli Shu, Jiongxiao Wang, Chen Zhu, Jonas Geiping, Chaowei Xiao, and Tom Goldstein. On the exploitability of instruction tuning. *Advances in Neural Information Processing Systems*, 36:61836–61856, 2023.

56. Ying Zhao, Junjun Chen, Jiale Zhang, Di Wu, Michael Blumenstein, and Shui Yu. Detecting and mitigating poisoning attacks in federated learning using generative adversarial networks. *Concurrency and Computation: Practice and Experience*, 34(7):e5906, 2022.
57. Lucia Alonso and Mina Alishahi. Autoencoder for detecting malicious updates in differentially private federated learning. In *Proceedings of the 21st International Conference on Security and Cryptography - Volume 1: SECRYPT*, pages 467–474. INSTICC, SciTePress, 2024.
58. Thomas Ristenpart Matt Fredrikson, Somesh Jha. Model inversion attacks that exploit confidence information and basic countermeasures. In *Proceedings of the 22nd ACM SIGSAC Conference on Computer and Communications Security*, page 1322–1333. ACM, 2015.
59. Yuheng Zhang, Ruoxi Jia, Hengzhi Pei, Wenxiao Wang, Bo Li, and Dawn Song. The secret revealer: Generative model-inversion attacks against deep neural networks. In *Proceedings of the IEEE/CVF Conference on Computer Vision and Pattern Recognition*, pages 253–261, 2020.
60. Si Chen, Mostafa Kahla, Ruoxi Jia, and Guo-Jun Qi. Knowledge-enriched distributional model inversion attacks, 2021.
61. Lukas Struppek, Dominik Hintersdorf, Antonio De Almeida Correia, Antonia Adler, and Kristian Kersting. Plug & play attacks: Towards robust and flexible model inversion attacks. *arXiv preprint arXiv:2201.12179*, 2022.
62. Ngoc-Bao Nguyen, Keshigeyan Chandrasegaran, Milad Abdollahzadeh, and Ngai-Man Cheung. Re-thinking model inversion attacks against deep neural networks. In *Proceedings of the IEEE/CVF Conference on Computer Vision and Pattern Recognition (CVPR)*, pages 16384–16393, June 2023.
63. Xueluan Gong, Ziyao Wang, Shuaike Li, Yanjiao Chen, and Qian Wang. A gan-based defense framework against model inversion attacks. *IEEE Transactions on Information Forensics and Security*, 18:4475–4487, 2023.
64. Reza Shokri, Marco Stronati, Congzheng Song, and Vitaly Shmatikov. Membership inference attacks against machine learning models, 2017.
65. Yunhui Long, Lei Wang, Diyue Bu, Vincent Bindschaedler, Xiaofeng Wang, Haixu Tang, Carl A. Gunter, and Kai Chen. A pragmatic approach to membership inferences on machine learning models. In *2020 IEEE European Symposium on Security and Privacy (EuroS&P)*, pages 521–534, 2020.
66. Jamie Hayes, Luca Melis, George Danezis, and Emiliano De Cristofaro. Logan: Membership inference attacks against generative models, 2018.
67. Benjamin Hilprecht, Martin Härterich, and Daniel Bernau. Reconstruction and membership inference attacks against generative models. , 2019.
68. Hui Sun, Tianqing Zhu, Jie Li, Shoulin Ji, and Wanlei Zhou. Attribute-based membership inference attacks and defenses on gans. *IEEE Transactions on Dependable and Secure Computing*, 21(4):2376–2393, 2024.
69. Filippo Galli, Luca Melis, and Tommaso Cucinotta. Noisy neighbors: Efficient membership inference attacks against llms. *arXiv preprint arXiv:2406.16565*, 2024.
70. Michael Duan, Anshuman Suri, Niloofar Mireshghallah, Sewon Min, Weijia Shi, Luke Zettlemoyer, Yulia Tsvetkov, Yejin Choi, David Evans, and Hannaneh Hajishirzi. Do membership inference attacks work on large language models? *arXiv preprint arXiv:2402.07841*, 2024.

71. Rui Wen, Zheng Li, Michael Backes, and Yang Zhang. Membership inference attacks against in-context learning. *arXiv preprint arXiv:2409.01380*, 2024.
72. Yi Shi, Yalin E. Sagduyu, Kemal Davaslioglu, and Jason H. Li. Generative adversarial networks for black-box api attacks with limited training data. In *2018 IEEE International Symposium on Signal Processing and Information Technology (ISSPIT)*, pages 453–458, 2018.
73. Tao Bai, Jinqi Luo, Jun Zhao, Bihan Wen, and Qian Wang. Recent advances in adversarial training for adversarial robustness. *arXiv preprint arXiv:2102.01356*, 2021.
74. Uri Shaham, Yutaro Yamada, and Sahand Negahban. Understanding adversarial training: Increasing local stability of supervised models through robust optimization. *Neurocomputing*, 307:195–204, September 2018.
75. Chunchuan Lyu, Kaizhu Huang, and Hai-Ning Liang. A unified gradient regularization family for adversarial examples. *2015 IEEE International Conference on Data Mining*, pp. 301-309, 2015.
76. Florian Tramèr, Alexey Kurakin, Nicolas Papernot, Ian Goodfellow, Dan Boneh, and Patrick McDaniel. Ensemble adversarial training: Attacks and defenses. *arXiv preprint arXiv:1705.07204*, 2017.
77. Nina Narodytska and Shiva Prasad Kasiviswanathan. Simple black-box adversarial perturbations for deep networks. *arXiv preprint arXiv:1612.06299*, 2016.
78. Geoffrey Hinton, Oriol Vinyals, and Jeff Dean. Distilling the knowledge in a neural network. *arXiv preprint arXiv:1503.02531*, 2015.
79. Cristian Buciluǎ, Rich Caruana, and Alexandru Niculescu-Mizil. Model compression. In *Proceedings of the 12th ACM SIGKDD International Conference on Knowledge Discovery and Data Mining*, pages 535–541, 2006.
80. Nicolas Papernot, Patrick McDaniel, Xi Wu, Somesh Jha, and Ananthram Swami. Distillation as a defense to adversarial perturbations against deep neural networks. In *2016 IEEE Symposium on Security and Privacy (SP)*, pages 582–597, 2016.
81. Qiaochu Gao, Zhiwei Cao, and Dou Li. Defensive distillation based end-to-end autoencoder communication system. In *2021 7th International Conference on Computer and Communications (ICCC)*, pages 109–114, 2021.
82. Raza Imam, Ibrahim Almakky, Salma Alrashdi, Baketah Alrashdi, and Mohammad Yaqub. Seda: Self-ensembling vit with defensive distillation and adversarial training for robust chest x-rays classification. In *MICCAI Workshop on Domain Adaptation and Representation Transfer*, pages 126–135. Springer, 2023.
83. Hossein Hosseini, Yize Chen, Sreeram Kannan, Baosen Zhang, and Radha Poovendran. Blocking transferability of adversarial examples in black-box learning systems. *arXiv preprint arXiv:1703.04318*, 2017.
84. Akshay Agarwal, Richa Singh, Mayank Vatsa, and Nalini Ratha. Image transformation-based defense against adversarial perturbation on deep learning models. *IEEE Transactions on Dependable and Secure Computing*, 18(5):2106–2121, 2021.
85. Andrew McCarthy, Essam Ghadafi, Panagiotis Andriotis, and Phil Legg. Defending against adversarial machine learning attacks using hierarchical learning: A case study on network traffic attack classification. *Journal of Information Security and Applications*, 72:103398, 2023.
86. Han Qiu, Yi Zeng, Qinkai Zheng, Tianwei Zhang, Meikang Qiu, and Gerard Memmi. Mitigating advanced adversarial attacks with more advanced gradient obfuscation techniques. *arXiv preprint arXiv:2005.13712*, 2020.

87. Weilin Xu, David Evans, and Yanjun Qi. Feature squeezing: Detecting adversarial examples in deep neural networks. In *Proceedings 2018 Network and Distributed System Security Symposium*. Internet Society, 2018.
88. Ishai Rosenberg, Asaf Shabtai, Yuval Elovici, and Lior Rokach. Sequence squeezing: A defense method against adversarial examples for api call-based rnn variants. In *2021 International Joint Conference on Neural Networks (IJCNN)*, pages 1–10, 2021.
89. Yu Zheng and Senem Velipasalar. Part-based feature squeezing to detect adversarial examples in person re-identification networks. In *2021 IEEE International Conference on Image Processing (ICIP)*, pages 844–848, 2021.
90. Pouya Samangouei, Maya Kabkab, and Rama Chellappa. Defense-gan: Protecting classifiers against adversarial attacks using generative models. *arXiv preprint arXiv:1805.06605*, 2018
91. Dongyu Meng and Hao Chen. Magnet: a two-pronged defense against adversarial examples. In *Proceedings of the 2017 ACM SIGSAC Conference on Computer and Communications Security*, pages 135–147, 2017.

8 Case Studies and Detailed Implementation/Experiments

This chapter presents detailed case studies that demonstrate the practical applications of generative AI in different cybersecurity domains. Each case study provides the comprehensive description of the problem statement, the data sources utilized, and the specific generative AI approach employed. It further presents the experimental setup, evaluation metrics, and results obtained. The case studies cover a wide range of cybersecurity scenarios, including Intrusion detection system, Data Synthesis, Malware classification, Traffic analysis, and Deepfake and Counter-measurements. They showcase how generative AI techniques can effectively enhance the accuracy and efficiency of existing cybersecurity systems, leading to improved threat intelligence and proactive defense mechanisms. In each case study, we outline four key steps for building a generative model based on data: defining the problem, preprocessing the data, designing the model, and training and testing the model.

8.1 INTRUSION DETECTION SYSTEM

8.1.1 PROBLEM STATEMENT

Intrusion Detection Systems (IDSs) are essential tools in cybersecurity, designed to monitor network traffic and detect unauthorized access. They analyze data to find potential security threats, using known attack patterns or spotting deviations from normal network behavior. IDSs play a key role in protecting network systems, ensuring timely detection of malicious activity, and reducing the risks of cyberattacks.

Autoencoder (AE) is a type of neural network that can learn and reconstruct normal patterns in data through unsupervised learning. In IDSs, AEs are valuable because they can identify anomalies that differ from usual network behavior, which may indicate intrusions. Since AEs do not rely on predefined attack signatures, they can detect new or unknown threats, making them especially useful in various network environments. AEs also help reduce false positives and simplify data processing, improving anomaly detection accuracy in IDSs.

8.1.2 DATA PREPROCESSING

Data preprocessing is crucial for preparing datasets for machine learning models by ensuring they are clean and well-structured. Raw datasets often contain inconsistencies, missing values, or features with varying scales, which can negatively impact

DOI: 10.1201/9781003502531-8

model performance. For this case study, we utilize the UNSW dataset [1], a widely recognized benchmark in intrusion detection research. Other popular datasets, such as NSLKDD [2] and CICIDS2017 [3], can be used similarly, applying comparable preprocessing techniques to maintain data consistency.

```
1 import pandas as pd
2 from sklearn.preprocessing import MinMaxScaler
3 import joblib
4 import os
5 def prepare_data(dir_path, data_type="train"):
6     data = pd.read_csv(f"{dir_path}/{data_type}.csv", sep=",",
    header=None)
7     data.dropna(inplace=True)
8     X, y = data.iloc[:, :-1].to_numpy(), (data.iloc[:, -1] > 0).
    astype(int)
9     if data_type == "train":
10        X, y = X[y == 0], y[y == 0]
11    scaler_path = os.path.join("checkpoints", "scaler.pkl")
12    scaler = MinMaxScaler()
13    if data_type == "train":
14        X = scaler.fit_transform(X)
15        joblib.dump(scaler, scaler_path)
16    else:
17        scaler = joblib.load(scaler_path)
18        X = scaler.transform(X)
19    return X, y
```

Listing 8.1: Data preprocessing for training and testing.

In this table data, e.g., the UNSW dataset, data preprocessing involves five key steps as in Listing 8.1, i.e., cleaning data, extracting features and labels, preprocessing labels, and normalizing feature values. The first step is to clean the dataset by removing missing values and None (NaN) values. In the line 6 in Listing 8.1, the dataset is loaded using the `pandas` library, and missing data points are eliminated in the line 7 using the `dropna()` function. This helps the dataset to be free from incomplete entries. The data features and labels are extracted as in the line 8. Here, the data features are represented by `X` while data labels y are taken from the last column of the dataset file. The labels are preprocessed as follows. The label values greater than zero are anomalies and vice versa. Due to the AE model be trained with only normal data, only normal data samples, i.e., the data samples with label values as 0, are retained for training data as shown in lines 9 and 10, while abnormal data is filtered out according to the `data_type` argument. Finally, the data is normalized for consistent scaling across all features. The code in line 13 to line 15 applies a `MinMaxScaler` function as a scaler object on the training data, and this scaler object is saved for future use. In the testing process, the code in the line 16 to the line 18 shows that the saved scaler object is loaded to normalize the testing data. The preprocessed feature set `X` and corresponding labels y are returned at the end of this preprocessing process.

8.1.3 DESIGNING AE ARCHITECTURE

The AE model is implemented by using the `nn.Module` class from PyTorch, as shown in Listing 8.2. The architecture of this model composes of two primary components, i.e., an encoder and a decoder.

The encoder is designed to map input data into a latent representation space. This process is outlined from line 6 to 11 in Listing 8.2. The input vector consists of n_{features} neurons, while the latent vector has $\sqrt{n_{\text{features}}}+1$ neurons. These numbers of neurons can be adjusted for different experiments and datasets. The encoder has three Linear layers followed by the Tanh activation function. Readers can modify the number of layer or use other activation functions. However, Tanh is a common activation function used in AE trained on the table structure data. This encoder reduces gradually the input dimensionality from the input layer with n_{features} neurons to the latent representation layer with $\sqrt{n_{\text{features}}}+1$ neurons as in [18, 8]. The readers can modify these hyperparameters to suitable with different configurations.

The decoder aims to reconstruct the original input data from its latent representation. The method to create the decoder is detailed from line 12 to line 17 in Listing 8.2. The latent vector, which has $\sqrt{n_{\text{features}}}+1$ neurons, is the input of the decoder part, while the output of the decoder has the same number of neurons with the input of the encoder part, i.e., n_{features}. Similar to the encoder, the decoder has three layers where each layer is followed by the activation function, e.g., `Tanh`. In an AE, each layer of the decoder is usually designed to have the same number of neurons as the corresponding layer of the encoder. This symmetry ensures that the decoder can effectively learn to reconstruct the input data from the latent representation, maintaining alignment between the features extracted by the encoder and those generated by the decoder.

The `forward` function, defined from line 20 to line 22 in Listing 8.2, is integral to call the AE model. This model takes the input data and passes it through its encoder component, which compresses the data into a lower-dimensional representation known as the latent representation, latent˙z. This latent representation captures the essential features of the input while reducing its dimensionality. Subsequently, this latent˙z vector is fed into the decoder component, which reconstructs the original input data from this latent vector. The `forward` function returns both the latent vector and the reconstructed data, which can be used for further adaptation of the AE model.

```
1 import torch.nn as nn
2 from math import sqrt
3 class AutoEncoder(nn.Module):
4     def __init__(self, n_features):
5         super(AutoEncoder, self).__init__()
6         self.encoder = nn.Sequential(
7             nn.Linear(n_features, round(n_features * 0.75)),
8             nn.Tanh(),
9             nn.Linear(round(n_features * 0.75), round(n_features
    * 0.5)),
10            nn.Tanh(),
11            nn.Linear(round(n_features * 0.5), round(sqrt(
```

```
    n_features)) + 1))
        self.decoder = nn.Sequential(
            nn.Linear(round(sqrt(n_features)) + 1, round(
    n_features * 0.5)),
            nn.Tanh(),
            nn.Linear(round(n_features * 0.5), round(n_features *
     0.75)),
            nn.Tanh(),
            nn.Linear(round(n_features * 0.75), n_features))
    def forward(self, x):
        latent_z = self.encoder(x)
        x_reconstruction = self.decoder(latent_z)
        return latent_z, x_reconstruction
```

Listing 8.2: The Autoencoder architecture for Intrusion Detection System.

8.1.4 TRAINING AND TESTING PROCESS

8.1.4.1 Training Process

To train the AE, it's essential to define hyperparameters, e.g., learning_rate, batch_size, and n_epochs. Details about these settings are provided in Listing 8.3. The batch_size (line 2) specifies the number of data samples used simultaneously to calculate the loss and update the model's weights. We can change this batch size for suitable memory usage. The learning_rate (line 3) controls the adjustment size of the model's weights, directly influencing the speed and effectiveness of the learning process. Finally, the n_epochs (line 4) indicates how many complete passes the model makes through the training data, affecting how well it learns from the data.

```
1 cfg = {
2     "batch_size": 8,
3     "learning_rate": 1e-3,
4     "n_epochs": 100}
```

Listing 8.3: Configuration settings for training the AE in an Intrusion Detection System.

Sequently, we need to define the training function to load training data, define a loss function, and define an optimization method. Specifically, this function can be drawn as in Listing 8.4. The training process for the AE model involves the following steps. First, the procedure starts by loading the dataset for training. This is executed by using the prepare_data function (line 3), which reads data from the specified directory (dataset_dir). The dataset is then divided into training and validation sets based on the validation_split hyperparameter (lines 5–7) and further broken down into mini-batches according to the batch_size (lines 8–9). Based on the computing power, the model is initialized and moved to the appropriate device, either GPU or CPU (lines 10–12). For each batch, the input data goes through the model (line 26). The encoder compresses the data into a latent representation, while the decoder reconstructs the original input.

Next step is defining the loss function. Training the AE model aims to minimize the difference between the input and output vectors. Note that, these vectors are in the FLOAT type. Thus, we can use a type of loss function suitable for measuring the difference between two these vectors, e.g., Mean Squared Error (MSE)[1] (line 14). The MSE loss measures how well the model captures the data patterns by comparing between the input vector and its reconstruction, i.e., the output vector.

Third, we need to define an optimization algorithms for training. The aim of an optimization algorithm in training a neural network is to adjust the model's weights and biases to minimize the loss function, thereby improving its accuracy and performance. There are many types of optimization algorithm to train neural networks[2]. Here, we choose a popular algorithm, i.e., Adam, as in the line 13.

Each training epoch consists of multiple steps, as shown in the line 23. In each step, the model performs a forward pass, where data flows from the input layer to the output layer, followed by a backward pass to calculate gradients and update the model's weights. This iterative process enables the model to gradually reduce the loss values until it reaches convergence (lines 19–32). At the end of each epoch of training, the model's weights are saved to the specified directory (`save_dir`) (line 47) as checkpoints, and the training losses for the epoch are recorded in a text file for future visualization observation (line 44).

After each training epoch, we evaluate the model's performance using a validation dataset by switching to evaluation mode, as shown in line 33. This process allows us to assess and select the best model checkpoint after training. Various metrics, such as accuracy, precision, recall, and loss values, can be used to validate model checkpoints. In this experiment, we measure the quality of model checkpoints based on their loss values on the validation set (lines 36–43). These validation loss values are recorded using the TensorBoard library, as shown in line 45, enabling further assessment and analysis.

```
1  def train(cfg):
2      device = torch.device("cuda" if cfg["gpu_usage"] else "cpu")
3      X_train, _ = prepare_data(dir_path=cfg["dataset_dir"], data_type="train")
4      X_train = torch.tensor(X_train, dtype=torch.float32)
5      val_size = int(cfg["validation_split"] * len(X_train))
6      train_size = len(X_train) - val_size
7      train_dataset, val_dataset = random_split(X_train, [train_size, val_size])
8      train_loader = DataLoader(train_dataset, batch_size=cfg["train_batch_size"], shuffle=True, num_workers=cfg["num_workers"])
9      val_loader = DataLoader(val_dataset, batch_size=cfg["train_batch_size"], shuffle=False, num_workers=cfg["num_workers"])
10     model = AutoEncoder(n_features=X_train.shape[1])
11     model = model.to(device)
12     model.train()
13     optimizer = Adam(model.parameters(), lr=cfg["learning_rate"])
14     loss_fn = nn.MSELoss()
```

[1] https://pytorch.org/docs/stable/generated/torch.nn.MSELoss.html.

[2] https://pytorch.org/docs/stable/optim.html.

```
15    writer = SummaryWriter(log_dir=cfg["log_dir"])
16    n_epochs = cfg["n_epochs"]
17    epoch_losses = []
18    for epoch in range(n_epochs):
19        model.train()
20        pbar = tqdm(train_loader, ncols=80, desc="Training")
21        running_loss = 0.0
22        train_num = 0
23        for step, x in enumerate(pbar):
24            x = x.to(device)
25            optimizer.zero_grad()
26            _, x_reconstruction = model(x)
27            loss = loss_fn(x, x_reconstruction)
28            loss.backward()
29            optimizer.step()
30            running_loss += loss.item() * x.shape[0]
31            train_num += x.shape[0]
32        train_loss = running_loss / train_num
33        model.eval()
34        val_running_loss = 0.0
35        val_num = 0
36        with torch.no_grad():
37            for x in val_loader:
38                x = x[0].to(device)
39                _, x_reconstruction = model(x)
40                val_loss = loss_fn(x, x_reconstruction)
41                val_running_loss += val_loss.item() * x.shape[0]
42                val_num += x.shape[0]
43        val_loss = val_running_loss / val_num
44        writer.add_scalar("Loss/Train_epoch", train_loss, epoch)
45        writer.add_scalar("Loss/Validation_epoch", val_loss,
    epoch)
46        epoch_losses.append(train_loss)
47        torch.save(model.state_dict(), f"{cfg["save_dir"]}/model_
    {epoch}.pth")
48    writer.close()
49    with open(f"{cfg["save_dir"]}/epoch_losses.txt", "w") as f:
50        for loss in epoch_losses:
51             f.write(f"{loss}\n")
```

Listing 8.4: Training procedure for training the AE in an IDS.

Visualizing the training log is essential for monitoring the model's progress. This log, recorded throughout the training process (as indicated in the lines 44-45 of Listing 8.4), provides valuable insights into the model's performance. For example, in Figure 8.1, we can see a steady decrease in the training loss value across the training steps, indicating that the model is learning accurately. To access this log, the readers can use the command `tensorboard --logdir=log_dir`, which will open the default web browser of the operating system to display the log. Note that, if the training loss is still decreasing but the validation loss starts increasing, it indicates the overfitting problem of machine learning. In such cases, we need to adjust the training process by changing data pre-processing, model architecture, or hyperparameters to improve the model's performance.

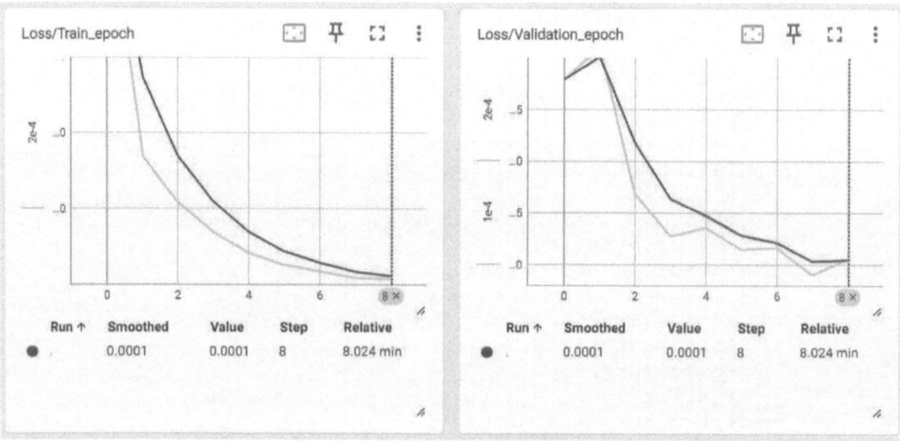

Figure 8.1 Visualization of training and validation loss for IDS using AE in TensorBoard.

8.1.4.2 Testing Process

```
def calculate_metrics(labels, predictions):
    accuracy = metrics.accuracy_score(labels, predictions)
    precision = metrics.precision_score(labels, predictions)
    recall = metrics.recall_score(labels, predictions)
    f1_score = metrics.f1_score(labels, predictions)
    confusion_matrix = metrics.confusion_matrix(labels, predictions)
    plt.figure(figsize=(8, 6))
    sns.heatmap(confusion_matrix, annot=True, fmt="d", cmap="Blues", cbar=False)
    plt.title("Confusion Matrix")
    plt.xlabel("Predicted Labels")
    plt.ylabel("True Labels")
    plt.savefig("confusion_matrix.pdf")
    plt.show()
    return {
        "accuracy": accuracy,
        "precision": precision,
        "recall": recall,
        "f1_score": f1_score
    }
```

Listing 8.5: Functions for calculating evaluation metrics for model performance.

The primary goal of the testing phase is to evaluate how effectively the AE model detects anomalies by analyzing its performance on unseen data. The process described in Listing 8.6 begins by loading the testing dataset and converting it into PyTorch tensors using the `prepare_data` function (lines 9–11). These tensors are then wrapped in the TensorDataset object and loaded into a DataLoader with a specified batch size (lines 12–13).

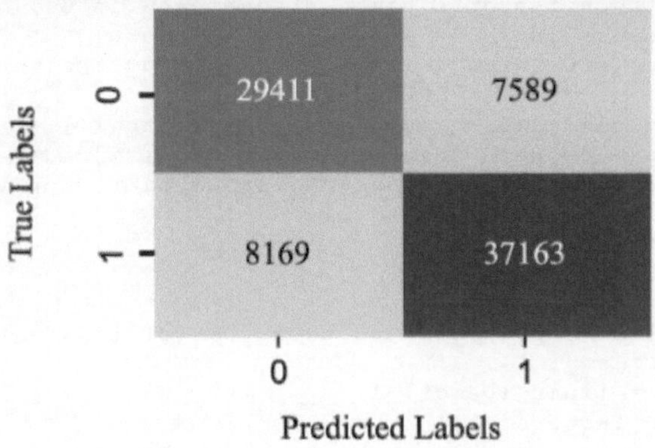

Figure 8.2 Confusion matrix generated by test procedure

Once the data is prepared, the trained AE model is restored from a checkpoint and set to evaluation mode to ensure that its parameters remain static during testing (lines 14–18). Each input sample is fed into the model, and the testing loss is calculated using the MSE loss function (line 19). This loss is then compared to a predefined threshold called as an anomaly score, allowing the model to classify the data as either normal or abnormal. Specifically, if one testing data sample outputs the MSE value larger than the anomaly score, it will be indicated as abnormal data sample or attacked data sample in the IDS dataset and vice versa.

To facilitate a comprehensive evaluation of the model's performance, we define a function called `calculate_metrics` (as shown in Listing 8.5). This function takes two inputs, i.e., labeled (ground-truth) value and the predicted value. This function computes four classification metrics, i.e., accuracy, precision, recall, and F1-score. Additionally, the function generates a confusion matrix to illustrate how well the model's predictions align with the actual classes. This confusion matrix is displayed as a heatmap, providing a clear visualization of the model's performance, as shown in Figure 8.2. Specifically, the anomaly score of each testing data sample is calculated as in the lines from 33 to 40 in Listing 8.6. Then, the model's performance in detecting anomalies is evaluated using the `calculate_metrics` function, providing a detailed analysis of its capabilities (line 32).

```
1 cfg = {
2     "gpu_usage": True,
3     "seed": 42,
4     "dataset_dir": "data/UNSW",
5     "batch_size": 8,
6     "save_dir": "checkpoints",
7 }
8 def test(cfg):
9     X_test, y_test = prepare_data(dir_path=cfg["dataset_dir"],
      data_type="test")
10    X_test = torch.tensor(X_test, dtype=torch.float32)
```

```
    y_test = torch.tensor(y_test, dtype=torch.int64)
    test_dataset = TensorDataset(X_test, y_test)
    test_loader = DataLoader(test_dataset, batch_size=cfg["
    test_batch_size"], shuffle=False)
    model = AutoEncoder(n_features=X_test.shape[1])
    model_path = os.path.join(cfg["save_dir"], "model_last.pth")
    model.load_state_dict(torch.load(model_path, map_location=
    device))
    model = model.to(device)
    model.eval()
    loss_fn = nn.MSELoss(reduction="none")
    train_losses = np.loadtxt(os.path.join(cfg["save_dir"], "
    epoch_losses.txt"))
    threshold = train_losses[-1]
    comparison_factors = []
    labels = []
    with torch.no_grad():
        for x_batch, y_batch in tqdm(test_loader, desc="Testing")
    :
            x_batch = x_batch.to(device)
            _, x_reconstruction = model(x_batch)
            cmp_factor = loss_fn(x_batch, x_reconstruction).mean(
    dim=1)
            comparison_factors.extend(cmp_factor.cpu().tolist())
            labels.extend(y_batch.tolist())
    predictions = [int(factor > threshold) for factor in
    comparison_factors]
    performance_metrics = calculate_metrics(labels, predictions)
    return performance_metrics
```

Listing 8.6: Testing proceduce for testing the AE in an IDS.

Readers can download the complete implementation code for this section from here[3].

8.2 DATA SYNTHESIS

8.2.1 PROBLEM STATEMENT

Data synthesis plays a vital role in overcoming the difficulties of acquiring large volumes of labeled or high-quality data in the field of cybersecurity. Creating synthetic data that closely mimics real-world data, helps to bridge the gap when actual datasets are scarce. This synthetic data can be integrated with existing datasets, enhancing the training of machine learning and deep learning models for threat detection. Furthermore, synthetic data can be tailored to simulate various types of cyberattacks, making it a versatile and effective resource for testing and refining cybersecurity systems. This approach not only improves the performance of models but also helps organizations better prepare for potential threats.

In this study, we utilize the Conditional Deep Convolutional Generative Adversarial Networks (cDCGAN), which is a common generative models introduced in[3],

[3]https://github.com/lyvt/genai˙cybersecurity/tree/main/01-intrusion-detection-system

for synthesizing malware data samples. By using cDCGAN, we can create synthetic malware that mimics the behavior of real malware, thereby enhancing the diversity and quantity of training data available. This approach not only alleviates the shortage of malware data but also enables the exploration of various types of malware, facilitating a more comprehensive understanding of potential threats. Moreover, cDCGAN allows for the generation of malware samples with specific characteristics, which provides researchers with the opportunity to investigate different attack scenarios and refine malware detection methods.

8.2.2 DATA PREPROCESSING

In this experiment, we use the Microsoft Malware Classification Challenge (BIG 2015) dataset, which can be downloaded from [4]. The dataset contains malware data samples structured as Portable Executable (PE) files in binary format, representing nine different types of malware. These malware samples are converted into fixed-size grayscale images, each measuring 256x256 pixels, as illustrated in Figure 8.3. The converting process is executed by the function `binary_to_image` in Listing 8.7. The function begins by creating a blank 256x256 pixel image filled with zero pixels using `np.zeros()` (line 3 in Listing 8.7). It then opens the PE file in the read-mode (line 4) and fills the content of the PE file into a variable called `byte_data` (line 5). To focus on relevant data, the function starts processing from the 9th byte (line 6).

The binary data is processed in chunks of 47 bytes (line 7). Within each chunk, the function extracts 5 bytes at a time (line 11). The first two bytes are used to determine the x-coordinate (line 12), while the last two bytes are converted into the y-coordinate (line 13) using the hexadecimal-to-decimal conversion. Both coordinates are clamped to a maximum value of 255 (lines 14–15) to ensure fitting within the image dimensions.

This process continues until all the values in the PE file are processed (lines 7–18). Finally, the function `binary_to_image` returns the grayscale image (line 19).

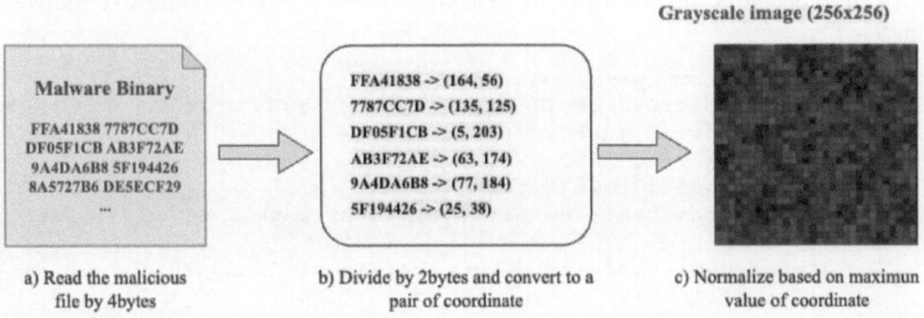

Figure 8.3 Diagram illustrating the conversion of a malware binary file into an image.

By converting malware in binary files into images, this method provides us with the visualizations of malware patterns. It allows us to effectively manage various types of malware and quickly adapt to new threats.

```
import numpy as np
def binary_to_image(binary_file):
    image = np.zeros((256, 256), dtype=np.uint8)
    with open(binary_file, "rb") as file:
        byte_data = file.read()
        i = 9
        while i < len(byte_data):
            data = byte_data[i:i + 47]
            j = 0
            while j < len(data):
                data_temp = data[j:j + 5]
                x = int(data_temp[:2], 16) if "?" not in data_temp[:2].decode("utf-8") else 0
                y = int(data_temp[3:], 16) if "?" not in data_temp[3:].decode("utf-8") else 0
                x1 = min(x, 255)
                y1 = min(y, 255)
                image[x1, y1] += 1
                j += 6
            i += 47 + 11
    return image
```

Listing 8.7: Procedure for converting malware binary data into images.

To convert all the training malware binary files in the dataset into images, we can use the `prepare_data` function in Listing 8.8. It starts by loading class labels from the file `trainLabels.csv` (line 2) and iterates through each entry (lines 3–5). It retrieves the `file_id` and corresponding `class_label` for each binary file. For each class, this function creates a dedicated folder to store the images (lines 6–7). The path to the binary file is then constructed using the `file_id` (line 8). The resulting image is saved as a PNG file in an appropriate class folder (lines 11–13). This approach ensures that all binary files are effectively transformed into images and organized by class, streamlining future analysis. After preprocessing steps, the data is fully prepared for use with the cDCGAN architecture to synthesize malware data.

```
1 def prepare_data(src_path, dest_path):
2     labels_df = pd.read_csv("trainLabels.csv")
3     for index, row in labels_df.iterrows():
4         file_id = row["Id"]
5         class_label = row["Class"]
6         class_directory = os.path.join(dest_path, str(class_label))
7         os.makedirs(class_directory, exist_ok=True)
8         binary_file_path = os.path.join(src_path, f"{file_id}.bytes")
9         if os.path.exists(binary_file_path):
10            image = binary_to_image(binary_file_path)
11            image_filename = os.path.join(class_directory, f"{file_id}.png")
```

```
            img = Image.fromarray(image)
            img.save(image_filename)
```

Listing 8.8: Data preprocessing workflow for training and testing with Data Synthesis.

8.2.3 DESIGNING THE ARCHITECTURE OF CDCGAN

The cDCGAN model is a specialized type of Generative Adversarial Network (GAN) that generates realistic images based on specific labels or conditions. This makes cDCGAN is particularly effective for tasks where image generation needs to align with predefined categories or classes.

The cDCGAN consists of two main components, i.e., the Generator and the Discriminator. The Generator generates images from noise values and label information, while the Discriminator evaluates images to determine if they are real, i.e., images from the dataset, or fake, i.e., images outputting from the Generator. By utilizing convolutional layers, cDCGAN enhances the ability to capture the spatial features of the images. This makes cDCGAN a powerful tool for synthesizing image-based data.

The Generator is designed to produce synthetic images from the noise vector (`z`) and label vector (`c`). The main goal is to generate realistic images that reflect the patterns learned from the training data (real images) with specific labels or classes. As outlined in Listing 8.9, the architecture of the Generator consists of multiple layers that progressively convert the input noise and label into a high-resolution image. It includes three key components:

- `hidden_layer1`: This layer processes the input noise vector (lines 14–15).
- `hidden_layer2`: This layer manages the label vector (lines 18–22).
- `hidden_layers`: This part combines both the noise and label inputs and passes them through a series of intermediate hidden layers (lines 26–30).

The intermediate hidden layers are made up of convolutional layers, each followed by batch normalization and ReLU activation functions. This combination gradually enhances the spatial resolution of the features being generated. Finally, the Generator concludes with an output layer that produces the final image with the shape (`output_dim`). The Tanh activation function is applied at the end to constrain the output values to a range between -1 and 1 (lines 31–36). This structure allows the Generator to effectively create high-quality synthetic images conditioned on the input labels. The reader can change the hyperparameters of these layers, e.g., kernel size, activation functions, etc., for further experiments.

```
1 class Generator(torch.nn.Module):
2     def __init__(self, input_dim, label_dim, num_filters,
    output_dim):
3         super(Generator, self).__init__()
4         self.hidden_layer1 = torch.nn.Sequential()
5         self.hidden_layer2 = torch.nn.Sequential()
```

```
        self.hidden_layers = torch.nn.Sequential()
        for i in range(len(num_filters)):
            if i == 0:
                input_deconv = torch.nn.ConvTranspose2d(input_dim, num_filters[i] // 2, kernel_size=4, stride=1,
                                                        padding=0)
                self.hidden_layer1.add_module("input_deconv", input_deconv)
                torch.nn.init.normal_(input_deconv.weight, mean=0.0, std=0.02)
                torch.nn.init.constant_(input_deconv.bias, 0.0)
                self.hidden_layer1.add_module("input_bn", torch.nn.BatchNorm2d(num_filters[i] // 2))
                self.hidden_layer1.add_module("input_act", torch.nn.ReLU())
                label_deconv = torch.nn.ConvTranspose2d(label_dim, num_filters[i] // 2, kernel_size=4, stride=1,
                                                        padding=0)
                self.hidden_layer2.add_module("label_deconv", label_deconv)
                torch.nn.init.normal_(label_deconv.weight, mean=0.0, std=0.02)
                torch.nn.init.constant_(label_deconv.bias, 0.0)
                self.hidden_layer2.add_module("label_bn", torch.nn.BatchNorm2d(num_filters[i] // 2))
                self.hidden_layer2.add_module("label_act", torch.nn.ReLU())
            else:
                deconv = torch.nn.ConvTranspose2d(num_filters[i - 1], num_filters[i], kernel_size=4, stride=2,
                                                  padding=1)
                self.hidden_layers.add_module(f"deconv_{i}", deconv)
                torch.nn.init.normal_(deconv.weight, mean=0.0, std=0.02)
                torch.nn.init.constant_(deconv.bias, 0.0)
                self.hidden_layers.add_module(f"bn_{i}", torch.nn.BatchNorm2d(num_filters[i]))
                self.hidden_layers.add_module(f"act_{i}", torch.nn.ReLU())
        self.output_layer = torch.nn.Sequential()
        out_deconv = torch.nn.ConvTranspose2d(num_filters[-1], output_dim, kernel_size=4, stride=2, padding=1)
        self.output_layer.add_module("output_deconv", out_deconv)
        torch.nn.init.normal_(out_deconv.weight, mean=0.0, std=0.02)
        torch.nn.init.constant_(out_deconv.bias, 0.0)
        self.output_layer.add_module("output_act", torch.nn.Tanh())
    def forward(self, z, c):
        h1 = self.hidden_layer1(z)
        h2 = self.hidden_layer2(c)
        x = torch.cat([h1, h2], dim=1)
        h = self.hidden_layers(x)
```

```
42         out = self.output_layer(h)
43         return out
```

Listing 8.9: The architecture of the cDCGAN 's Generator.

The forward method, outlined in the lines from 37 to 43 of Listing 8.9, details the data flow through the Generator. In this process, the noise and label vectors are processed separately. Specifically, the noise vector passes through a transpose convolutional layer in line 38, while the label vector undergoes similar processing in line 39. The outputs from these two branches are then concatenated along the channel dimension (line 40). Then, it is passed through additional transpose convolutional layer that progressively increases the spatial resolution (line 41).

```
class Discriminator(torch.nn.Module):
    def __init__(self, input_dim, label_dim, num_filters, output_dim):
        super(Discriminator, self).__init__()
        self.hidden_layer1 = torch.nn.Sequential()
        self.hidden_layer2 = torch.nn.Sequential()
        self.hidden_layers = torch.nn.Sequential()
        for i in range(len(num_filters)):
            if i == 0:
                input_conv = torch.nn.Conv2d(input_dim, num_filters[i] // 2, kernel_size=4, stride=2, padding=1)
                self.hidden_layer1.add_module("input_conv", input_conv)
                torch.nn.init.normal_(input_conv.weight, mean=0.0, std=0.02)
                torch.nn.init.constant_(input_conv.bias, 0.0)
                self.hidden_layer1.add_module("input_act", torch.nn.LeakyReLU(0.2))
                label_conv = torch.nn.Conv2d(label_dim, num_filters[i] // 2, kernel_size=4, stride=2, padding=1)
                self.hidden_layer2.add_module("label_conv", label_conv)
                torch.nn.init.normal_(label_conv.weight, mean=0.0, std=0.02)
                torch.nn.init.constant_(label_conv.bias, 0.0)
                self.hidden_layer2.add_module("label_act", torch.nn.LeakyReLU(0.2))
            else:
                conv = torch.nn.Conv2d(num_filters[i - 1], num_filters[i], kernel_size=4, stride=2, padding=1)
                self.hidden_layers.add_module(f"conv_{i}", conv)
                torch.nn.init.normal_(conv.weight, mean=0.0, std=0.02)
                torch.nn.init.constant_(conv.bias, 0.0)
                self.hidden_layers.add_module(f"bn_{i}", torch.nn.BatchNorm2d(num_filters[i]))
                self.hidden_layers.add_module(f"act_{i}", torch.nn.LeakyReLU(0.2))
        self.output_layer = torch.nn.Sequential()
        out_conv = torch.nn.Conv2d(num_filters[-1], output_dim, kernel_size=4, stride=1, padding=0)
```

```
28         self.output_layer.add_module("output_conv", out_conv)
29         torch.nn.init.normal_(out_conv.weight, mean=0.0, std
    =0.02)
30         torch.nn.init.constant_(out_conv.bias, 0.0)
31         self.output_layer.add_module("output_act", torch.nn.
    Sigmoid())
32     def forward(self, z, c):
33         h1 = self.hidden_layer1(z)
34         h2 = self.hidden_layer2(c)
35         x = torch.cat([h1, h2], dim=1)
36         h = self.hidden_layers(x)
37         out = self.output_layer(h)
38         return out
```

Listing 8.10: The architecture of the cDCGAN 's Discriminator.

The Discriminator is designed to distinguish between real images from the dataset and fake images generated by the Generator. As described in Listing 8.10, the Discriminator is composed of several convolutional layers that progressively reduce the spatial resolution of the input images while extracting useful features. The architecture of the Discriminator is organized into three main parts:

- `hidden_layer1` processes the input image using convolutional layers (line 13);
- `hidden_layer2` processes the label vector(line 18);
- `hidden_layers` combines the image and label features by concatenating them along the channel dimension and further puts them into other convolutional layers (lines 21–25).

To train the Discriminator, the input image and label data must be fed into it to classify whether the image is real or fake, as in Listing 8.10. Training the Discriminator is similar to a general classifier. Specifically, the image passes through the convolutional layer with a filter size of `num_filters[0]` (line 33), while the label undergoes a similar convolutional operation (line 34). The outputs from these two branches are concatenated along the channel dimension (line 35) and passed through other convolutional layers to extract high-level features. The final output layer (line 37) uses the Sigmoid activation function for classification purpose.

The integration of the Generator and Discriminator forms the core of the robust cDCGAN model. The Generator generates images from the noise vector, while the Discriminator distinguishes whether those images are real or fake. This adversarial process enables both these components to improve iteratively. The Generator learns to produce more realistic images while the Discriminator becomes better at distinguishing between real and generated images. Through this interaction, the cDCGAN effectively captures complex patterns of the real data, allowing it to generate high-quality images.

8.2.4 TRAINING AND TESTING PROCESS

8.2.4.1 Training Process

To effectively train the cDCGAN model, it is essential to define key components, including the loss function, optimization method, training procedure, and hyperparameters such as `learning_rate`, `batch_size`, and `n_epochs`. These hyperparameters are critical for optimizing the training process and ensuring efficient use of computational resources. The readers can also change these hyperparameters for further experiments.

The specific configuration for the Generator and Discriminator is detailed in Listing 8.11. The key hyperparameters in this configuration include `image_size` (line 5), `G_input_dim` (line 7), `G_output_dim` (line 8), `D_input_dim` (line 9), `D_output_dim` (line 10), `num_filters` (line 11). Specifically, `image_size` sets the resolution of the images processed by both the Generator and Discriminator to 256x256; `G_input_dim` defines the size of the latent space, typically represented as a random noise vector from which the Generator synthesizes images. `G_output_dim` indicates the number of channels in the generated images, corresponding to RGB color channels. `D_input_dim` reflects the RGB channels in the images fed into the Discriminator. `D_output_dim` indicates the output dimension of the Discriminator. Finally, `num_filters` determines the number of convolutional filters at each layer, starting from 1024 and decreasing to 32. This influences the model's ability to learn and capture intricate details.

```
1 cfg = {
2     "batch_size": 16,
3     "learning_rate": 1e-5,
4     "n_epochs": 20,
5     "image_size": 256,
6     "label_dim": 9,
7     "G_input_dim": 100,
8     "G_output_dim": 3,
9     "D_input_dim": 3,
10     "D_output_dim": 3,
11     "num_filters": [1024, 512, 256, 128, 64, 32],
12 }
```

Listing 8.11: Configuration settings for training the cDCGAN in Data Synthesis.

The training process of the cDCGAN model involves the following steps. First, the training data is loaded and prepared by resizing the images, converting them into tensors, and normalizing their values. A custom dataset named as `MalwareDataset` is created to allow for batch processing of training data samples using the `DataLoader` (lines 3–5). Next, both the Generator and Discriminator models are initialized with specific input and output dimensions. They are then moved to an appropriate computing device (lines 6–11). The Adam optimizer is used for training both Generator and Discriminator, with learning rates and other hyperparameters set in the configuration (lines 12–13). The loss function is Binary Cross-Entropy Loss, i.e., (`BCELoss`) in line 14. Additionally, we use `SummaryWriter` from TensorBoard to log training metrics for later visualization (line 15).

During training, the labels for the training samples are converted into one-hot encoding format (lines 16–20). The training loop runs for a number of epochs (lines 22–78). In each training iteration, the Discriminator is trained using both real images and fake images generated by the Generator (lines 50–55). Its loss is calculated based on how well it distinguishes real and fake images (line 52).

```
1 def train(cfg):
2     device = torch.device("cuda" if cfg["gpu_usage"] else "cpu")
3     transform_train = transforms.Compose([transforms.Resize((cfg[
      "image_size"], cfg["image_size"])), transforms.ToTensor(),
      transforms.Normalize(mean=(0.5, 0.5, 0.5), std=(0.5, 0.5,
      0.5))])
4     train_dataset = MalwareDataset(os.path.join(cfg["dataset_dir"
      ], "train"), transform_train)
5     train_loader = DataLoader(train_dataset, batch_size=cfg["
      train_batch_size"], shuffle=True, num_workers=cfg["
      num_workers"])
6     generator_model = Generator(cfg["G_input_dim"], cfg["
      label_dim"], cfg["num_filters"], cfg["G_output_dim"])
7     discriminator_model = Discriminator(cfg["D_input_dim"], cfg["
      label_dim"], cfg["num_filters"][::-1], cfg["D_output_dim"])
8     generator_model = generator_model.to(device)
9     discriminator_model = discriminator_model.to(device)
10    generator_model.train()
11    discriminator_model.train()
12    G_optimizer = Adam(generator_model.parameters(), lr=cfg["
      learning_rate"], betas=cfg["betas"])
13    D_optimizer = Adam(discriminator_model.parameters(), lr=cfg["
      learning_rate"] * 0.3, betas=cfg["betas"])
14    loss_fn = torch.nn.BCELoss()
15    writer = SummaryWriter(log_dir=cfg["log_dir"])
16    onehot = torch.zeros(cfg["label_dim"], cfg["label_dim"])
17    onehot = onehot.scatter_(1, torch.LongTensor([0, 1, 2, 3, 4,
      5, 6, 7, 8]).view(cfg["label_dim"], 1), 1).view(cfg["
      label_dim"], cfg["label_dim"], 1, 1)
18    fill = torch.zeros([cfg["label_dim"], cfg["label_dim"], cfg["
      image_size"], cfg["image_size"]])
19    for dim in range(cfg["label_dim"]):
20        fill[dim, dim, :, :] = 1
21    n_epochs = cfg["n_epochs"]
22    for epoch in range(n_epochs):
23        if epoch == 5 or epoch == 10:
24            G_optimizer.param_groups[0]["lr"] /= 2
25            D_optimizer.param_groups[0]["lr"] /= 2
26        pbar = tqdm(train_loader, ncols=80, desc="Training")
27        D_running_loss = 0.0
28        G_running_loss = 0.0
29        train_num = 0
30        for step, data in enumerate(pbar):
31            images, labels = data["image"], data["label"]
32            images = images.to(device)
33            labels = labels.to(device)
34            mini_batch = images.size()[0]
35            labels_fill_ = fill[labels].to(device)
36            D_real_decision = discriminator_model(images,
      labels_fill_).squeeze()
```

```
        y_real_ = torch.ones_like(D_real_decision).to(device)
        y_fake_ = torch.zeros_like(D_real_decision).to(device)
        D_real_loss = loss_fn(D_real_decision, y_real_)
        z_ = torch.randn(mini_batch, cfg["G_input_dim"]).view(-1, cfg["G_input_dim"], 1, 1).to(device)
        c_ = (torch.rand(mini_batch, 1) * cfg["label_dim"]).type(torch.LongTensor).squeeze()
        if len(c_.shape) == 0:
            c_ = c_.unsqueeze(0)
        c_onehot_ = onehot[c_].to(device)
        if epoch == 0:
            fixed_noise = z_
            fixed_label = c_onehot_
        gen_image = generator_model(z_, c_onehot_)
        c_fill_ = fill[c_].cuda()
        D_fake_decision = discriminator_model(gen_image, c_fill_).squeeze()
        D_fake_loss = loss_fn(D_fake_decision, y_fake_)
        D_loss = D_real_loss + D_fake_loss
        D_optimizer.zero_grad()
        D_loss.backward()
        D_optimizer.step()
        z_ = torch.randn(mini_batch, cfg["G_input_dim"]).view(-1, cfg["G_input_dim"], 1, 1).to(device)
        c_ = (torch.rand(mini_batch, 1) * cfg["label_dim"]).type(torch.LongTensor).squeeze()
        if len(c_.shape) == 0:
            c_ = c_.unsqueeze(0)
        c_onehot_ = onehot[c_].to(device)
        gen_image = generator_model(z_, c_onehot_)
        c_fill_ = fill[c_].cuda()
        D_fake_decision = discriminator_model(gen_image, c_fill_).squeeze()
        G_loss = loss_fn(D_fake_decision, y_real_)
        G_optimizer.zero_grad()
        G_loss.backward()
        G_optimizer.step()
        D_running_loss += D_loss.item() * mini_batch
        G_running_loss += G_loss.item() * mini_batch
        train_num += mini_batch
    D_train_loss = D_running_loss / train_num
    G_train_loss = G_running_loss / train_num
    print(f"\tG train loss: {G_train_loss:.4f}, D_train_loss: {D_train_loss:.4f}")
    writer.add_scalar("Discriminator Loss/Train_epoch", D_train_loss, epoch)
    writer.add_scalar("Generator Loss/Train_epoch", G_train_loss, epoch)
    torch.save(generator_model.state_dict(), f"{cfg["save_dir"]}/G_model_{epoch}.pth")
    torch.save(discriminator_model.state_dict(), f"{cfg["save_dir"]}/D_model_{epoch}.pth")
writer.close()
```

Listing 8.12: Training procedure for training the cDCGAN in Data Synthesis.

We record the training log to monitor the model's progress and performance (as illustrated in lines 76-77 of Listing 8.12). Figure 8.4 demonstrates that balancing the losses of the Discriminator and Generator can be challenging. The readers can adapt this training by adjusting learning rates, using gradient clipping, label smoothing, etc., to stabilize the training process.

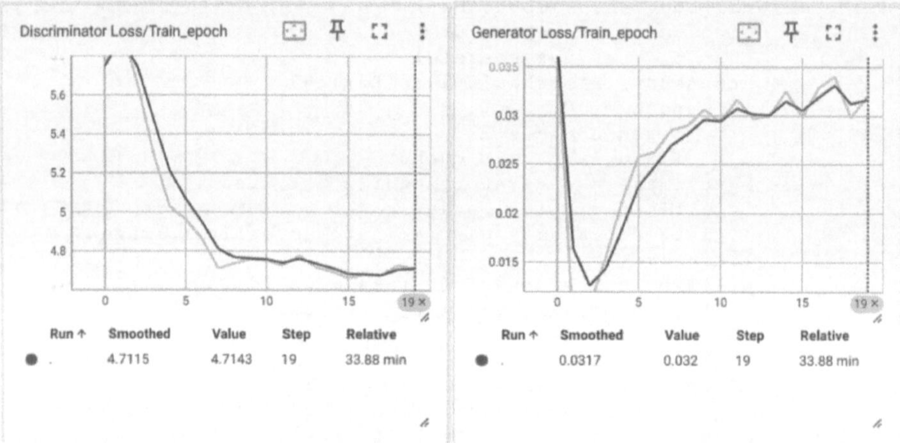

Figure 8.4 Visualization of training the Discriminator and Generator losses for Data Synthesis using cDCGAN in Tensorboard

8.2.4.2 Testing Process

```
def test(cfg):
    device = torch.device("cuda" if cfg["gpu_usage"] else "cpu")
    generator = Generator(cfg["G_input_dim"], cfg["label_dim"], cfg["num_filters"], cfg["G_output_dim"]).to(device)
    num_epoch = 19
    generator_path = os.path.join(cfg["save_dir"], f"G_model_{num_epoch}.pth")
    if os.path.exists(generator_path):
        generator.load_state_dict(torch.load(generator_path, map_location=device))
    generator.eval()
    onehot = torch.eye(cfg["label_dim"]).view(cfg["label_dim"], cfg["label_dim"], 1, 1)
    noise = torch.randn(cfg["test_batch_size"], cfg["G_input_dim"], 1, 1).to(device)
    c_ = (torch.rand(cfg["test_batch_size"]) * cfg["label_dim"]).long()
    label = onehot[c_].to(device)
    with torch.no_grad():
        gen_image = generator(noise, label)
        gen_image = denorm(gen_image)
    n_samples = noise.size(0)
    n_rows = int(np.ceil(np.sqrt(n_samples)))
    n_cols = int(np.ceil(n_samples / n_rows))
```

```
19     fig, axes = plt.subplots(n_rows, n_cols, figsize=(5, 5))
20     for ax, img in zip(axes.flatten(), gen_image):
21         ax.axis("off")
22         img = img.cpu().numpy().transpose(1, 2, 0) and transpose
23         img = (img - img.min()) * 255 / (img.max() - img.min())
24         ax.imshow(img.astype(np.uint8), aspect="equal")
25     plt.subplots_adjust(wspace=0, hspace=0)
26     plt.title(f"Epoch {num_epoch+1}")
27     plt.show()
```

Listing 8.13: Testing procedure for testing the cDCGAN in Data Synthesis.

The testing procedure for the cDCGAN model is detailed in Listing 8.13. Initially, the testing configuration is set up to align with the training process, which includes loading the trained Generator and its weights from a checkpoint and switching to evaluation mode (lines 2–8). The process then creates random noise vectors and one-hot encoded labels to facilitate the generation of new data samples (lines 9–12). Within a `torch.no_grad()` context (line 20), the Generator produces images from the noise and labels. Finally, the generated images are organized into a grid format for evaluation purposes (lines 17–27), allowing for a comprehensive assessment of the generator's output quality. Figure. 8.5 shows generated data samples.

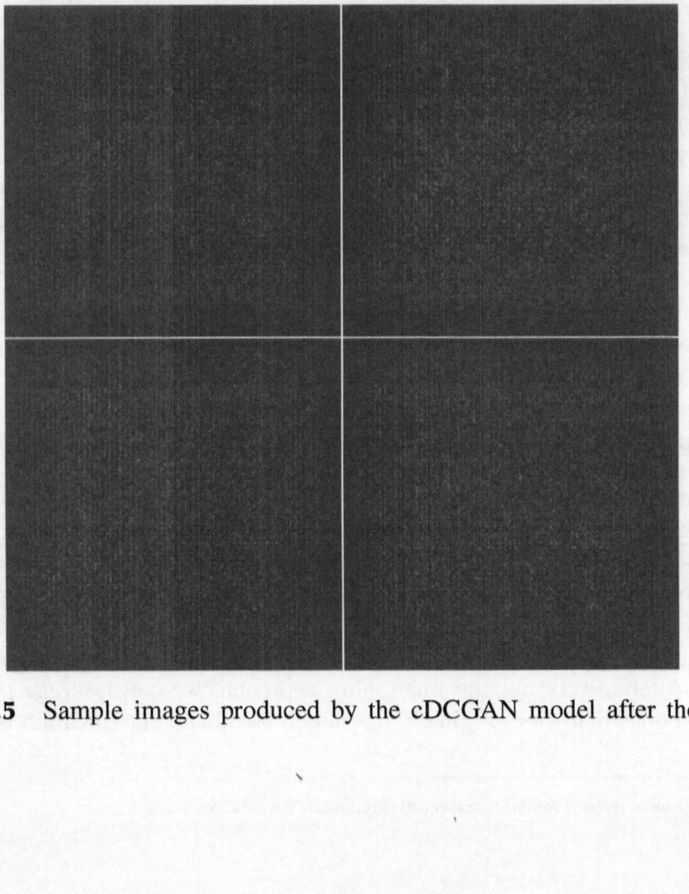

Figure 8.5 Sample images produced by the cDCGAN model after the training phase.

Readers can download the complete implementation code for this section from here[4].

8.3 MALWARE CLASSIFICATION

8.3.1 PROBLEM STATEMENT

This experimental study focuses on developing a practical and effective method for malware classification—an essential task in cybersecurity that enables early threat identification and timely incident response. Accurate classification not only helps security systems understand the type and behavior of malware but also supports the development of targeted defense strategies. In this case study, we explore an innovative approach by converting malware binaries into image representations, which reveal structural patterns that are often difficult to detect in raw binary format. These visual cues can significantly improve the accuracy and robustness of classification models. By integrating this technique with generative AI, such as GAN-based models, students and practitioners can build efficient, scalable, and real-world-ready solutions for malware detection, gaining hands-on experience in applying generative models to pressing cybersecurity challenges.

8.3.2 DATA PREPROCESSING

The dataset used in this experiment is consistent with the one employed in the Data Synthesis case study, allowing readers to follow the same preprocessing steps outlined in Section 8.2.2. After preprocessing, each malware sample is converted into a grayscale image, enabling visual analysis and deep learning-based classification. This image-based representation captures structural patterns within the malware, which can enhance the performance of generative models. However, readers are encouraged to explore alternative formats of malware data—such as raw binaries, opcode sequences, or byte-level representations—to evaluate the adaptability and effectiveness of generative approaches across different data types. This flexibility provides valuable opportunities to deepen understanding and assess the broader applicability of deep neural network models in malware analysis.

8.3.3 DESIGNING THE ARCHITECTURE OF A DEEP NEURAL NETWORK

In this section, we apply a Deep Neural Network (DNN) model to perform malware classification. Two implementation strategies are available: training a model from scratch or fine-tuning a pre-trained model. For the first approach, we design a custom DNN using PyTorch, following architectural guidelines found in prior literature [5]. Readers can refer to Section 8.2.3 for guidance on constructing neural network models, particularly the discriminator, which can be adapted for classification purposes. Alternatively, for the fine-tuning approach, we can leverage pre-trained DNN backbones that were originally developed for image classification tasks, such

[4] https://github.com/lyvt/genai˙cybersecurity/tree/main/02-data˙synthesis

as InceptionV3 or GoogLeNet [6]. These pre-trained models provide a solid foundation for transfer learning, enabling more efficient training and improved performance on malware image datasets.

In this case study, we demonstrate how to implement a DNN model for malware classification using GoogLeNet, as detailed in Listing 8.14. The process begins by initializing the model with pre-trained weights from GoogLeNet (line 3), which have been learned from a large-scale image classification dataset. Next, we modify the model by replacing the top fully connected layer with output predictions for our specific set of nine malware classes (line 4). This approach leverages transfer learning, allowing us to fine-tune the model for the task of malware classification while benefiting from the general image features learned during pre-training. While GoogLeNet is used in this example, readers are encouraged to experiment with other popular CNN backbones such as ResNet, InceptionV3, or EfficientNet to explore and compare their performance in malware classification.

```
1 import torch
2 import torchvision.models as models
3 googlenet = models.googlenet(pretrained=True)
4 googlenet.fc = torch.nn.Linear(googlenet.fc.in_features, 9)
```

Listing 8.14: Customizing GoogleNet architecture for Malware Classification Tasks

8.3.4 TRAINING AND TESTING PROCESS

8.3.4.1 Training Process

The training process for the DNN model is designed to iteratively improve its ability to classify malware by minimizing prediction errors. As illustrated in Listing 8.15, the workflow begins with loading and preprocessing the training and validation datasets, which includes transformations such as random resizing, cropping, and normalization to enhance generalization (lines 3–10). These datasets are then wrapped in `DataLoader` objects to enable efficient batch processing. The GoogLeNet model is initialized, moved to the appropriate computation device (CPU or GPU), and set to training mode (lines 11–12). During each training batch, input data is passed through the model to generate predictions (line 25), and the cross-entropy loss is computed by comparing these predictions with the ground-truth labels (line 26). This loss indicates the model's classification error, guiding the optimization process. The backpropagation algorithm computes the gradients (line 27), and the Adam optimizer updates the model's weights to reduce the loss (line 28). After each epoch, the model is evaluated on the validation dataset by switching to evaluation mode, allowing the calculation of validation loss and helping to monitor performance and detect overfitting (lines 32–42).

```
1 def train(cfg):
2     device = torch.device("cuda" if cfg["gpu_usage"] else "cpu")
3     transform_train = transforms.Compose([
4         transforms.RandomResizedCrop(224)...])
5     transform_val = transforms.Compose([
6         transforms.ToTensor(), ...])
```

```
    train_datasetset = MalwareDataset(os.path.join("dataset", "
    train"), transform_train)
    train_loader = torch.utils.data.DataLoader(train_datasetset,
    batch_size=cfg["train_batch_size"], shuffle=True, num_workers
    =cfg["num_workers"])
    val_dataset = MalwareDataset(os.path.join("dataset", "test"),
     transform_val)
    val_loader = torch.utils.data.DataLoader(val_dataset,
    batch_size=cfg["train_batch_size"], shuffle=False,
    num_workers=cfg["num_workers"])
    model = googlenet.to(device)
    model.train()
    optimizer = SGD(model.parameters(), lr=cfg["learning_rate"],
    momentum=0.9, weight_decay=5e-4)
    loss_fn = nn.CrossEntropyLoss()
    writer = SummaryWriter(log_dir=cfg["log_dir"])
    n_epochs = cfg["n_epochs"]
    for epoch in range(n_epochs):
        model.train()
        pbar = tqdm(train_loader, ncols=80, desc="Training")
        running_loss = 0.0
        train_num = 0
        for step, data in enumerate(pbar):
            inputs, targets = data["image"].to(device), data["
    label"].to(device)
            optimizer.zero_grad()
            outputs = model(inputs)
            loss = loss_fn(outputs, targets)
            loss.backward()
            optimizer.step()
            running_loss += loss.item() * inputs.shape[0]
            train_num += inputs.shape[0]
        train_loss = running_loss / train_num
        model.eval()
        val_running_loss = 0.0
        val_num = 0
        with torch.no_grad():
            for inputs in val_loader:
                inputs, targets = data["image"].to(device), data[
    "label"].to(device)
                outputs = model(inputs)
                val_loss = loss_fn(outputs, targets)
                val_running_loss += val_loss.item() * inputs.
    shape[0]
                val_num += inputs.shape[0]
        val_loss = val_running_loss / val_num
        writer.add_scalar("Loss/Train_epoch", train_loss, epoch)
        writer.add_scalar("Loss/Validation_epoch", val_loss,
    epoch)
        torch.save(model.state_dict(), f"{cfg["save_dir"]}/model_
    {epoch}.pth")
    writer.close()
```

Listing 8.15: Training proceduce for training the DNN model in Malware Classification.

Visualizing the training log is crucial for tracking the model's convergence during the training process. The training log (as shown in the lines 44-45 of Listing 8.4), allows us to monitor the training's progress and model's performance. To access this log, use the command `tensorboard --logdir=log_dir`, which opens the default browser. If we observe the overfitting problem presented in the log visualization, we need to adjust the data pre-process hyper-parameter set to enhance performance.

Visualizing the training log is essential for monitoring the model's learning behavior and ensuring proper convergence throughout the training process. As shown in lines 44–45 of Listing 8.4, the training log records important metrics such as training and validation loss over time, helping to track the model's performance. Readers can visualize these metrics using TensorBoard by running the command `tensorboard --logdir=log_dir`, which launches an interactive dashboard in your default web browser. This visualization enables early detection of issues like overfitting—where the model performs well on training data but poorly on validation data. If overfitting or other training issues are observed, readers are encouraged to refine the preprocessing steps, adjust hyperparameters, or apply regularization techniques to improve model generalization.

8.3.4.2 Testing Process

The primary objective of the testing phase is to evaluate how well the trained DNN model can classify previously unseen malware samples. This is accomplished using the function shown in Listing 8.5, which computes standard classification performance metrics such as accuracy, precision, recall, and F1-score. As detailed in Listing 8.16, the testing process begins by preparing the test dataset: loading malware images from the designated directory using the `MalwareDataset` class, applying the necessary transformations, and organizing the data into batches via a `DataLoader` (lines 3–5). The previously trained model is then restored from a checkpoint and switched to evaluation mode to disable gradient computation and dropout layers (lines 6–11). During testing, the model processes each batch, predicts class labels by selecting the highest probability scores (lines 16–18), and stores both the predicted and true labels (lines 19–20). Finally, the `calculate_metrics` function is called (line 21) to evaluate the overall performance of the model, providing a clear and quantitative summary of its classification effectiveness.

```
1 def test(cfg):
2     device = torch.device("cuda" if cfg["gpu_usage"] else "cpu")
3     transform_test = transforms.Compose([transforms.ToTensor(),
      ...)
4     test_dataset = MalwareDataset(os.path.join("dataset", "test")
      , transform_test)
5     test_loader = torch.utils.data.DataLoader(test_dataset,
      batch_size=cfg["train_batch_size"], shuffle=False,
      num_workers=cfg["num_workers"])
6     model = googlenet.to(device)
7     model_path = os.path.join(cfg["save_dir"], "model_9.pth")
8     if os.path.exists(model_path):
9         model.load_state_dict(torch.load(model_path))
```

```
    model = model.to(device)
    model.eval()
    labels = []
    predictions = []
    with torch.no_grad():
        for x_batch, y_batch in tqdm(test_loader):
            x_batch = x_batch.to(device)
            outputs = model(x_batch)
            _, y_predict = outputs.max(1)
            labels.extend(y_batch.tolist())
            predictions.extend(y_predict.cpu().tolist())
    performance_metrics = calculate_metrics(labels, predictions)
    return performance_metrics
```

Listing 8.16: Testing procedure for testing the DNN model in Malware Classification.

Readers can download the complete implementation code for this section from here [5].

8.4 TRAFFIC ANALYSIS

8.4.1 PROBLEM STATEMENT

Network traffic classification plays a vital role in managing and securing modern networks by identifying the types of traffic generated by various applications. This process not only ensures smooth network performance but also aids in detecting and mitigating potential security threats. However, with the growing adoption of encryption to protect user privacy, accurately classifying traffic has become increasingly challenging. Encrypted traffic conceals the payload, making it difficult for conventional methods to inspect and analyze content.

Traditional approaches such as Deep Packet Inspection (DPI), which depend on examining patterns within packet payloads, become ineffective when encryption is present. Moreover, the rapid evolution of encryption protocols further complicates classification efforts, rendering many existing techniques obsolete or less effective. As a result, developing models that can recognize subtle and robust patterns in encrypted traffic is now crucial for maintaining strong network security.

To address these challenges, recent advancements in pre-trained deep learning models have shown great promise. These models, which have revolutionized fields like natural language processing and computer vision, are trained on large volumes of unlabeled data to extract generalizable features. They can then be fine-tuned on smaller, task-specific datasets to achieve high performance with minimal labeled data. In this experiment, we adopt the Encrypted Traffic Bidirectional Encoder Representations from Transformers (ET-BERT) architecture [7] to learn meaningful representations of encrypted traffic. By leveraging ET-BERT's ability to capture contextual information in network flows, this approach provides a powerful solution for handling encrypted traffic classification with improved accuracy and adaptability.

[5] https://github.com/lyvt/genai˙cybersecurity/tree/main/03-malware-image-classification

8.4.2 DATA PREPROCESSING

In this experiment, we utilize a real-world TLS 1.3 encrypted traffic dataset, collected between March and July 2021 on the China Science and Technology Network (CSTNET) [7]. This dataset captures traffic from various applications under real network conditions, making it highly suitable for evaluating encrypted traffic classification models. The dataset is preprocessed using the `prepare_data()` function, as illustrated in Listing 8.17, which formats the raw input for BERT-based classification by parsing and tokenizing each traffic sample.

Effective data preprocessing is a critical step in preparing datasets for machine learning models, ensuring the input data is clean, consistent, and appropriately structured. In this case, since the input traffic is represented in text format, preprocessing involves parsing CSV-formatted records and converting them into tokenized sequences suitable for BERT-based models. The `prepare_data()` function begins by initializing an empty list called `dataset` to store preprocessed records and a `columns` dictionary to map column names to their respective indices (line 2). It reads the CSV file line by line, maps column headers, and extracts relevant fields, including the target classification label from the `label` column (lines 3–9).

Each traffic sample is then tokenized with BERT-specific tokens such as `[CLS]` and `[SEP]` to mark the beginning and separation of input segments. Segment embeddings are assigned (1 for the first segment, 2 for the second) to help the model differentiate between parts of the input (line 23). To maintain a consistent input length for the model, sequences are either truncated if they exceed the maximum length (lines 24–26) or padded with special tokens if they are shorter (lines 27–29). Finally, each processed sequence, along with its corresponding label, is added to the `dataset` list (lines 30–33), resulting in a structured and tokenized dataset ready for training and evaluation.

```
1 def prepare_data(args, path):
2     dataset, columns = [], {}
3     with open(path, mode="r", encoding="utf-8") as f:
4         for line_id, line in enumerate(f):
5             if line_id == 0:
6                 for i, column_name in enumerate(line.strip().
    split("\t")):
7                     columns[column_name] = i
8                 continue
9             line = line[:-1].split("\t")
10            tgt = int(line[columns["label"]])
11            if args.soft_targets and "logits" in columns.keys():
12                soft_tgt = [float(value) for value in the line[
    columns["logits"]].split(" ")]
13            if "text_b" not in columns:
14                text_a = line[columns["text_a"]]
15                src = args.tokenizer.convert_tokens_to_ids([
    CLS_TOKEN] + args.tokenizer.tokenize(text_a))
16                seg = [1] * len(src)
17            else:
18                text_a, text_b = line[columns["text_a"]], line[
    columns["text_b"]]
```

```
                src_a = args.tokenizer.convert_tokens_to_ids(
                    [CLS_TOKEN] + args.tokenizer.tokenize(text_a)
 + [SEP_TOKEN])
                src_b = args.tokenizer.convert_tokens_to_ids(args
.tokenizer.tokenize(text_b) + [SEP_TOKEN])
                src = src_a + src_b
                seg = [1] * len(src_a) + [2] * len(src_b)
            if len(src) > args.seq_length:
                src = src[: args.seq_length]
                seg = seg[: args.seq_length]
            while len(src) < args.seq_length:
                src.append(0)
                seg.append(0)
            if args.soft_targets and "logits" in columns.keys():
                dataset.append((src, tgt, seg, soft_tgt))
            else:
                dataset.append((src, tgt, seg))
    return dataset
```

Listing 8.17: Data preprocessing workflow for Traffic Analysis.

8.4.3 DESIGNING THE ARCHITECTURE OF BERT

For network traffic classification, we use a classifier architecture based on the pre-trained BERT model to efficiently classify encrypted traffic protocols. The architecture has three main parts, i.e., the embedding layer, the encoder layer, and the output layer. Each part has a specific role in helping the model accurately recognize traffic types in encrypted environments, as described in Listing 8.18.

```
class Classifier(nn.Module):
    def __init__(self, args):
        super(Classifier, self).__init__()
        self.embedding = WordPosSegEmbedding(args, len(args.
    tokenizer.vocab))
        self.encoder = TransformerEncoder(args)
        self.labels_num = args.labels_num
        self.soft_targets = args.soft_targets
        self.soft_alpha = args.soft_alpha
        self.output_layer_1 = nn.Linear(args.hidden_size, args.
    hidden_size)
        self.output_layer_2 = nn.Linear(args.hidden_size, self.
    labels_num)
    def forward(self, src, tgt, seg):
        emb = self.embedding(src, seg)
        output = self.encoder(emb, seg)[:, 0, :]
        output = torch.tanh(self.output_layer_1(output))
        logits = self.output_layer_2(output)
        if tgt is not None:
            loss = nn.N Los()(nn.LogSoftmax(dim=-1)(logits), tgt.
    view(-1))
            return loss, logits
        else:
            return None, logits
```

Listing 8.18: The classifier architecture for traffic analysis.

The embedding layer, described in line 4 of Listing 8.18, plays a crucial role in converting input traffic data into meaningful token embeddings. It combines word embeddings with positional and segmentation embeddings to effectively capture the structural and sequential features of encrypted traffic. In this architecture, the `WordPosSegEmbedding` layer is utilized for generating embeddings, as demonstrated in Listing 8.19. The `WordPosSegEmbedding` class integrates three types of word-embeddings, positional embedding, and segmentation to create the comprehensive input representations.

Word embeddings, which map each token in the vocabulary to a fixed-size dense vector defined by `args.emb_size` (line 7), capture the meaning of the input. Positional embeddings indicate the order of the sequence, specifying the position of each token within the input (line 8). Segmentation embeddings provide details about specific segments, with the parameter 3 representing three possible segments (line 9). To generate token embeddings from the input, the data is fed into the embedding layer, and the output is retrieved. The `forward` function, defined in the architecture, starts by computing the word embeddings in line 13, where `src` represents the input token indices. In the line 14, positional embeddings are generated by creating a tensor of sequential indices corresponding to the length of the input sequence. This tensor is then passed to the positional embedding layer to obtain the respective embeddings. Next, the final embedding is the summarization of the word, positional, and segmentation embeddings in the line 18. This summation ensures that each token representation encompasses both its meaning and its position within the sequence.

```
1 class WordPosSegEmbedding(nn.Module):
2     def __init__(self, args, vocab_size):
3         super(WordPosSegEmbedding, self).__init__()
4         self.remove_embedding_layernorm = args.
    remove_embedding_layernorm
5         self.dropout = nn.Dropout(args.dropout)
6         self.max_seq_length = args.max_seq_length
7         self.word_embedding = nn.Embedding(vocab_size, args.
    emb_size)
8         self.position_embedding = nn.Embedding(self.
    max_seq_length, args.emb_size)
9         self.segment_embedding = nn.Embedding(3, args.emb_size)
10        if not self.remove_embedding_layernorm:
11            self.layer_norm = LayerNorm(args.emb_size)
12    def forward(self, src, seg):
13        word_emb = self.word_embedding(src)
14        pos_emb = self.position_embedding(
15            torch.arange(0, word_emb.size(1), device=word_emb.
    device, dtype=torch.long)
16            .unsqueeze(0).repeat(word_emb.size(0), 1))
17        seg_emb = self.segment_embedding(seg)
18        emb = word_emb + pos_emb + seg_emb
19        if not self.remove_embedding_layernorm:
20            emb = self.layer_norm(emb)
21        emb = self.dropout(emb)
22        return emb
```

Listing 8.19: The WordPosSegEmbedding layer architecture for Traffic Analysis.

In the BERT architecture, the `TransformerEncoder` is utilized as the encoder layer, as detailed in Listing 8.20. The `TransformerEncoder` class initializes several key components within its `__init__` method. It first establishes the mask and the number of layers based on parameters defined in `args` (lines 4–5). The line 6 creates `self.transformer`, which initializes a list of transformer layers using the `TransformerLayer` class, repeated for the specified number of layers. The line 7 initializes `relative_pos_emb`, which incorporates relative positional information to enhance the representation of token relationships. The `forward` function begins by determining the size of the input embeddings, which is stored in the variables `batch_size` and `seq_length` (line 12). A mask is then generated using the segment input, designed to remove tokens during processing (lines 13–15). The position bias is calculated using the `RelativePositionEmbedding` class, incorporating relative positional information between tokens (line 17). Core processing occurs within a loop that iterates for the specified number of layers. During each iteration, the hidden states are updated by passing them through the transformer layers. The final output `hidden` of the encoder represents the transformed embeddings, ready for further processing in the model (lines 18–19).

```
1 class TransformerEncoder(nn.Module):
2     def __init__(self, args):
3         super(TransformerEncoder, self).__init__()
4         self.mask = args.mask
5         self.layers_num = args.layers_num
6         self.transformer = TransformerLayer(args)
7         self.relative_pos_emb = RelativePositionEmbedding(
    bidirectional=True, heads_num=args.heads_num, num_buckets=
    args.relative_attention_buckets_num)
8     def forward(self, emb, seg):
9         batch_size, seq_length, _ = emb.size()
10        mask = (seg > 0).unsqueeze(1).repeat(1, seq_length, 1).
    unsqueeze(1)
11        mask = mask.float()
12        mask = (1.0 - mask) * -10000.0
13        hidden = emb
14        position_bias = self.relative_pos_emb(hidden, hidden)
15        for i in range(self.layers_num):
16            hidden = self.transformer(hidden, mask, position_bias
    =position_bias)
17        return hidden
```

Listing 8.20: The Transformer Encoder layer architecture for Traffic Analysis.

The output layers, defined in lines 9-10 of Listing 8.18, transform the final hidden states into classification logits. This process is composed of two fully connected layers. The first layer performs a non-linear transformation using the `Tanh` activation function (line 16), which introduces non-linearity to capture complex patterns in the traffic data. The second layer maps the transformed hidden states to the label space, producing the classification logits that serve as the model's output for prediction. To train the classifier, the processed traffic data is passed through the model, and the corresponding output is generated. This process is defined in the `forward()` function, starting at line 11 of Listing 8.18.

8.4.4 TRAINING AND TESTING PROCESS

8.4.4.1 Training Process

The training process for the traffic analysis model involves feeding input data through the model, calculating the loss, and updating the model's weights based on the gradients. As outlined in Listing 8.21, this first loads the training and validation datasets, configuring various paths and hyperparameters via command-line arguments (lines 2–6). The datasets are processed to create tensors for the source, target, and segment inputs. The dataset is then shuffled to introduce randomness (lines 12–18). The `BertTokenizer` is initialized (line 7), followed by the `Classifier` model (line 9). After loading or initializing the model parameters (line 10), the training setup proceeds by calculating the total number of training steps based on the size of the datasets, the number of epochs, and the batch size (line 19). An optimizer and a learning rate schedulers are defined to manage the learning process. The main training loop iterates through each epoch, with the model set to training mode (line 24). For each batch, the gradients are reset (line 28), and the input data is transferred to the appropriate device (lines 29–31).

Subsequently, calculating the loss and updating the model's weights are executed as follows. The loss is computed by performing forward propagation through the model (line 32). The loss value, computed by using the classification logits and the true labels, is backpropagated to update the model's parameters (line 33). The learning rate is then adjusted dynamically (line 35). After completing each epoch, the average loss is recorded (line 39). Once the training is finalized, the model's state dictionary is saved to an output path (line 40). This model leverages pre-trained BERT model and fine-tunes on the traffic data allowing the encoder to effectively capture the relationships between tokens in the input data, as discussed in [7].

```
1 def train():
2     args = parser.parse_args()
3     args.train_path = "data/cstnet-tls-1.3/fine-tuning/packet/
      train_dataset.tsv"
4     args.dev_path = "data/cstnet-tls-1.3/fine-tuning/packet/
      valid_dataset.tsv"
5     args.vocab_path = "models/encryptd_vocab.txt"
6     ...
7     args.tokenizer = BertTokenizer(args)
8     args.device = torch.device("cuda:0" if torch.cuda.
      is_available() else "cpu")
9     model = Classifier(args)
10    load_or_initialize_parameters(args, model)
11    model = model.to(args.device)
12    train_data = prepare_data(args, args.train_path)
13    random.shuffle(train_data)
14    instances_num = len(train_data)
15    batch_size = args.batch_size
16    src = torch.LongTensor([example[0] for example in train_data
      ])
17    tgt = torch.LongTensor([example[1] for example in train_data
      ])
```

```
    seg = torch.LongTensor([example[2] for example in train_data])
    args.train_steps = int(instances_num * args.epochs_num / batch_size) + 1
    optimizer, scheduler = build_optimizer(args, model)
    args.model = model
    writer = SummaryWriter(log_dir="runs")
    for epoch in tqdm.tqdm(range(1, args.epochs_num + 1)):
        model.train()
        running_loss = 0.0
        train_num = 0
        for i, (src_batch, tgt_batch, seg_batch) in enumerate(batch_loader(batch_size, src, tgt, seg)):
            model.zero_grad()
            src_batch = src_batch.to(args.device)
            tgt_batch = tgt_batch.to(args.device)
            seg_batch = seg_batch.to(args.device)
            loss, _ = model(src_batch, tgt_batch, seg_batch)
            loss.backward()
            optimizer.step()
            scheduler.step()
            running_loss += loss.item() * src_batch.shape[0]
            train_num += src_batch.shape[0]
        train_loss = running_loss / train_num
        writer.add_scalar("Loss/Train_epoch", train_loss, epoch)
        torch.save(model.state_dict(), args.output_model_path)
    writer.close()
```

Listing 8.21: Training procedure for a training classifier for Traffic Analysis.

8.4.4.2 Testing Process

The main objective of the testing phase is to assess how well the classifier model performs on unseen data by evaluating its ability to classify samples accurately. The testing procedure for the BERT model in network traffic classification involves feeding input data into the model to generate predictions. As detailed in Listing 8.22, the process begins by configuring the argument parser, which sets the hyperparameters, dataset paths, and additional options such as the tokenizer and batch size (lines 2–9). Next, the model's hyperparameters are loaded, and the number of labels in the dataset is determined (lines 10–11). A `BertTokenizer` is initialized (line 12), and the model is restored from a pre-trained checkpoint (lines 14–16). The test dataset is prepared by converting it into tensors representing the source, target, and segment inputs (lines 17–21).

We need to switch to the evaluation mode and disable gradients of the model's weights to test data in batches (line 23). For each batch, the inputs are transferred to the appropriate device (CPU or GPU) (lines 27–29). Then, the model predicts the output logits (line 31), which are converted into label predictions by applying a softmax function followed by an argmax operation (line 32). Finally, the `calculate_metrics()` function computes performance metrics to provide a comprehensive summary of the model's performance on the test dataset (line 36).

```
def test():
    parser = argparse.ArgumentParser(formatter_class=argparse.ArgumentDefaultsHelpFormatter)
    finetune_opts(parser)
    args = parser.parse_args()
    args.test_path = "data/cstnet-tls-1.3/fine-tuning/packet/test_dataset.tsv"
    args.train_path = "data/cstnet-tls-1.3/fine-tuning/packet/train_dataset.tsv"
    args.vocab_path = "models/encryptd_vocab.txt"
    ...
    args.batch_size = 1
    args = load_hyperparam(args)
    args.labels_num = count_labels_num(args.train_path)
    args.tokenizer = BertTokenizer(args)
    args.device = torch.device("cuda:0" if torch.cuda.is_available() else "cpu")
    model = Classifier(args)
    model.load_state_dict(torch.load(args.output_model_path))
    model = model.to(args.device)
    test_data = prepare_data(args, args.test_path)
    batch_size = args.batch_size
    src = torch.LongTensor([example[0] for example in test_data])
    tgt = torch.LongTensor([example[1] for example in test_data])
    seg = torch.LongTensor([example[2] for example in test_data])
    args.model = model
    model.eval()
    labels = []
    predictions = []
    for i, (src_batch, tgt_batch, seg_batch) in enumerate(batch_loader(batch_size, src, tgt, seg)):
        src_batch = src_batch.to(args.device)
        tgt_batch = tgt_batch.to(args.device)
        seg_batch = seg_batch.to(args.device)
        with torch.no_grad():
            _, logits = args.model(src_batch, tgt_batch, seg_batch)
        y_pred = torch.argmax(nn.Softmax(dim=1)(logits), dim=1)
        y_gt = tgt_batch
        predictions.extend(y_pred.cpu().tolist())
        labels.extend(y_gt.cpu().tolist())
    performance_metrics = calculate_metrics(labels, predictions)
    return performance_metrics
```

Listing 8.22: Testing procedure for testing classifier for Traffic Analysis.

Readers can download the complete implementation code for this section from here [6].

[6]https://github.com/lyvt/genai˙cybersecurity/tree/main/04-traffic-analysis

8.5 DEEPFAKE COUNTER-MEASUREMENTS

8.5.1 PROBLEM STATEMENT

Deepfakes are a technology that uses advanced deep learning techniques, like Autoencoders and Generative Adversarial Networks (GANs), to change or swap faces in images or videos, resulting in very realistic but fake content. While this technology can create impressive media, it also brings serious security and privacy risks. Malicious individuals can misuse deepfakes to create fake videos of well-known figures.

Variational Autoencoders (VAEs) are powerful Deep Learning models that can be used for various applications, including Deepfake Detection. In this section, we explore the potential of VAEs in addressing deepfake detection as an anomaly detection problem. Specifically, real images are classified as "normal," while deepfake images are treated as anomalies. This approach allows us to effectively identify and differentiate between real and fake content, leveraging the unique capabilities of VAEs to capture the underlying distribution of real images.

8.5.2 DATA PREPROCESSING

In this experiment, we utilize the DFFD dataset [8] which comprises several publicly accessible collections, including both real and synthetic images generated by various methods. These images are different in resolution and quality. Specifically, we focus on the Flickr-Faces-HQ (FFHQ) subset of the DFFD dataset for real images, while the synthetic images are generated using StyleGAN-based models trained on the FFHQ dataset. From the original data, we separate the images into two folders: one for real images and another for fake images. Samples from each class are visualized in Fig 8.6.

Figure 8.6 Examples of real and fake images used for training and evaluation, with real images sourced from the FFHQ dataset and fake images generated by StyleGAN-based models.

8.5.3 DESIGNING THE ARCHITECTURE OF VAE

The Variational AutoEncoder (VAE) architecture is implemented by the `nn.Module` class, as shown in Listing 8.23. This architecture is composed of two primary components, i.e., the encoder and the decoder.

```
class VAE_CNN(nn.Module):
    def __init__(self):
        super(VAE_CNN, self).__init__()
        self.encoder = nn.Sequential(
            nn.Conv2d(3, 16, kernel_size=3, stride=1, padding=1,
    bias=False),
            ...
            nn.ReLU())
        self.fc1 = nn.Linear(25 * 25 * 16, 1024)
        self.fc_bn1 = nn.BatchNorm1d(1024)
        self.fc_mu = nn.Linear(1024, 1024)
        self.fc_logvar = nn.Linear(1024, 1024)

        self.fc3 = nn.Linear(1024, 1024)
        self.fc_bn3 = nn.BatchNorm1d(1024)
        self.fc4 = nn.Linear(1024, 25 * 25 * 16)
        self.decoder = nn.Sequential(
            nn.ConvTranspose2d(16, 64, kernel_size=3, stride=1,
    padding=1, bias=False),
            ...
            nn.ReLU(),
            nn.ConvTranspose2d(16, 3, kernel_size=3, stride=2,
            padding=1, output_padding=1, bias=False))
```

Listing 8.23: The VAE architecture for Deepfake Counter-measurements.

```
    def reparameterize(self, mu, logvar):
        std = torch.exp(0.5 * logvar)
        eps = torch.randn_like(std)
        return mu + eps * std
```

Listing 8.24: The VAE architecture for Deepfake Counter-measurements: components

```
    def encode(self, x):
        conv_out = self.encoder(x)
        conv_out = conv_out.view(-1, 25 * 25 * 16)
        fc_out = self.relu(self.fc_bn1(self.fc1(conv_out)))
        mu = self.fc_mu(fc_out)
        logvar = self.fc_logvar(fc_out)
        return mu, logvar

    def decode(self, z):
        fc_out = self.relu(self.fc_bn3(self.fc3(z)))
        fc_out = self.fc4(fc_out).view(-1, 16, 25, 25)
        decoded = self.decoder(fc_out)
        return decoded.view(-1, 3, 100, 100)

    def forward(self, x):
```

```
        mu, logvar = self.encode(x)
        z = self.reparameterize(mu, logvar)
        return self.decode(z), mu, logvar
```

Listing 8.25: The VAE architecture for Deepfake Counter-measurements: components

In the VAE, the encoder is designed to map input data into a latent representation space and consists of several convolutional layers followed by fully connected layers. Specifically, the encoder includes four convolutional layers (the lines 4-11 in Listing 8.23) that change the input from three channels (RGB images) to 16 feature map channels. This helps to effectively extract features at various resolutions. Each convolution is followed by the batch normalization operation and the ReLU activation function. The final output of the encoder is flattened and passed through two fully connected layers, which produce the mean and log variance of the latent representation.

The decoder aims to reconstruct the original input data from its latent representation. The decoder is set up similarly to the encoder but uses transposed convolutional layers (the lines 20-28 in Listing 8.23). This will increase the latent representation back to the original image size. It has three transposed convolutional layers that gradually increase the number of channels from 16 channel of feature maps back to 3 channels. Each transposed convolutional layer also uses batch normalization and ReLU activations ensures the output stays within a suitable range for generating images. The reparameterization trick (lines 38–41 in Listing 8.23) is a key part of VAEs, allowing the model to sample from the learned distribution while keeping the process differentiable. This trick adds randomness by generating a sample from a Gaussian distribution defined by the mean and standard deviation (derived from the log variance).

To train the VAE, it is essential to input data into the model and obtain the corresponding output. The `forward` method, defined in lines 49-52 of Listing 8.23, outlines the data flow through the VAE. The encoder part computes the mean (`mu`) and log variance (`logvar`) of the latent space using the `encode` function (line 50). It then samples from this latent space using the reparameterization trick to obtain `z` (line 51). Finally, `z` is decoded back into image space to produce a reconstructed output (line 52). The method returns three outputs, i.e., the reconstructed data, the mean `mu`, and the log variance `logvar` (line 52). This step-by-step process not only enables the model to generate new images but also ensures it to learn important features from the training data

8.5.4 TRAINING AND TESTING PROCESS

8.5.4.1 Training Process

To train the Variational Autoencoder (VAE), similar to other training neuron network models in Pytorch, we need to define the loss function, optimization method, training procedure, and hyperparameters. Listing 8.26 outlines a custom loss function and the training process for the VAE model using PyTorch. The `customLoss` class (lines 1–8) inherits from the `nn.Module` class that combines the Mean Squared Error (MSE) loss and the Kullback-Leibler Divergence (KLD) loss. The `init` method (lines 2–4) initializes an MSE loss object, while the `forward` method (lines 5–8) calculates the total loss. This total loss includes the MSE between the reconstructed output (`x_recon`) and the original input (`x`). Besides, the KLD loss as well as the KLD is calculated based on the encoder's output parameters (`mu` and `logvar`).

```
1 class customLoss(nn.Module):
2     def __init__(self):
3         super(customLoss, self).__init__()
4         self.mse_loss = nn.MSELoss(reduction="sum")
5     def forward(self, x_recon, x, mu, logvar):
6         loss_MSE = self.mse_loss(x_recon, x)
7         loss_KLD = -0.5 * torch.sum(1 + logvar - mu.pow(2) -
    logvar.exp())
8         return loss_MSE + loss_KLD
```

Listing 8.26: Custom loss function implementation for Variational Autoencoder.

The training procedure for the VAE model involves feeding input data through the model, calculating the loss, and updating the model weights accordingly. As described in Listing 8.27, the process begins by creating data loaders, i.e., `train_loader` and `val_loader` (lines 3–12), using the PyTorch `DataLoader` class. This loads the batch data from the specified directories, applies transformations, e.g., resizing and normalization. The model is initialized from the `VAE_CNN` class (lines 13–14) and transferred to the appropriate device for training (lines 13–14). For each mini-batch, the input data passes through the model to generate predictions (line 27). The loss function is computed to assess the model's performance (line 28). The optimizer, e.g., Adam, then adjusts the model weights to minimize this loss (line 30). Each training epoch consists of multiple forward passes (from input to output) and backward passes (for gradient calculation and weight updates) (lines 28–30). After training on the training set, the model's performance is evaluated on the validation set by switching to the evaluation mode. In this assessment, the validation loss is computed without updating the model's weights (lines 35–45). At the end of each epoch, the model's state is saved as checkpoints in the directory specified by `save_dir` (line 48).

```
1 def train(cfg):
2     device = torch.device("cuda" if cfg["gpu_usage"] else "cpu")
3     transform_train = transforms.Compose([
4         transforms.Resize(100),
5         ...
6     ])
7     transform_val = transforms.Compose([
8         transforms.Resize(100),
9         ...
10    ])
11    train_loader = DataLoader(datasets.ImageFolder(cfg["
    train_dataset_dir"], transform=transform_train), batch_size=
    cfg["batch_size"], shuffle=True)
```

```
    val_loader = DataLoader(datasets.ImageFolder(cfg["val_dataset_dir"], transform=transform_val), batch_size=cfg["batch_size"])
    model = VAE_CNN().to(device)
    model.train()
    optimizer = Adam(model.parameters(), lr=cfg["learning_rate"])
    loss_fn = customLoss()
    writer = SummaryWriter(log_dir=cfg["log_dir"])
    n_epochs = cfg["n_epochs"]
    for epoch in range(n_epochs):
        model.train()
        pbar = tqdm(train_loader, ncols=80, desc="Training")
        running_loss = 0.0
        train_num = 0
        for step, x in enumerate(pbar):
            x = x[0].to(device)
            optimizer.zero_grad()
            recon_batch, mu, logvar = model(x)
            loss = loss_fn(recon_batch, x, mu, logvar)
            loss.backward()
            optimizer.step()
            running_loss += loss.item() * x.shape[0]
            train_num += x.shape[0]
        train_loss = running_loss / train_num

        model.eval()
        val_running_loss = 0.0
        val_num = 0
        with torch.no_grad():
            for x in val_loader:
                x = x[0].to(device)
                x_reconstruction, mu, logvar = model(x)
                val_loss = loss_fn(x, x_reconstruction, mu, logvar)
                val_running_loss += val_loss.item() * x.shape[0]
                val_num += x.shape[0]
        val_loss = val_running_loss / val_num
        writer.add_scalar("Loss/Train_epoch", train_loss, epoch)
        writer.add_scalar("Loss/Validation_epoch", val_loss, epoch)
        torch.save(model.state_dict(), f"{cfg["save_dir"]}/model_{epoch}.pth")
    writer.close()
```

Listing 8.27: Training procedure for training the VAE for Deepfake Countermeasurements.

8.5.4.2 Testing Process

The primary goal of the testing phase is to assess the effectiveness of the VAE model in distinguishing between real and fake images by evaluating its performance on unseen data. In the context of Deepfake detection, the testing procedure involves feeding input data into the trained model, calculating the reconstruction loss, and comparing it to a predefined threshold to classify each sample as either real or fake.

As described in Listing 8.28, the process begins by loading the dataset and creating the data loader, `train_loader` and `test_loader`, (lines 4–5), using PyTorch's `DataLoader` class. The trained VAE model is then restored from a checkpoint and switched to evaluation mode to ensure that the model's parameters remain static during testing (lines 6–10). The MSE loss function is calculated (line 11) to compare with the anomaly score (lines 13–19). The interquartile range (IQR) is used to identify the 80th percentile (T80) of this error distribution, which is anomaly score for classifying images as real or fake, following the approach in [9]. During the testing phase (lines 22–28), the reconstruction losses are computed for each data sample. These losses are then compared to the anomaly score to make predictions (line 29). Finally, performance metrics are computed using the `calculate_metrics` function (line 30). These metrics provide a comprehensive summary of the model's performance during the testing phase.

```
1 def test(cfg):
2     device = torch.device("cuda" if cfg["gpu_usage"] else "cpu")
3     transform = transforms.Compose([transforms.Resize(100), transforms.ToTensor(), transforms.Normalize([0.485, 0.456, 0.406], [0.229, 0.224, 0.225])])
4     train_loader = DataLoader(datasets.ImageFolder(cfg["train_dataset_dir"],transform=transform), batch_size=cfg["batch_size"], shuffle=True)
5     test_loader = DataLoader(datasets.ImageFolder(cfg["test_dataset_dir"],transform=transform), batch_size=cfg["batch_size"])
6     model = VAE_CNN()
7     model_path = os.path.join(cfg["save_dir"], cfg.get("model_checkpoint"))
8     if os.path.exists(model_path):
9         model.load_state_dict(torch.load(model_path, map_location=device))
10    model.to(device).eval()
11    loss_fn = nn.MSELoss(reduction="none")
12    loss_list = []
13    with torch.no_grad():
14        for x_batch, _ in tqdm(train_loader, desc="Processing training data"):
15            x_batch = x_batch.to(device)
16            x_reconstruction, _, _ = model(x_batch)
17            cmp_factor = loss_fn(x_batch, x_reconstruction).mean(dim=[1])
18            loss_list.extend(cmp_factor.cpu().numpy())
19    threshold = np.quantile(loss_list, 0.8)
20    comparison_factors = []
21    labels = []
22    with torch.no_grad():
23        for x_batch, y_batch in tqdm(test_loader, desc="Testing"):
24            x_batch = x_batch.to(device)
25            x_reconstruction, _, _ = model(x_batch)
26            cmp_factor = loss_fn(x_batch, x_reconstruction).mean(dim=[1])
27            comparison_factors.extend(cmp_factor.cpu().numpy())
```

```
            labels.extend(y_batch.numpy())
    predictions = (np.array(comparison_factors) > threshold).
    astype(int)
    return calculate_metrics(labels, predictions)
```

Listing 8.28: Testing procedure for testing the VAE for Deepfake Countermeasurements.

Readers can download the complete implementation code for this section from here [7].

8.6 SUMMARY

This chapter guides readers, especially students, in implementing generative models for various cybersecurity applications. Through detailed case studies, it demonstrates how generative AI techniques—such as GANs, VAEs, and AAE-based models—can be effectively applied in areas like intrusion detection, malware classification, traffic analysis, data synthesis, and deepfake countermeasures. Each case study follows a structured approach, covering problem definition, data preprocessing, model design, and model training and evaluation, to help readers understand the end-to-end process of building generative AI solutions. Overall, this chapter serves as a practical roadmap for students and practitioners to apply generative AI in solving real-world cybersecurity challenges.

1. Nour Moustafa and Jill Slay. Unsw-nb15: a comprehensive data set for network intrusion detection systems (unsw-nb15 network data set). *2015 Military Communications and Information Systems Conference (MilCIS)*, pages 1–6, 2015.
2. Mahbod Tavallaee, Ebrahim Bagheri, Wei Lu, and Ali A. Ghorbani. A detailed analysis of the kdd cup 99 data set. *2009 IEEE Symposium on Computational Intelligence for Security and Defense Applications*, pages 1–6, 2009.
3. Iman Sharafaldin, Arash Habibi Lashkari, and Ali A Ghorbani. Toward generating a new intrusion detection dataset and intrusion traffic characterization. *ICISSp*, 1:108–116, 2018.
4. Royi Ronen, Marian Radu, Corina Feuerstein, Elad Yom-Tov, and Mansour Ahmadi. Microsoft malware classification challenge. *arXiv preprint arXiv:1802.10135*, 2018.
5. Google. Defining a Neural Network in PyTorch. `https://pytorch.org/tutorials/recipes/recipes/defining_a_neural_network.html`, 2023. [Online; accessed 20-May-2024].
6. Google. Models and pre-trained weights. `https://pytorch.org/vision/stable/models.html`, 2023. [Online; accessed 20-May-2024].
7. Xinjie Lin, Gang Xiong, Gaopeng Gou, Zhen Li, Junzheng Shi, and Jing Yu. Et-bert: A contextualized datagram representation with pre-training transformers for encrypted traffic classification. In *Proceedings of the ACM Web Conference 2022*, pages 633–642, 2022.
8. Joel Stehouwer, Hao Dang, Feng Liu, Xiaoming Liu, and Anil K. Jain. On the detection of digital face manipulation. *2020 IEEE/CVF Conference on Computer Vision and Pattern Recognition (CVPR)*, pages 5780–5789, 2019.
9. Matthew Groh, Ziv Epstein, Chaz Firestone, and Rosalind Picard. Deepfake detection by human crowds, machines, and machine-informed crowds. *Proceedings of the National Academy of Sciences*, 119(1):e2110013119, 2022.

---

[7] https://github.com/lyvt/genai˙cybersecurity/tree/main/05-deepfake-counter-measurements

Part III

Conclusion and Future Perspectives

In Part III, the book discusses critical issues and challenges associated with generative AI applications to cybersecurity, such as training data availability, privacy, and security concerns, and required computational complexity. The book will then explore potential solutions to these challenges, along with highlighting future research directions in the field. Part III has 2 chapters as follows.

9 Emerging Topics of Generative AI in Cybersecurity

This chapter provides an exploration of the emerging topics of generative AI in cybersecurity. By investigating areas such as deepfakes, privacy-preserving techniques, proactive threat hunting, and adversarial settings, the chapter highlights the potential of generative AI to address evolving cybersecurity challenges. It is a valuable resource for researchers, practitioners, and policymakers seeking to stay abreast of the latest advancements and emerging trends in generative AI for cybersecurity. Moreover, this chapter discusses issues such as detecting sophisticated deepfake attacks, the ethical implications of using generative models for offensive purposes, and the trade-offs between privacy and utility in privacy-preserving generative AI techniques.

9.1 DEEPFAKES AND THEIR IMPACT ON CYBERSECURITY

9.1.1 USING AI TO TRACE AND VERIFY DEEPFAKE ORIGINS

To trace and track the origin of documents, such as footage, images, audio, and videos, the authors in [4] aim to use Ethereum smart contracts to prove the history of documents, even if they are copied multiple times. This technique uses hash functions applied to the documents and stores their hash values to check whether a document has been modified in the future. Therefore, any document modification can be detected by comparing the stored hash value with the hash value of the checked document.

Watermarking techniques are widely applied to AI-generated image verification. The authors in [5] first generate subtle watermarks and then embed them into digital objects, such as images and audio files. Note that subtle watermarks can be text, images, or codes. Watermark embedding refers to adding a watermark to a digital object, including an image, to assert ownership or provide protection against unauthorized use or distribution. After embedding the watermark into the document, it must be hidden using preprocessing techniques such as compression and noise addition. Subsequently, the document can be shared or distributed, and its integrity can be checked for modifications. This can be done by extracting the watermark from the document and comparing it with the original watermark. If they match, the document is considered protected; if they do not, the document has been modified or faked.

Machine learning combined with metadata visualization is also applied to check and verify digital material. The authors in [6] proposed a web-based image

DOI: 10.1201/9781003502531-9

verification platform based on feature image tampering detection and metadata visualization. First, metadata extraction aims to list the metadata in the image and display any embedded thumbnails. If GPS metadata is detected, the location of the image can be shown. The second step includes six tampering localization maps using different forensic algorithms, which target capturing different potential tampering traces. The third step links to Google Image Search to check the image using a browser, such as Firefox or Chrome. Some image forensics services can be used, such as ELA, Thumbnail, Metadata, and Geotagging[1234].

9.1.2 DETECTION AND MITIGATION OF DEEPFAKE ATTACKS

Deepfake detection plays a crucial role in protecting information from fabricated content. Initially, deepfake detection began as a classification problem, where objects—such as news, images, audio, and videos—are categorized as either real or generated by models. Fake image detection emerged in early 2017 with the application of handcrafted feature-based techniques [1]. For example, image preprocessing techniques, such as Gaussian blur and noise addition, can increase the pixel-level statistical similarity between real and fake images. However, this method may struggle to detect fake images synthesized by GANs, which learn the distribution of complex input data and generate images that closely mimic that distribution.

In [2], the authors demonstrated that for low-resolution images generated by GANs, the distance between the synthesized image distribution (referred to as the oracle error) and that of the real images can be used to detect fakes. This is because an increase in the oracle error indicates that the GAN model is generating images less accurately. However, as the resolution increases, applying the oracle error becomes more challenging. A deep feature-based method has been applied to detect fake images. The authors in [3] aim to extract features from fake and real images using pairwise information. The features of two images are labelled as 1 if both images are of the same type, i.e., either both real or both fake. Otherwise, the features extracted will be zero for a pair consisting of one real and one fake image. Note that the fake images are generated by a generative model, such as a GAN. The extracted features are then fed into classifiers for further analysis.

In addition, the self-consistency of the local source can be applied to detect fake images. This is because an edited image may have different sources in different regions. Therefore, using a pairwise self-consistency technique can help extract distinctive representations of fake inputs. This is achieved by penalizing pairs of features within the same image to have a low similarity score while increasing the similarity score of pairs from different image inputs. The above methods can detect fake images where only local regions or parts of an image have been manipulated. However,

[1] http://fotoforensics.com/

[2] https://29a.ch/photo-forensics/

[3] https://www.getghiro.org/

[4] https://github.com/MKLab-ITI/image-forensics

feature-based methods may struggle to detect fake images when a generative model generates the entire image.

Fake video detection is also a contemporary topic because detection methods applied to images may fail when applied to videos. Temporal features across video frames can be used for fake video detection. The authors in [7] utilize both convolutional neural networks (CNN) and Long Short-Term Memory (LSTM) networks to detect fake videos. Initially, the CNN extracts features from the video frames, creating a sequence input that is subsequently processed by the LSTM for classification.

Interestingly, physiological signals, such as eye blinking, can also be used to detect fake videos. For example, an adult human typically blinks every 2 to 10 seconds, with each blink lasting between 0.2 and 0.4 seconds. Based on this observation, deepfake models struggle to replicate natural eye blinking patterns in a way that matches real human videos. Consequently, blinking rates in deepfake videos are often much lower than those in real videos. To detect fake videos, the authors in [8] attempt to crop the eye areas in the video and use a CNN to extract features before feeding them into an LSTM for classification.

9.1.3 ADVANCEMENTS IN REAL-TIME DEEPFAKE DETECTION USING AI

Real-time fake detection plays a crucial role in ensuring trust in online content. The authors in [9] propose a novel approach to detecting fake images using Binary Neural Networks (BNNs) for fast inference. They combine Fast Fourier Transform (FFT) and Local Binary Pattern (LBP) to extract additional features for the BNN model. The FFT magnitude is used to highlight anomalies by extracting the anomaly spectrum from the image. The LBP, on the other hand, is designed to capture the spatial structure of the image by comparing each pixel with its neighbors. Finally, three channel formats—original input, FFT, and LBP—are combined and fed into the BNN for classification.

Real-time deepfake detection using local regions is also applied to detect fake images. The authors in [11] use a single 9×9 image patch to generate a deepfake score, which can be used to detect fake images. The deepfake score for an image is given as: $S(I) = \frac{1}{N}\sum_{i=1}^{N} \varphi(p_i)$, where II is the image that needs to be checked, and $\varphi(p_i)$ is the deepfake score of patch p_i. Note that, to calculate the deepfake score of patch p_i, the patch must be passed through two models: a pre-trained model such as ResNet50 and a neural network with 4 layers. The outputs of these two models are then compared by measuring the distance between them using the mean squared error (MSE), which serves as the deepfake score.

Real-time video-conferencing fake detection poses challenges for current methods. The authors in [10] make the assumption that during a video call, a user is often positioned in front of a light source, such as the light from the computer screen. Therefore, any changes in the appearance of the user's face may increase deviations from the expected changes, which could indicate the presence of a fake video. To address this, an active illumination source is modeled as follows: $H(t) = 0.1307 \times \cos\left(\frac{t}{8}\right)$, where $t \in [0, 16]$ is a hue value ranging from yellow to magenta (i.e., [-0.1307, 0.1307]). Note that a face is extracted from each video frame using the Dlib library. To counterattack models that might predict the illumination pattern, the authors randomly inject blank frames into the temporal illumination sequence.

Combining spatial and temporal information is also applied to real-time deepfake video detection. The authors in [12] use a transfer learning technique with a pre-trained model, which is employed for spatial feature extraction to reduce the amount of training data before feeding it into a Long Short-Term Memory (LSTM) network for classification. Specifically, video frames are collected from the video and passed through the VGG16 model to extract features. These features are then processed by global average pooling to obtain feature matrices. Next, the feature matrices are input into the LSTM, followed by a fully connected network for classification.

9.1.4 CROSS-MODAL DEEPFAKE DETECTION (E.G., COMBINING AUDIO-VISUAL CUES)

Traditional deepfake detection methods require forensic expertise to analyze manipulated images. Therefore, there is a growing need for a framework that enables both experts and non-experts to combat the spread of fake information. The authors in [13] introduced a Multi-Model GAN Guard (MMGANGuard), which integrates four models for detecting fake images. Additionally, transfer learning is applied to enhance the accuracy of the detection model. For example, GramNet, the Co-Occurrence Matrix, ResNet-V2, and DenseNet-201 are used simultaneously to analyze an input, with each model assigned a weight representing its contribution to the final detection result.

Learning cross-modal correlation is also applied to enhance the generalizability of detection accuracy. The authors in [14] proposed a cross-modal distillation approach to leverage cross-modal correlations based on content information. To achieve this, both image frames and audio are simultaneously fed into the model to determine whether they are fake or real in the first detection branch. In a second branch, features from audio and image frames are extracted and synchronized. These synchronized features are then processed using correlation distillation and contrastive learning. Contrastive learning seeks to develop meaningful representations by maximizing the similarity between similar (positive) samples and minimizing the similarity between dissimilar (negative) samples. Meanwhile, correlation distillation transfers relational information between data points from a larger (teacher) model to a smaller (student) model. Instead of focusing solely on individual sample predictions, it preserves pairwise or structural relationships among features, embeddings, or outputs during training. Ultimately, the distillation branch is responsible for learning cross-modal correlations.

A multi-modal attention framework can effectively leverage contextual information for detecting fake audio. The authors in [15] employed attention mechanisms in a multi-modal, multi-sequence setting to extract useful features from sequences of image frames, lip movements, and audio. This approach enables the model to determine whether the input is real or fake. Notably, the model can also identify the start and end timestamps of the manipulated audio segments, contributing to the precise localization of fake segments within a video.

Leveraging multi-modal information to bridge the representation gap between different modalities has proven to be an effective strategy for deepfake detection. In [16], the authors utilized multimodal contrastive learning (MCL) to capture both intra-modal and cross-modal relationships. Specifically, cross-modal contrastive learning aims to generate embeddings that integrate information from various modalities. For instance, the model employs three types of modality features, i.e., audio, frames, and video, for feature extraction. Notably, audio and frame features are processed separately from video features to create cross-modal representations. By contrasting these cross-modal features with intra-modal features, the method effectively reduces the cross-modal gap. Additionally, noise-based feature augmentation is applied to adaptively perturb the features, which enhances the model's generalization performance in deepfake detection.

9.2 PRIVACY-PRESERVING IMPLICATIONS ON GENERATIVE AI

9.2.1 PRIVACY RISKS FROM AI-GENERATED DATA

Generative AI plays a crucial role in medicine by transforming medical diagnostics, treatment planning, and patient care. However, securing data-intensive health information presents significant challenges. The authors in [17] aim to identify security and privacy threats related to data collection, model development, and application implementation when generative AI is applied. Training GenAI models require sensitive patient data, which can be exposed by malicious software. This can lead to serious consequences, such as data breaches, erosion of patients' trust, and compromised reliability of medical treatment methods. Another challenge arises from the bias in GenAI models, which may result in inaccuracies in the methods applied to treatment. Additionally, another risk comes from the generator of the GenAI model, such as GAN. For example, when using GAN to generate data from imbalanced classes collected from images of cancer patients, the GAN's generator can produce images that resemble real ones. If attackers compromise the generator model, they could generate images of patients and use them for blackmail.

Rendering images from the gradients of the model can pose a risk of information leakage. The authors in [19] introduced tools to visualize the activation functions at each layer of a neural network, allowing for the visualization of features learned by the network at different layers. The main idea is to maximize the activations of specific neurons in a deep neural network, revealing the internal representations that the network has learned. For example, the results of reproducing an image from gradients are shown in Figure 9.1. The specific code is provided below:

Figure 9.1 Reproduce image from the gradient of the model.

```
import tensorflow as tf
import numpy as np
import matplotlib.pyplot as plt

# Load the target image
target_image_path = 'dog.png'  # Replace with the actual image
    path
target_img = tf.keras.preprocessing.image.load_img(
    target_image_path, target_size=(224, 224))
target_array = tf.keras.preprocessing.image.img_to_array(
    target_img)
target_array = tf.convert_to_tensor(target_array, dtype=tf.
    float32)
# Normalize the target image to [0, 1]
target_array /= 255.0
# Create a random image (initialized with noise)
random_img = tf.Variable(np.random.rand(224, 224, 3), dtype=tf.
    float32)
# Define the loss function (Mean Squared Error)
def compute_loss(input_image, target_image):
    return tf.reduce_mean(tf.square(input_image - target_image))

# Define the optimizer (Gradient Descent)
optimizer = tf.optimizers.Adam(learning_rate=0.1)
# Store images for visualization
images_to_plot = []

# Gradient descent loop
iterations = 1000
interval = iterations // 5  # 5 images after 1000 epochs (one
    every 200 epochs)
for i in range(iterations):
    with tf.GradientTape() as tape:
        # Compute the loss between the random image and the
    target image
        loss = compute_loss(random_img, target_array)
    # Compute the gradients with respect to the random image
    gradients = tape.gradient(loss, random_img)

    # Update the image using the gradients
    optimizer.apply_gradients([(gradients, random_img)])
    # Store the image every interval epochs (200, 400, 600, 800,
    1000)
    if i % interval == 0:       images_to_plot.append(np.clip(
    random_img.numpy(), 0.0, 1.0))  # Append the image for
    display
        print(f"Iteration {i}, Loss: {loss.numpy()}")

# Create a figure to display the images
fig, axes = plt.subplots(1, len(images_to_plot), figsize=(15, 5))

# Plot each image in a separate subplot
for ax, img in zip(axes, images_to_plot):
    ax.imshow(img)
    ax.axis('off')

```

```
# Adjust layout for better spacing
plt.tight_layout()
plt.show()
```

Listing 9.1: Code for reproducing an image using gradient

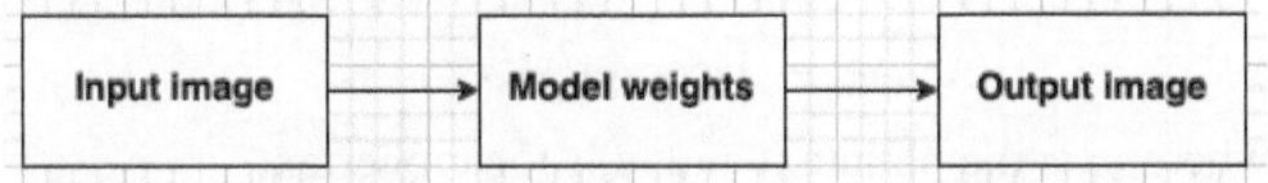

Figure 9.2 Diagram of reproducing input data from weight model.

Training AI-generated models are highly susceptible to privacy leakage because private data, such as a user's information, can be memorized by the model, specifically in the weight matrix of the neural network. An attacker can potentially reproduce the input data from the model weights. The authors in [18] proposed a post-hoc visual explanation method based on class activation. Instead of using gradients, this method leverages the weights of each activation to map to the target class and then obtains a linear combination of weights and activation maps. Finally, it can be used to reconstruct and display the input image, as observed in Figure 9.2.

9.2.2 FUNDAMENTALS OF PRIVACY-PRESERVING GENERATIVE MODELS

GenAI models, including large language models (LLMs), raise security concerns due to their outsourcing to third-party clouds and their ability to retain previously trained input within the models. Fully homomorphic encryption (FHE) presents a natural solution to this issue. By encrypting the input data before sending it to the GenAI models for processing, only clients holding the secret key can decrypt the plaintext, as observed in Figure 9.3. However, applying FHE to GenAI models presents challenges due to unsupported active functions in LLMs and the high complexity of both LLMs and FHE. The authors in [20] address this challenge by applying FHE to GPT-2, achieving performance 150 times faster than the CPU baseline when deploying on a GPU.

Homomorphic Encryption (HE) begins with the key generation algorithm:

$$(sk, pk, evk) \leftarrow \text{KeyGen}(1^{\lambda})$$

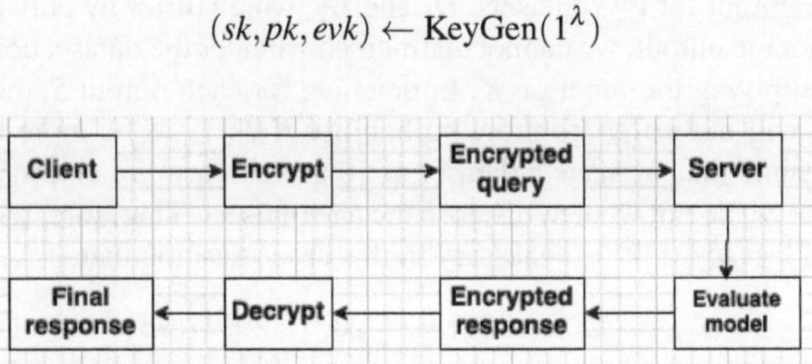

Figure 9.3 Diagram of homomorphic encryption to GenAI model.

where λ is the security parameter. sk is the secret key, used for decrypting the input data, while pk is the public key, used for encryption. evk is the evaluation key, which enables homomorphic computations over encrypted data. Two procedures are used for encrypting plaintext and decrypting encrypted data:

$$ct \leftarrow \text{Encrypt}(pk, m),$$

$$m' \leftarrow \text{Decrypt}(sk, ct')$$

respectively, where m is the plaintext and ct is the ciphertext.

$$\text{ctf} \leftarrow \text{Eval}(\text{evk}, \text{ct}, f)$$

is a homomorphic evaluation procedure that uses the evaluation key evk, ciphertext ct, and function f. The function $\text{Eval}(\text{evk}, \text{ct}, f)$ evaluates the function f on the encrypted data ct using the evaluation key evk, and the result is an encrypted version of the output of f. For example, when applied in the context of HE, if we need to perform operations like addition or multiplication on encrypted numbers without decrypting them, the function Eval allows us to do that.

Differential Privacy (DP) ensures privacy when sharing data for Generative AI (GenAI) models. DP guarantees that the inclusion or exclusion of a single data point does not significantly change the outcome of an analysis, thereby preserving classification accuracy. To achieve this, noise is added to the data, making it more obscure and difficult for third parties to extract sensitive information. However, adding excessive noise can reduce accuracy. The ε parameter, known as the privacy budget, is used to balance privacy and accuracy.

The fundamental background starts with the randomized function K, an algorithm used in information release [21]. We define K as follows: A randomized function K ensures ε-differential privacy for all datasets D_1 and D_2 that differ by at most one element, for all $S \in \text{Range}(K)$, and

$$\Pr[K(D_1) \in S] \leq \exp(\varepsilon) \times \Pr[K(D_2) \in S]$$

Note that D_1 and D_2 differing by at most one element means that one is a subset of the other, and the larger dataset contains just one additional row. Specifically, we assume that the function K adds noise to the data, ensuring that the probability of the output of K is similar for two datasets, D_1 and D_2, which differ by only one element. Therefore, from the output, we cannot distinguish whether the dataset comes from D_1 or D_2 before applying the function K. In practice, for each output S, the probability of $K(D_1) \in S$ will not differ from the probability $K(D_2) \in S$ by more than a factor of $exp(\varepsilon)$. Assume that we have a dataset of 10 users' ages. We aim to calculate the average age but we do not want to disclose the age of users. Therefore, the application of Differential Privacy (DP) is necessary in this case. Assume that

$$D_1 = \{25, 30, 22, 28, 34, 26, 31, 29, 27, 33\}$$

and

$$D_2 = \{27, 30, 22, 28, 34, 26, 31, 29, 27, 33\}$$

where D_2 differs from D_1 by the first element, i.e., 25 vs 27. As a result, D_1 and D_2 differ by one element. We aim to calculate the average age of D_1 and D_2, while ensuring that the outputs do not reveal whether the data comes from D_1 or D_2. To address this, we apply a randomized function K which adds noise to the output to ensure differential privacy. Without noise, the average of D_1 is:

$$\frac{25+30+22+28+34+26+31+29+27+33}{10} = 28.5,$$

and without noise, the average of D_2 is:

$$\frac{27+30+22+28+34+26+31+29+27+33}{10} = 28.6$$

Now, let's add stronger Laplace noise with $\varepsilon = 0.1$. Suppose the Laplace noise generated for D_1 and D_2 are $\text{noise}_1 = -0.5$ and $\text{noise}_2 = 0.3$, respectively. Then, the privatized averages are: For D_1:

$$\text{Privatized Average for } D_1 = 28.5 + (-0.5) = 28.0$$

For D_2:

$$\text{Privatized Average for } D_2 = 28.6 + 0.3 = 28.9$$

As can observed from the private average ages for D_1 and D_2 are 28.0 and 28.9, respectively. It gets difficult to guess which dataset gives this output. For the ε, if its value is high, it can add less noise. In contrast, a lower value of ε adds more noise, providing stronger privacy.

Next, we discuss the sensitivity of the function f, which is defined as follow.
Definition 2. For $f : D \to \mathbb{R}^k$, the sensitivity of f is

$$\Delta f = \max_{D_1, D_2 \in D, \|D_1 - D_2\|_1 \leq 1} \|f(D_1) - f(D_2)\|_1,$$

for all D_1 and D_2 which distinguishing at most one element. The sensitivity is used to capture how great difference between two value of f could be on two database differing at most one element.

Integrating differential privacy (DP) into Generative AI (GenAI) can protect the privacy of data used to train the GenAI model. The authors in [22] introduced a hybrid method that combines DP with a Generative Adversarial Network (GAN) model to preserve sensitive data in the Industrial Internet of Things (IIoT). To implement DP, the Laplace distribution is used to add noise to both the training and testing datasets. The experimental results show that this method effectively preserves privacy with minimal accuracy loss.

9.2.3 TRADE-OFFS AND CHALLENGES

9.2.3.1 Balancing Privacy with Utility

The trade-off between privacy and utility is challenging when applying differential privacy to GenAI. Adding noise to data helps protect sensitive information,

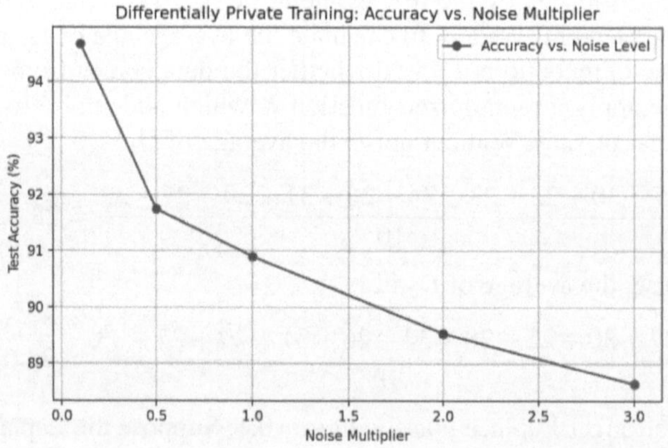

Figure 9.4 Dependency of privacy and accuracy of classification model.

but it also reduces the utility of the data, making it harder to derive useful insights or accurate statistical information. Specifically, when noise is added to the data, the classification accuracy of classifiers may decrease. For example, the trade-off between accuracy and the rate of noise used is illustrated in Figure 9.4. To generate Figure 9.4, we added noise at different levels to the training sets, i.e., *noise˙levels* $= [0.1, 0.5, 1.0, 2.0, 3.0]$, while classifying the MNIST dataset using simple neural networks, as follows.

```
import torch
import torch.nn as nn
import torch.optim as optim
from opacus import PrivacyEngine

class SimpleNN(nn.Module):
    def __init__(self):
        super(SimpleNN, self).__init__()
        self.fc1 = nn.Linear(28 * 28, 128)
        self.relu = nn.ReLU()
        self.fc2 = nn.Linear(128, 10)
    def forward(self, x):
        x = x.view(-1, 28 * 28)
        x = self.fc1(x)
        x = self.relu(x)
        x = self.fc2(x)
        return x  # No softmax (handled in loss function)

# Function to train model with differential privacy
def train_with_dp(train_loader, noise_multiplier, epochs=10):
    model = SimpleNN().to(device)
    optimizer = optim.Adam(model.parameters(), lr=0.001)
    criterion = nn.CrossEntropyLoss()
    # Attach PrivacyEngine for differential privacy
    privacy_engine = PrivacyEngine()
```

```
    private_model, private_optimizer, private_train_loader = privacy_engine.make_private(
        module=model,
        optimizer=optimizer,
        data_loader=train_loader,
        noise_multiplier=noise_multiplier,
        max_grad_norm=1.0,  # Gradient clipping
    )
    # Training loop
    for epoch in range(epochs):
        private_model.train()
        for images, labels in private_train_loader:  # Use the new private loader
            images, labels = images.to(device), labels.to(device)
            private_optimizer.zero_grad()
            outputs = private_model(images)
            loss = criterion(outputs, labels)
            loss.backward()
            private_optimizer.step()
    # Evaluate model
    private_model.eval()
    correct = 0
    total = 0
    with torch.no_grad():
        for images, labels in test_loader:
            images, labels = images.to(device), labels.to(device)
            outputs = private_model(images)
            _, predicted = torch.max(outputs, 1)
            total += labels.size(0)
            correct += (predicted == labels).sum().item()
    accuracy = 100 * correct / total
    return accuracy
```

Listing 9.2: Code for a simple neural network with differential privacy

As observed in Figure 9.4, as the noise rate in the training sets increases, the classification accuracy on the test set decreases. For example, when the noise rate is 0.1, the classification accuracy achieved by the model is around 94.7%. In contrast, when the noise rate is 3.0, the classification accuracy drops to below 89%.

The authors in [23] formulate the trade-off between privacy and utility as a minimax information leakage problem. The privacy-preserving attack and defense game framework is introduced to provide an effective method compared to other ε-privacy methods.

9.2.3.2 Technical Challenges and Limitations

Integrating DP into GenAI faces numerous challenges. The difficulties stem from the need for a suitable computing environment, the quality of data, the difficulty in accounting for users' confidential data, the challenge of setting the value of the privacy-loss parameter ε, and the lack of a framework to verify the DP implementation [24].

Although differential privacy has been proposed for a decade, the existing methods are primarily used for scientific applications and not effective for official use.

For scientific purposes, the hierarchical methods used in the 2020 census are limited to small groups and small geographies. This requires the Census Bureau to develop a new method to optimize the accuracy of many queries simultaneously. In addition, there are too many metrics for assessing the quality of a published dataset. Furthermore, the setting of the value of ε is a key question. The debate over whether policymakers or scientists should set the value of ε has garnered attention.

Operational issues stem from a shortage of experts employed by agencies. The Bipartisan Commission on Evidence-Based Policymaking noted this scarcity of expertise [25]. Furthermore, there is a significant absence of toolkits designed for conducting data practices to verify the accuracy of specific implementations.

9.2.3.3 Future Directions in Privacy-Preserving GenAI

Implementing Differential Privacy (DP) in the Industrial Internet of Things (IIoT) plays an essential role in protecting user data privacy [26]. However, several obstacles hinder its implementation. First, edge devices in IIoT handle sensitive data, which may lead to privacy breaches. Real-time privacy is a major challenge due to the limited computational resources and memory constraints of IoT devices. Additionally, the trade-off between privacy and utility poses a significant challenge, requiring careful consideration of both policy decisions and technical approaches. Budget optimization is crucial when controlling the privacy parameter ε in DP. A higher ε improves model accuracy but weakens privacy, making it essential to balance these factors effectively. Moreover, IoT devices have limited resources, making it difficult to implement computationally complex DP algorithms. Ensuring secure data handling during model training and storage is another critical challenge for maintaining data privacy. Finally, IoT deployment and scalability become more difficult when applying DP, as it can impact the overall performance and efficiency of the IoT network. These challenges are illustrated in Figure 9.5.

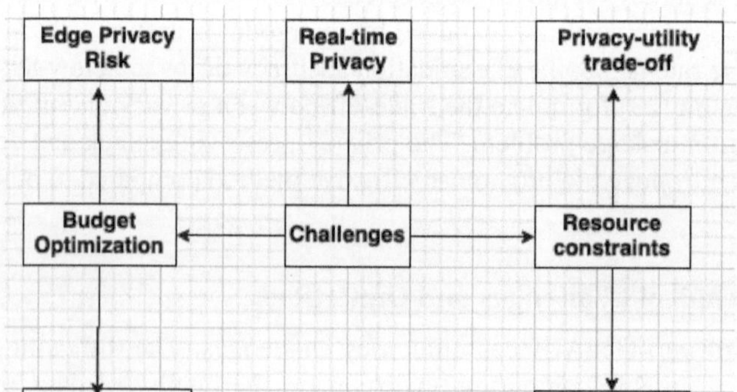

Figure 9.5 Challenges for DP in IIoT.

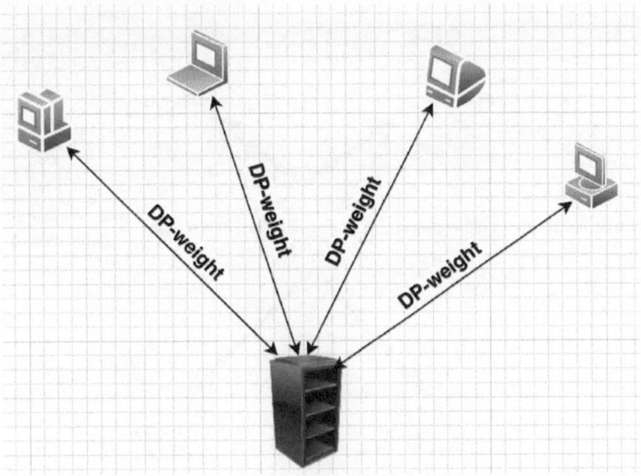

Figure 9.6 Diagram of DP applied to federated learning.

DP is widely applied to federated learning, where training data remains on the client side before sending model weights for aggregation. This is because data may be leaked through the model's weights, and DP is an effective way to prevent such leakage. The authors in [27] propose a framework based on DP, called NbAFL, which adds noise to client weights before sending them to the server for model aggregation, as observed in Figure 9.6. To achieve this, NbAFL establishes a theoretical convergence bound on the loss of the trained FL model. This bound ensures a trade-off between convergence performance and privacy protection. Additionally, as the number of clients increases, the privacy level remains ensured.

The implementation of differential privacy (DP) in Generative AI faces many challenges. First, training GAN variants is often prone to mode collapse, where the generator struggles to produce fake data that closely resembles real data. Adding noise to the input data increases the complexity of the GAN models, making it harder to avoid mode collapse. Ensuring data privacy in GAN models becomes difficult due to the structure of GAN variants. This is because the generator must ensure it does not disclose individual data points, while the discriminator must ensure it does not rely too heavily on any single data point. As a result, it increases the complexity of ensuring privacy for both the generator and the discriminator. Additionally, adding gradient noise for DP can contribute to the model's collapse in GAN variants. Finally, achieving a balance between the GAN's performance—aiming to generate fake data that closely resembles real data—and ensuring data privacy is a difficult trade-off. The diagram of challenges when applying DP to GAN variants is shown in Figure 9.7.

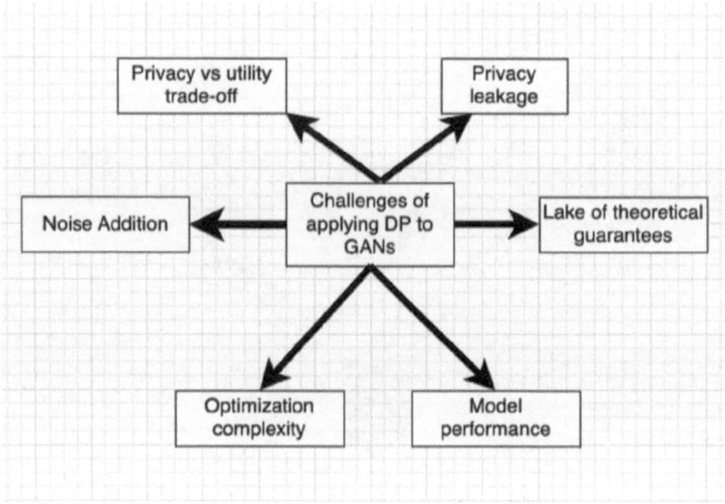

Figure 9.7 Challenges of applying DP to GAN variants.

9.3 PROACTIVE THREAT HUNTING AND CYBER THREAT INTELLIGENCE WITH GENERATIVE AI

9.3.1 AI-POWERED CYBER DECEPTION TECHNIQUES

9.3.1.1 Honeypots and Deceptive AI: Using GenAI to create deceptive traps for cyber attackers

Using Generative AI (GenAI) to create honeypots has become a growing trend in recent years. Traditional honeypots are typically designed and implemented once but rarely updated, leading to obsolescence and a reduced ability to attract malicious software from attackers.

The authors in [28] explore the use of GenAI, including large language models (LLMs), to simulate honeypots without the risk of an environment breakout. Traditional honeypot methods have limitations that may expose their deceptive nature to attackers. In contrast, GenAI can better control inputs and outputs, reducing non-deterministic responses and limiting token variations, thereby enhancing honeypot effectiveness.

As shown in Figure 9.8, the methodology consists of five steps: (1) receiving a user query, (2) customizing the input, (3) querying the model, (4) customizing the output, and (5) returning the response to the user.

For example, an LLM-based honeypot can engage with attackers, as illustrated in Figure 9.9. When an attacker submits a query through the front-end interface, the system modifies and appends the query to create an appropriate prompt before sending it to the LLM. After receiving the model's response, the system appends it to a history log, refines the final output, and then delivers it back to the attacker.

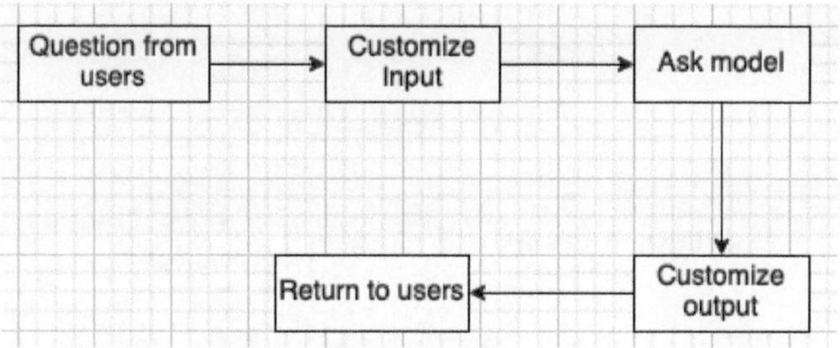

Figure 9.8 Methodology outline of honeypot

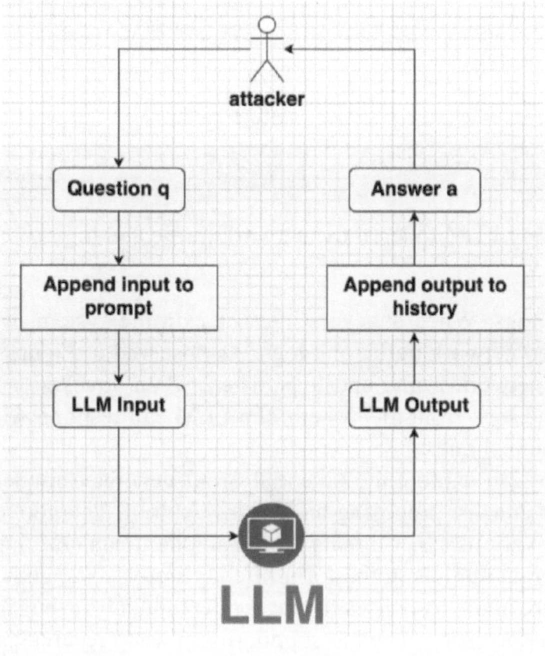

Figure 9.9 LLM honeypot

As a result, the attacker remains unaware that they are interacting with an LLM rather than a real system. Additionally, analyzing attacker interactions can help identify attack patterns and improve future intrusion detection mechanisms.

9.3.1.2 Mimicking Attack Patterns to Confuse Hackers: AI-generated fake network traffic and logs to mislead attackers

GenAI can be used to generate fake network traffic and logs to mislead attackers. Using TCP dump, we can capture network traffic and obtain logs from systems and software. Next, we can use LLMs to learn from this data and generate new network

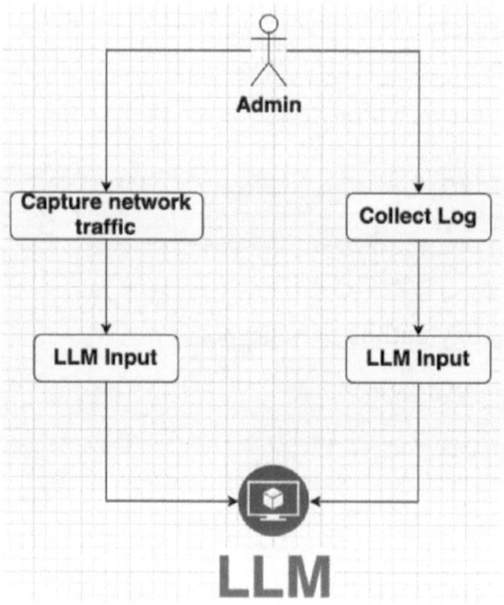

Figure 9.10 Using LLM to learn the network traffic and logs.

traffic as well as logs. When attackers interact with the system, we can then send them the generated network traffic and logs. As observed in Figure 9.10, LLMs can be used to learn network traffic and logs. Then, when attackers attempt to collect network traffic and logs from the system, the LLM can respond with the generated ones instead of the real ones.

The authors in [29] explore the potential of AI-generated models being used as weapons in cyber attacks, proposing an Occupy AI tool. This tool can generate executable code for various cyber threats. For example, it can craft inputs that prompt ChatGPT to generate outputs in the form of code scripts. The authors carefully designed the structure of the responses to prevent the model from engaging in unethical behavior. The interface is designed to support users when receiving responses.

9.3.1.3 Dynamic Attack Surface Management: AI-driven deception to make attack surfaces unpredictable

AI-driven deception potentially reduces the attack surface of the system and then brings a challenge to the attacker. An attack surface refers to a system's total set of potential vulnerabilities or entry points that an attacker could exploit to gain unauthorized access or cause harm. This includes both the software and hardware components of the system and user interactions, APIs, network connections, and other interfaces that might be exposed to potential threats. The authors in [30] indicated that by using cyber deception, organizations can reduce their attack surface, thereby limiting the chances for attackers to identify and exploit vulnerabilities in the system.

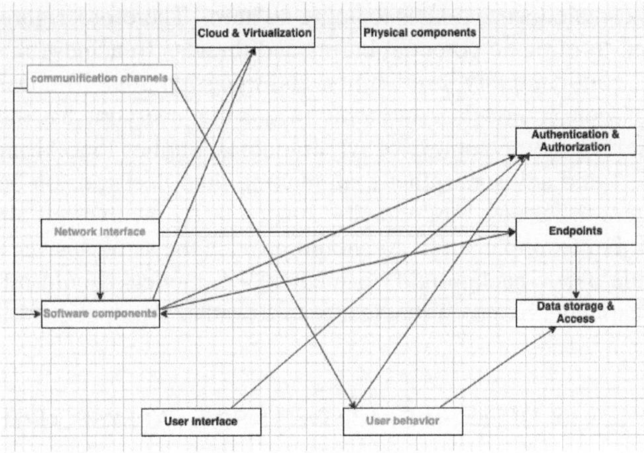

Figure 9.11 Attack surface of the system

For example, organizations can leverage LLMs to automatically generate honeypots—fake systems or services that appear vulnerable to attackers. These fake systems can lure attackers away from actual critical assets. Additionally, LLMs can be used to generate fake network traffic that mimics real, legitimate communication patterns, drawing attackers' attention and making them target the decoy systems rather than the real network. Furthermore, fake logs can be generated and fed to attackers to mislead them into thinking they have successfully compromised the system, causing them to focus on the decoys instead of finding actual vulnerabilities. By using these deceptive techniques, the organization reduces the effective attack surface, protecting real assets while deceiving attackers into engaging with fake, non-critical components. As observed in Figure 9.11, one may use GenAI to make the red components more transparent to the attacker, thereby reducing the system's vulnerability to future attacks.

9.4 ETHICAL AND REGULATORY CHALLENGES OF GENERATIVE AI IN CYBERSECURITY

With the rise of concerns about cybersecurity, many recent works have discussed the risks of GenAI, such as ChatGPT, Gemini, DeepSeek, and others. These risks include data breaches, unauthorized access, user interactions, and employees inadvertently sharing sensitive company information with LLMs. The authors in [31] discussed the cybersecurity risks of GenAI models and the ethical obligations of organizations. To analyze the ethical concerns associated with GenAI risks, four moral principles are considered: beneficence, non-maleficence, autonomy, and justice. Beneficence refers to well-being or doing good. The implementation of GenAI should aim to promote human well-being. Although GenAI models may pose risks, such as data breaches, data leakage, and the disclosure of users' sensitive information, they can also enhance productivity and free humans from tedious tasks and labor costs. Non-maleficence relates to doing no harm. Therefore, rigorous measures must be applied to mitigate potential harms and threats. To achieve this, three core principles, i.e., confidentiality, availability, and integrity, must be considered in the development of GenAI models. Autonomy means that everyone has the right to decide for themselves. In the context of AI, autonomy implies that AI models should act as supporters and advisors rather than making final decisions on behalf of people. Justice refers to fairness and equality. Justice in AI involves eliminating unfair discrimination. For example, an LLM should provide the same answer to a question regardless of the user's background. In medical AI, models should offer advice or recommend treatments based on medical records rather than patients' demographic profiles.

9.4.1 AI ETHICS IN OFFENSIVE AND DEFENSIVE CYBERSECURITY

9.4.1.1 Ethical dilemmas of AI-assisted offensive cybersecurity

There are existing dilemmas in AI-assisted offensive cybersecurity. GenAI-based models and cybersecurity have a multi-dimensional relationship with many interdependencies. GenAI models are generally designed to generate human-like content, enhancing productivity, reducing labor costs, supporting decision-making, and improving human well-being. However, as GenAI models continue to advance, their applications spark debates in both offensive and defensive cybersecurity.

From an offensive perspective, attackers can exploit the advancements of GenAI models to discover system vulnerabilities and compromise systems for malicious activities such as blackmail and stealing sensitive user information (e.g., credit card numbers, online banking credentials). For example, ChatGPT-3.5 can easily generate scripts that mimic an organization's website or imitate emails from social platforms such as Facebook, Amazon, and Netflix. These techniques are then used in phishing campaigns to extract sensitive user information.

On the other hand, GenAI models play an essential role in cybersecurity defense. Thanks to advancements in GenAI, security tasks can be implemented to protect systems from both internal and external threats, including malicious software. By learning data distributions from training data, GenAI models can generate synthetic data that closely resembles real data. For example, a GenAI model such as GAN can be used to learn the distribution of commonly used passwords. When deployed in a real system, the model can then suggest stronger passwords to help users avoid vulnerabilities. Another example is GenAI's ability to address data imbalance in cybersecurity datasets. By learning the distribution of underrepresented classes, a GenAI model can generate new data samples for skewed classes, ultimately improving the classification accuracy of detection engines in network intrusion detection systems. Furthermore, GenAI models can learn to replicate network traffic patterns. The synthetic traffic they generate can be used in honeypot systems to attract malicious software, thereby reducing the attack surface and strengthening system security against external threats and attacks.

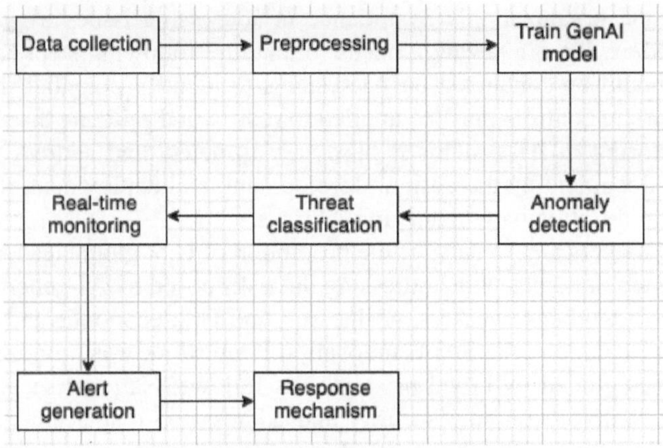

Figure 9.12 The process of intrusion detection systems is used to monitor malicious attacks.

9.4.1.2 AI-powered surveillance and its implications

AI-powered surveillance is one of the most significant benefits of AI models, enhancing productivity and reducing labor costs. Intrusion detection systems (IDSs) are designed to detect malicious attacks in real time, helping humans protect systems from unauthorized access. Without IDS support, identifying threats and attacks becomes challenging when an intruder gains access without the administrator's permission. GenAI models can directly classify a user's transaction as legitimate or illegitimate by comparing it to the learned distribution of normal user behavior. The authors in [32] introduced a neural network model based on an autoencoder that distinguishes attacks from normal data samples. By leveraging unsupervised learning techniques, the detection engine can identify malicious software attacks, thereby improving the security of IoT networks, as observed in Figure 9.12. Additionally, the authors in [33] implemented an AI-powered surveillance system using real-time facial recognition technology.

9.4.2 REGULATORY FRAMEWORKS FOR AI-GENERATED CYBER THREATS

Due to the benefits and risks associated with AI-generated models, many governments and organizations worldwide recognize the need for comprehensive AI regulations. The National Institute of Standards and Technology (NIST) has proposed the AI Risk Management Framework (RMF) to address the risks posed by AI models. The NIST AI RMF contains several key components. First, the governance structure is designed to monitor the development of AI systems. The operations of AI systems must be understandable and transparent. AI systems should clearly define their roles to ensure accountability and ethics. Regarding privacy, there is a need to protect sensitive and personal data. Additionally, AI models must ensure equitable outcomes for users from diverse backgrounds.

To ensure the organization's compliance in addressing issues triggered by AI-generated models, the organization needs to take the following steps. A cyber risk assessment should be conducted on AI systems to identify potential risks and impacts. Appropriate controls must be applied to reduce and mitigate these risks. Continuous monitoring of AI systems is necessary to detect and address future risks. Additionally, documenting the design, development, and deployment of the AI system is essential for effectively managing AI risks.

U.S. policymakers aim to balance the promotion of U.S. global leadership in critical emerging technologies with addressing the risks posed by AI-generated models, which impact individuals, workers, businesses, and U.S. economic and national security [34]. For example, the SAFE Innovation Framework was proposed to outline specific policies guided by broad principles as follows: National security must be protected to promote economic security and address job displacement. It ensures the transparency and accountability of AI systems and holds responsible those who spread misinformation, bias, or infringe on intellectual property. Algorithms should be developed to protect demographics and promote U.S. values, such as liberty, civil rights, and justice. AI systems, including data and content, need to ensure transparency and disclosure. Finally, innovation regulations must promote U.S. global technology leadership.

9.4.3 HUMAN-AI COLLABORATION IN CYBERSECURITY

Human oversight in AI-driven security operations plays an essential role in addressing the risks posed by AI-generated models. The authors in [35] discussed the role of human oversight in AI governance and evaluated how GenAI models, including LLMs, can undermine human understanding and capabilities. In specific domains, such as biological and biotechnological applications, human oversight is crucial to ensure trustworthy results that protect human health. For example, when AI-generated models provide advice or instructions for patient treatments, these suggestions must be verified by doctors to ensure their accuracy, preventing serious issues related to human health, environmental safety, and regulatory compliance. In addition, explainable AI is mentioned as a means to enable human oversight of AI-generated models. Current techniques, including GenAI models and LLMs, struggle to explain their decision-making, and they are often considered black-box models. Therefore, the need for human oversight of GenAI models has become more practical in today's context.

Balancing AI efficiency and human judgment has become a contemporary topic. Due to the rapid development of GenAI models, including LLMs, these models have come closer to mimicking human mindsets in several aspects. For example, a chess model can easily defeat the best human players in the world in face-to-face games. These models also provide valuable statistics that help chess players improve their abilities. In mathematics, LLMs can solve about 80% of the problems in competitive math contests compared to humans. However, trust in GenAI models needs to be verified in specific domains, such as medicine, autonomous driving, and others. Therefore, there is a need for a combination of human and AI capabilities to produce more robust and effective results while preventing the risks posed by AI models. Rather than replacing the human workforce, the combination of human expertise and AI can more effectively identify system threats and malicious attacks. For example, we can consider an AI model as the first line of defense, automatically monitoring the system and flagging the most critical issues before sending them for human analysis. This way, the human cyber workforce can focus solely on analyzing the threats flagged by the AI model, uncovering deeper threats, and protecting the system.

Enhancing the cybersecurity workforce's efficiency with AI augmentation is essential to prevent AI systems from potential risks. Since AI cannot cover every aspect of a system, the role of the human workforce remains crucial. For example, intrusion detection systems implemented by AI models can protect the system from harmful activities and notify the admin when a threat enters the system. However, determining whether incoming traffic is legitimate or malicious must be verified by cybersecurity experts, who can deeply analyze specific cases. The decision of cybersecurity experts may not only be based on the current parameters of user traffic but also on the history of the user's access or their profile. The detection engine used in IDSs may find it difficult to analyze and propose a suitable decision in such cases.

9.5 SUMMARY

This chapter explored the emerging topic of GenAI in cybersecurity. We discussed key aspects of GenAI, such as deepfakes and privacy-preserving techniques, including homomorphic encryption and differential privacy, which were applied to GenAI models to prevent sensitive user information from being leaked. We also addressed proactive threat hunting and cyber threat intelligence using GenAI models. Finally, we examined the ethical and regulatory challenges associated with GenAI applications and reviewed the relevant regulatory frameworks.

1. Nguyen, Thanh Thi, et al. "Deep learning for deepfakes creation and detection: A survey." Computer Vision and Image Understanding 223 (2022): 103525.
2. Agarwal, Shruti, Farid, Hany, Gu, Yuming, He, Mingming, Nagano, Koki, Li, Hao, 2019. Protecting world leaders against deepfakes. In: Computer Vision and Pattern Recognition Workshops, vol. 1, pp. 38–45.
3. Hsu, Chih-Chung, Zhuang, Yi-Xiu, Lee, Chia-Yen, 2020. Deepfake image detection based on pairwise learning. Appl. Sci. 10 (1), 370.
4. Hasan, Haya R., and Khaled Salah. "Combating deepfake videos using blockchain and smart contracts." IEEE Access 7 (2019): 41596-41606
5. Thakkar, Jinal Jagdishkumar, and Arashdeep Kaur. "From Deepfakes to Digital Truths: The Role of Watermarking in AI-Generated Image Verification." 2024 47th International Conference on Telecommunications and Signal Processing (TSP). IEEE, 2024.
6. Zampoglou, Markos, et al. "Web and social media image forensics for news professionals." Proceedings of the International AAAI Conference on Web and Social Media. Vol. 10. No. 2. 2016.
7. Sabir, Ekraam, Cheng, Jiaxin, Jaiswal, Ayush, AbdAlmageed, Wael, Masi, Iacopo, Natarajan, Prem, 2019. Recurrent convolutional strategies for face manipulation detection in videos. In: Proceedings of the IEEE Conference on Computer Vision and Pattern Recognition Workshops, vol. 3, (1), pp. 80–87

8. Li, Yuezun, and Siwei Lyu. "Exposing DeepFake Videos By Detecting Face Warping Artifacts." 2018.
9. Lanzino, Romeo, et al. "Faster Than Lies: Real-time Deepfake Detection using Binary Neural Networks." Proceedings of the IEEE/CVF Conference on Computer Vision and Pattern Recognition. 2024.
10. Gerstner, Candice R., and Hany Farid. "Detecting real-time deep-fake videos using active illumination." Proceedings of the IEEE/CVF Conference on Computer Vision and Pattern Recognition. 2022.
11. Cavia, Bar, et al. "Real-Time Deepfake Detection in the Real-World." arXiv preprint arXiv:2406.09398 (2024).
12. Masud, Umar, et al. "LW-DeepFakeNet: a lightweight time distributed CNN-LSTM network for real-time DeepFake video detection." Signal, Image and Video Processing 17.8 (2023): 4029-4037
13. Raza, Syed Ali, et al. "MMGANGuard: A Robust Approach for Detecting Fake Images Generated by GANs using Multi-Model Techniques." IEEE Access (2024).
14. Yu, C., Jia, S., Fu, X., Liu, J., Tian, J., Dai, & Han, J. (2024). Explicit Correlation Learning for Generalizable Cross-Modal Deepfake Detection. arXiv preprint arXiv:2404.19171.
15. Katamneni, Vinaya Sree, and Ajita Rattani. "Contextual cross-modal attention for audio-visual deepfake detection and localization." 2024 IEEE International Joint Conference on Biometrics (IJCB). IEEE, 2024.
16. Liu, Xiaolong, et al. "Mcl: multimodal contrastive learning for deepfake detection." IEEE Transactions on Circuits and Systems for Video Technology (2023).
17. Chen, Yan, and Pouyan Esmaeilzadeh. "Generative AI in medical practice: in-depth exploration of privacy and security challenges." Journal of Medical Internet Research 26 (2024): e53008.
18. Wang, Haofan, et al. "Score-CAM: Score-weighted visual explanations for convolutional neural networks." Proceedings of the IEEE/CVF conference on computer vision and pattern recognition workshops. 2020.
19. Yosinski, Jason, et al. "Understanding neural networks through deep visualization." arXiv preprint arXiv:1506.06579 (2015).
20. de Castro, Leo, Antigoni Polychroniadou, and Daniel Escudero. "Privacy-Preserving Large Language Model Inference via GPU-Accelerated Fully Homomorphic Encryption." Neurips Safe Generative AI Workshop 2024. 2024.
21. Dwork, Cynthia. "Differential privacy: A survey of results." International conference on theory and applications of models of computation. Berlin, Heidelberg: Springer Berlin Heidelberg, 2008.
22. Hindistan, Yavuz Selim, and E. Fatih Yetkin. "A hybrid approach with GAN and DP for privacy preservation of IIoT data." IEEE Access 11 (2023): 5837-5849.
23. Wu, Ningbo, Changgen Peng, and Kun Niu. "A privacy-preserving game model for local differential privacy by using information-theoretic approach." IEEE Access 8 (2020): 216741-216751.
24. Garfinkel, Simson L., John M. Abowd, and Sarah Powazek. "Issues encountered deploying differential privacy." Proceedings of the 2018 Workshop on Privacy in the Electronic Society. 2018.
25. Katherine, et al, 2017, "The Promise of Evidence-Based Policymaking. Commission on Evidence-Based Policymaking," Washington, DC. https://www.cep.gov/cep-final-report.html

26. Jiang, Bin, et al. "Differential privacy for industrial internet of things: Opportunities, applications, and challenges." IEEE Internet of Things Journal 8.13 (2021): 10430-10451.
27. K. Wei et al., "Federated Learning With Differential Privacy: Algorithms and Performance Analysis," in IEEE Transactions on Information Forensics and Security, vol. 15, pp. 3454-3469, 2020.
28. J. Ragsdale and R. V. Boppana, "On Designing Low-Risk Honeypots Using Generative Pre-Trained Transformer Models With Curated Inputs," in IEEE Access, vol. 11, pp. 117528-117545, 2023.
29. Usman, Yusuf, et al. "Is generative AI the next tactical cyber weapon for threat actors? unforeseen implications of AI generated cyber attacks." arXiv preprint arXiv:2408.12806 (2024).
30. López, Pedro Beltrán, Manuel Gil Pérez, and Pantaleone Nespoli. "Cyber Deception: State of the art, Trends and Open challenges." arXiv preprint arXiv:2409.07194 (2024).
31. Humphreys, D., Koay, A., Desmond, D., & Mealy, E. (2024). AI hype as a cyber security risk: The moral responsibility of implementing generative AI in business. AI and Ethics, 4(3), 791-804.
32. Phai, V. D., Diep Nguyen, D. T. Hoang, N. Q. Uy, S. P. Bao and E. Dutkiewicz, A Deep Learning Approach for Outlier Detection in Heterogeneous/Non-IID Data, GLOBECOM 2024 IEEE Global Communications Conference, Cape Town, South Africa, 2024, pp. 1215-1220
33. Fontes, C., Hohma, E., Corrigan, C. C., & Lütge, C. (2022). AI-powered public surveillance systems: why we (might) need them and how we want them. Technology in Society, 71, 102137.
34. Covington & Burling LLP. (2023, October). U.S. artificial intelligence policy: Legislative and regulatory developments. Covington & Burling LLP. `https://www.cov.com/en/news-and-insights/insights/2023/10/us-artificial-intelligence-policy-legislative-and-regulatory-developments`
35. Holzinger, Andreas, Kurt Zatloukal, and Heimo Müller. "Is Human Oversight to AI Systems Still Possible?" New Biotechnology (2024).

10 Future Directions and Open Challenges

This chapter discusses the future directions and open challenges, including the role of GenAI in post-quantum cryptography and AI-assisted quantum attacks. It serves as a valuable resource for researchers, practitioners, and policymakers seeking to understand the current challenges of using quantum computers and developing algorithms to apply security to systems. Moreover, this chapter discusses trustworthy AI in cybersecurity by specifically addressing the current explainable AI framework in security applications. Last but not least, it covers self-adaptive AI-driven cybersecurity solutions and highlights the weaknesses of security operations centers (SOCs). This can help organizations, researchers, and partners focus on the key factors in SOCs to address their weaknesses in the future.

10.1 AI AND QUANTUM CYBERSECURITY

10.1.1 THE ROLE OF GENERATIVE AI IN POST-QUANTUM CRYPTOGRAPHY

A quantum computer is a machine that can exploit the full complexity of quantum wavefunctions to address computational challenges [1]. Unlike classical computers, which use bits (0 and 1) as the basic unit of information, quantum computers use quantum bits (qubits) to encode information. Qubits can exist in a superposition of both 0 and 1 simultaneously, which can exponentially increase computing power for certain problems, as observed in Figure 10.1. Furthermore, unlike classical computers that process tasks sequentially, quantum computers leverage superposition to perform multiple calculations simultaneously, enabling significant speedups for specific problems. This inherent parallelism goes beyond classical multiprocessing techniques, offering advantages in cryptography, optimization, and complex simulations. In the future, quantum computers are expected to provide enough computational power to break current cryptographic algorithms. At that time, information shared via the Internet will be at risk, even if the length of private keys is increased or encryption algorithms significantly improve in complexity.

Quantum encryption techniques play an essential role in defending against adversarial attacks on the GenAI model. The authors in [2] introduced hybrid-quantum encryption to strengthen protection against generative reconstruction attacks. To achieve this, quantum noise is added to each training sample before training the federated learning model. For instance, when applying quantum noise to the MNIST dataset, the quantum encryption scheme includes eight encrypted values, i.e., $[q1, q2, q3, q1n, q2n, q3n, z, z, z]$. These values, consisting of $q1, q2, q3$, represent random quantum gates, while $q1n, q2n$, and $q3n$ are quantum gates that fit specific

DOI: 10.1201/9781003502531-10

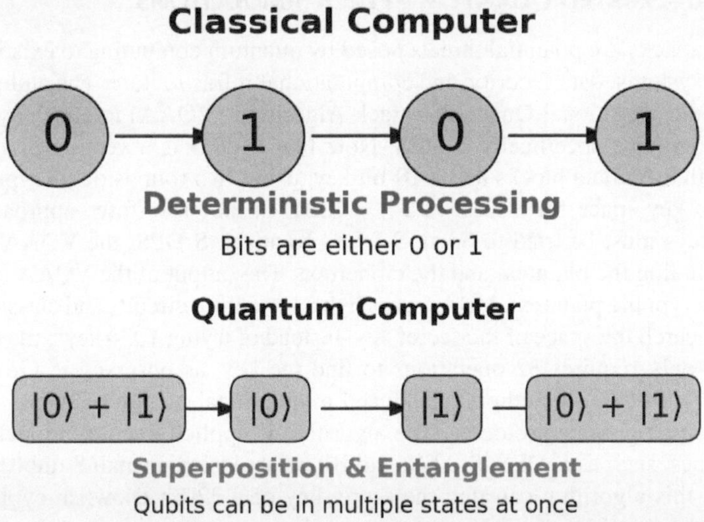

Figure 10.1 Comparison between classical and quantum computers.

spots. An attacker can exploit the model using the GAN model because the values received from the model contain noise compared to the original data.

Quantum computers have the potential to greatly enhance the performance of generative AI (GenAI) models. In [3], the authors improved anomaly detection by substituting the generator of Wasserstein GANs (WGANs) with a hybrid quantum-classical neural network. They specifically modified the WGAN generator by replacing its first component with a short-state preparation layer, $S(z)$, which consists of a parameterized quantum circuit $U_v(\theta)$ operating on N qubits. This circuit includes measurements of all qubits in the Z-basis and applies N single-qubit R_x rotations to prepare the quantum state. Additionally, noise circuits, utilizing nine qubits, were simulated using the TensorFlow Quantum simulator.

There are potential weaknesses of generative AI (GenAI) in the post-quantum era. Like many cryptographic algorithms, GenAI models could be vulnerable to quantum algorithms, such as Shor's algorithm, which can break RSA encryption or compromise the models and the data on which they are trained. If AI models depend on cryptographic methods to ensure data privacy or integrity, these quantum algorithms may undermine these protections in the future. Attackers could exploit quantum algorithms to manipulate the learning process or compromise training data in ways that traditional methods cannot prevent. Therefore, it is essential to develop new cryptographic protocols to safeguard the integrity of AI models against quantum threats. Additionally, integrating quantum-classical systems with GenAI models may present challenges, including inefficient algorithms and noise from quantum hardware. Furthermore, since GenAI models rely on probabilistic processes to generate creative outputs, the inherent randomness of quantum mechanics could introduce additional complications.

10.1.2 AI-ASSISTED QUANTUM ATTACK SIMULATIONS

Quantum attacks are potential threats posed by quantum computing to existing cryptographic systems, data security, and computational infrastructure. The authors in [6] proposed the Variational Quantum Attack Algorithm (VQAA) to attack symmetric key cryptography, specifically S-DES. Note that S-DES is a version of DES, designed with 8-bit data blocks and a 10-bit key, using two rounds of encryption. The size of the key space for S-DES is $2^{10} = 1024$. In the brute-force approach, 1024 possible keys must be tried to break S-DES. To attack S-DES, the VQAA requires inputs including the plaintext and the ciphertext. The output of the VQAA is the key used to encrypt the plaintext. VQAA uses hybrid quantum circuits and classical algorithms to search the space of the secret key. Instead of trying 1024 keys, the quantum algorithm only requires 32 operations to find the key, as observed in Grover's algorithm [7]. In [5], the authors introduced a variational quantum attack algorithm targeting cryptographic protocols. The algorithm is applied to conventional encryption schemes, such as S-DES, S-AES, and Blowfish. Using a small 8-qubit quantum computer, this algorithm can find the secret key of a 32-bit Blowfish cypher in 24 times fewer operations than a brute-force attack.

Quantum computing to break RSA encryption has gained attention in recent years. In [8], the authors devised an algorithm based on Shor's algorithm to crack RSA variants. RSA encryption relies on the fact that multiplying two large prime numbers is computationally easy, but factoring the resulting large number back into its two original primes is extremely difficult. Specifically, RSA encryption depends on the difficulty of factoring a composite number $N = p \times q$, where p and q are two prime numbers. As such, RSA is considered secure under classical computational assumptions but is vulnerable to quantum attacks. Shor's algorithm takes a composite number N as input and outputs two prime numbers p and q such that $N = p \times q$. The algorithm begins by selecting a random number a in the range $1 < a < N$ and then uses quantum parallelism to find the period r of the function $f(x) = a^x \mod N$. The period r is the smallest integer such that $a^r \equiv 1 \mod N$. Once the period r is found and is even, the algorithm computes the greatest common divisors (GCDs):

$$p = \gcd\left(a^{\frac{r}{2}} + 1, N\right),$$

$$q = \gcd\left(a^{\frac{r}{2}} - 1, N\right).$$

These GCDs, p and q, are the prime factors of N.

Quantum machine learning (QML) is widely applied to intrusion detection systems. In [4], the authors leveraged QML to classify denial-of-service attacks and normal traffic. To achieve this, the CIC-DDoS 2019 dataset is used before being input into the QML model. The data are encoded into quantum states using quantum circuits, where Hadamard gates are applied to create superpositions, and CNOT gates are used to generate entanglement. It is important to note that the CNOT gate is a two-qubit quantum gate that performs a conditional operation. After representing the datasets as quantum state vectors, QML algorithms are applied to exploit the parallelism of quantum computing, which enhances the efficiency of data processing. To utilize QML with a quantum neural network (QNN), the QNN is designed in blocks. The architecture of the QNN includes three main stages: data loading, data processing, and measurement. In the data-loading stage, classical data are encoded into quantum states using a feature map, preparing the qubits from $q0$ to $q6$. The data processing stage is represented by the Ansatz block, which refers to a specific component of a quantum circuit that defines a parameterized set of quantum gates or operations used to construct a trial quantum state. This stage aims to explore the quantum state space. Finally, the measurement stage is used to extract meaningful information, which is then utilized for classification.

10.2 TRUSTWORTHY AI IN CYBERSECURITY

10.2.1 DEVELOPING EXPLAINABLE AI (XAI) FOR SECURITY APPLICATIONS

Explainable AI (XAI) is an essential tool for explaining AI models, i.e., the decisions and actions of AI models to human users. This enables users to understand, appropriately trust, and effectively manage AI systems. For example, an AI-powered system may provide personalized health advice to a diabetic patient on how to manage their condition. The AI model might analyze the patient's health data, such as blood sugar levels, medication history, exercise habits, diet, and other factors, to provide recommendations. After that, the AI model can give advice to the patient, as follows: *"Take 5 units of insulin at 8 AM, eat a low-carb meal with 30 grams of carbohydrates, and walk for 30 minutes in the afternoon,"* as suggested by ChatGPT-3.5. The question is the reliability of this suggestion and whether the patient should follow this advice. Therefore, XAI is necessary to answer this question. The relationship between the performance of AI models and their explainability is shown in Figure 10.2. Although the superior performance of neural network models, especially deep learning and GenAI models, is notable, their explainability has some limitations. In contrast, conventional machine learning models, i.e., decision tree models, show better explainability in explaining most of the decision results.

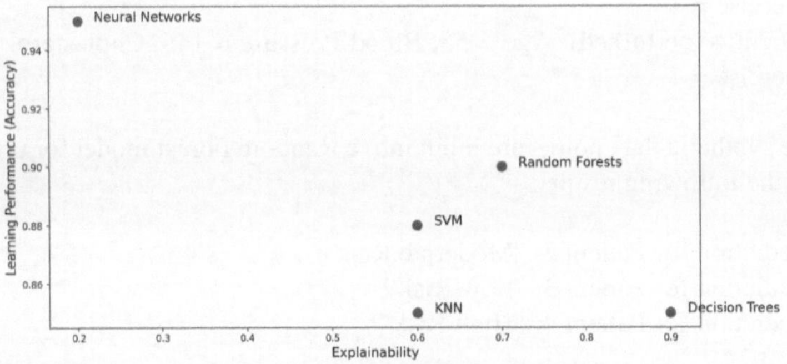

Figure 10.2 Learning Performance vs. explainability tradeoff for some learning models.

Shapley Additive Explanations (SHAP) and Local Interpretable Model-agnostic Explanations (LIME) are traditional methods for Explainable AI (XAI) [9]. SHAP helps to explain individual model predictions and provides insight into how each feature contributes to those predictions. Specifically, each feature, i.e., age, income, medical history, and so on, can be considered a "player." Next, Shapley values calculate how each feature contributes to the final prediction. For example, the Shapley value of each feature, i.e., age and blood pressure, are 0.25 and −0.1, respectively. These values tell us that the feature related to age contributes an increase of 0.25 to the prediction result, while the feature corresponding to blood pressure leads to a decrease of 0.1 in terms of prediction accuracy.

On the other hand, LIME aims to generate local approximations of the decision boundaries of the AI models. These approximations are tailored to interpret individual predictions, contributing to explaining the influence of the factors on the models' performance. To do this, LIME selects a specific instance to explain, i.e., a data sample that represents a patient. Next, LIME generates perturbed versions of the original data. The synthetic data points represent small changes compared to the original data. After that, these synthetic data points are input into a simple model to predict their results. This simple model, such as a Decision Tree or Linear Regression, is an interpretable model (surrogate) that is used to explain the synthetic data and predictions. Finally, the surrogate model's results are interpreted to understand the feature contributions.

Let's take an example where a patient has the following information: Age = 55, Blood Pressure = 140, Cholesterol = 230, and Exercise = 2 days per week. The model, i.e., the convolutional neural network (CNN), predicts a 'High Risk' of heart disease for this patient. Next, we need to use LIME to explain the result of the CNN model. To do this, LIME generates synthetic data as follows:

- Patient 1 (original data): Age = 55, Blood Pressure = 140, Cholesterol = 230, Exercise = 2
- Patient 2 (perturbed): Age = 56, Blood Pressure = 142, Cholesterol = 232, Exercise = 2
- Patient 3 (perturbed): Age = 54, Blood Pressure = 138, Cholesterol = 228, Exercise = 3
- Patient 4 (perturbed): Age = 55, Blood Pressure = 145, Cholesterol = 220, Exercise = 4

These synthetic data points are input into a Random Forest model for predictions, yielding the following results:

- Prediction for Patient 2: "Moderate Risk"
- Prediction for Patient 3: "Low Risk"
- Prediction for Patient 4: "High Risk"

LIME then fits a simple interpretable model, such as linear regression or a decision tree, to the synthetic data points and their corresponding predictions. After training the surrogate model, the results for the linear regression model are as follows:

- Age: Coefficient = 0.2 (Increasing age slightly increases the risk of heart disease)
- Blood Pressure: Coefficient = 0.5 (Higher blood pressure significantly increases the risk)
- Cholesterol: Coefficient = 0.1 (Cholesterol has a moderate positive influence on risk)
- Exercise: Coefficient = -0.3 (More exercise reduces the risk)

By analyzing the coefficients of the surrogate model, we can understand how each feature contributed to the model's prediction for the specific patient. For the patient with Age = 55, Blood Pressure = 140, Cholesterol = 230, and Exercise = 3 days per week, the prediction of "High Risk" of heart disease is primarily driven by high blood pressure (with the highest coefficient of 0.5), followed by the age feature (coefficient 0.2).

From the discussion of SHAP and LIME, we can observe that LIME relies on perturbing the original data point and training a simple, interpretable surrogate model. Therefore, LIME is more prone to the choice of perturbed samples. In addition, SHAP is based on Shapley values, which require calculating the contribution of each feature, leading to exponential time complexity as the number of features increases.

10.2.2 REDUCING BIAS IN AI-BASED SECURITY DECISION-MAKING

Bias is defined as *"an inclination of temperament or outlook,"* as observed in [11]. Bias in AI refers to systematic errors in AI models that lead to unfair or skewed outcomes, often reflecting existing biases in data or decision-making processes. The bias in AI-based security decision making comes from many aspects, i.e., human influence, data, and algorithms.

Regarding bias from human influence, the relationship between AI-based decision making and human experience is a crucial concern. The authors in [10] showed a hypothesis that *"those with the lowest levels of experience, knowledge, and/or familiarity (background) are relatively more averse to AI; people with middle levels of background are relatively over-reliant on AI, and those with the highest levels of background are relatively appropriately reliant on AI."* This suggests that when individuals have a high level of background in AI, they use AI applications appropriately. This contributes to increased effectiveness and efficiency in decision-making using AI models. In contrast, those with a moderate level of AI background tend to rely heavily on AI for decision-making. Meanwhile, individuals with little experience in AI are less likely to rely on AI for decision-making, as illustrated in Figure 10.3.

Bias in data occurs when the collected data does not represent the full distribution but only a subset. For example, when training a facial recognition system, if the data primarily includes adult faces while excluding children and elderly individuals, the system may struggle in real-world applications. Additionally, data bias can arise

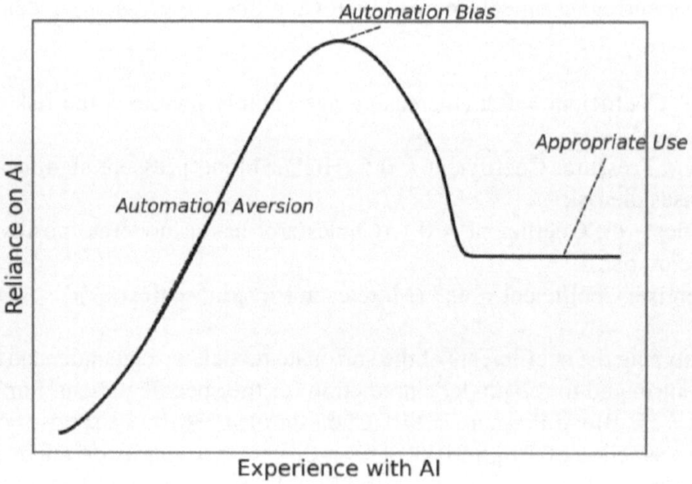

Figure 10.3 The relationship between human knowledge and their reliance on AI.

from historical inequalities or discrimination. For instance, if a bank's loan approval system is trained on past lending data, it may favor individuals who have previously received loans, thus reinforcing existing biases in financial decision-making. Furthermore, biases can result from incorrect data collection or labeling, leading to skewed or unfair AI outcomes.

Finally, several factors contribute to bias in AI models. AI models often operate based on assumptions that may not correspond with the actual distribution of the data. For instance, when training generative AI (GenAI) models under the assumption of a normal distribution, the reality of the data may not reflect this, resulting in flawed decision-making. Moreover, overfitting can occur when a model achieves high accuracy on training data but performs poorly on testing data. This situation arises when the model's complexity exceeds the simplicity of the training data. Data imbalance among classes also plays a significant role in model bias. AI models are prone to favoring the majority class while overlooking the minority class. For example, in a classification task that differentiates between normal and abnormal data, the model may frequently classify data as normal due to the training dataset containing a disproportionately larger amount of normal data than abnormal data. This imbalance skews the model's decision-making abilities, diminishing its effectiveness in detecting anomalies.

10.3 AUTONOMOUS SECURITY SYSTEMS USING GENERATIVE AI

10.3.1 SELF-ADAPTIVE AI-DRIVEN CYBERSECURITY SOLUTIONS

Real-time threat/attack detection systems are essential in the modern cybersecurity framework. With the rapid development of billions of malware variants in a short period, traditional offline defense systems are struggling to keep up with new types of attacks. Therefore, it is necessary to develop defense systems, such as real-time network intrusion detection/prevention systems, that can automatically update to detect and prevent emerging malware variants. The authors of [12] discussed the challenges of real-time cyber threat detection in cloud and network environments. One major challenge is detection accuracy, as real-time models must contend with new types of attacks that do not exist in training datasets. A high rate of false positive detections is also a weakness of current systems. This issue arises from the difficulty of collecting data on new types of attacks compared to normal traffic, leading to model bias in distinguishing between attacks and normal behavior. Additionally, real-time response presents challenges, as the detection engine requires significant computational resources to process input data. While the development of deep learning can improve detection accuracy, it demands more computational power due to the billions of model parameters involved. Privacy concerns in AI models also arise, as the data used for training may be exposed during or after the model training process. Therefore, it is essential to develop next-generation AI-based cybersecurity models that can protect systems online while maintaining high accuracy, low false alarm and miss detection rates, and fast response times to new threats and ensuring the confidentiality of data used in the AI models.

10.3.2 AI-POWERED SECURITY OPERATIONS CENTERS (SOCS)

AI-powered Security Operations Centers (SOCs) are advanced cybersecurity infrastructures that use AI and machine learning (ML) technologies to enhance the efficiency and effectiveness of security operations [13]. SOCs are the core of an organization's cybersecurity, providing monitoring, detection, and prevention to address threats from both external and internal activities. Traditional SOCs were designed to detect and respond to incidents in a proactive manner. This means that when attacks occur, the SOC analyzes the system and provides predictions based on detection engines, such as intrusion detection systems (IDS). SOCs can analyze a wide range of aspects within an organization's systems, including the physical layer, network traffic, system logs, software logs, and more.

The current SOCs have limitations, i.e., high rate of false alarm and missed detection, difficulty in detecting advanced persistent threats (APTs), data privacy concerns, lack of resources and complexity, lack of adaptability to new threats, overreliance on automation, limited real-time response, lack of transparency and explainability, and high cost of implementation, as observed in Figure 10.4. Among the drawbacks present in the current SOCs, some key features need to be prioritized. Reducing the rate of false alarms and missed detections is critical because the accuracy of SOCs should be prioritized first [14], [15], [16], [17]. The development of malicious software is becoming more complex each day. Instead of using a single attack on the system, attackers often conduct a series of attacks to gain control, utilizing various attack vectors to achieve root control of the system. As a result, APTs can cause severe damage to an organization, including data breaches, financial loss, and reputational harm. Improving adaptability to new threats is a crucial task for SOCs,

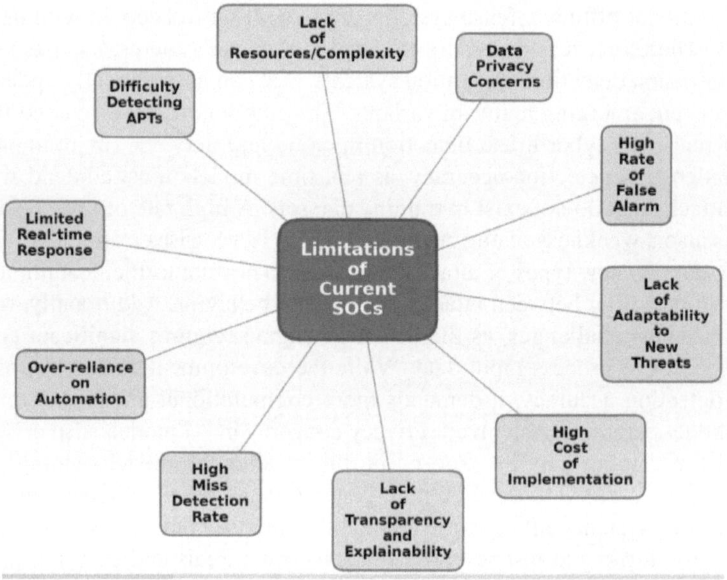

Figure 10.4 Limitations of Current SOCs.

as zero-day attacks with new malicious software variants change rapidly every day. Therefore, SOCs need to implement a flexible and scalable security framework and integrate the system in real time. Over-reliance on automatic defense mechanisms from the SOCs can lead to poor decision-making. This necessitates the development of explainable AI models to clarify the results of the detection.

10.3.3 GENAI IMPROVES CYBERSECURITY ATTACK DETECTION

GenAI models play an important role in improving the classification accuracy of classifiers by using data representations generated from these models. The authors in [16] proposed a constrained twin variational autoencoder (CTVAE) to enhance detection accuracy in intrusion detection systems (IDSs) for IoT systems. To achieve this, CTVAE's encoder aims to generate data from different classes, each originating from separate Gaussian distributions. Then, CTVAE's hermaphrodite maps these separated data points from the latent space into the input space. Finally, CTVAE's decoder learns the distribution of the various classes generated by the encoder. After training, only CTVAE's decoder is used to generate data representations for both the training and testing sets. Similarly, the authors in [18] proposed a twin auto-encoder (TAE) for representation learning. TAE addresses the issue of posterior collapse found in CTVAE, further improving the classification accuracy of classifiers using the data representations generated by TAE. Additionally, the authors in [17] leveraged the power of the twin architecture to propose a balanced twin auto-encoder (BTAE), which generates data samples for skewed attack labels before feeding them into the decoder. Since BTAE's decoder learns the data distribution of the skewed classes during training, it can generate data representations for both the training and testing sets without needing to increase the size of the training set. In contrast, traditional methods require generating data samples of skewed labels to balance the class distributions in the training set.

GenAI models are essential tools for detecting anomalies. They learn the distribution of normal data and can identify samples that significantly deviate from this distribution. The authors in [19] introduced an anomaly detection model called AnoGAN, which is based on generative adversarial networks (GANs). AnoGAN assigns an anomaly score to a data sample by using reconstruction errors from both the GAN's generator and its discriminator. A sample is classified as an anomaly if its reconstruction error greatly exceeds a predetermined threshold. Furthermore, the authors in [20] proposed a multiple-input variational autoencoder (MIVAE) for detecting anomalies in heterogeneous data. MIVAE employs an unsupervised learning approach, processing all feature subsets simultaneously to identify anomalies within each subset.

1. Ladd, T. D., Jelezko, F., Laflamme, R., Nakamura, Y., Monroe, C., & O'Brien, J. L. (2010). "Quantum computers." Nature, 464(7285), 45-53.
2. Moore, Ervin, Shabnam Rezapour, and M. Hadi Amini. "Quantum Encryption for Secure Federated Learning Against Generative Adversarial Network Attacks." Authorea Preprints.
3. Herr, Daniel, Benjamin Obert, and Matthias Rosenkranz. "Anomaly detection with variational quantum generative adversarial networks." Quantum Science and Technology 6.4 (2021): 045004.
4. Kücükkara, Muhammed Yusuf, Furkan Atban, and Cüneyt Bayılmıs. "Quantum-Neural Network Model for Platform Independent Ddos Attack Classification in Cyber Security." Advanced Quantum Technologies 7.10 (2024): 2400084.
5. Aizpurua, B., Bermejo, P., Etxezarreta Martínez, J., & Orús, R. (2025). Hacking cryptographic protocols with advanced variational quantum attacks. ACM Transactions on Quantum Computing, 6(2), 1-24.
6. Wang, Z., Wei, S., Long, G. L., & Hanzo, L. (2022). Variational quantum attacks threaten advanced encryption standard-based symmetric cryptography. Science China Information Sciences, 65(10), 200503.
7. Grover L K. A fast quantum mechanical algorithm for database search. In: Proceedings of the 28th Annual ACM Symposium on the Theory of Computing, 1996. 212–219
8. Sharma, M., Choudhary, V., Bhatia, R. S., Malik, S., Raina, A., & Khandelwal, H. (2021). Leveraging the power of quantum computing for breaking RSA encryption. Cyber-Physical Systems, 7(2), 73-92.
9. Thakur, A., Arunbalaji, C. G., Maddi, A., & Maheswari, B. U. (2024, May). Interpretable Predictive Modeling for Smoking and Drinking Behavior using SHAP and LIME. In 2024 International Conference on Current Trends in Advanced Computing (ICCTAC) (pp. 1-6). IEEE.
10. HOROWITZ, Michael C.; KAHN, Lauren. Bending the automation bias curve: a study of human and AI-based decision-making in national security contexts. International Studies Quarterly, 2024.

11. Kadiresan, Adheesh, Yuvraj Baweja, and Obi Ogbanufe. “Bias in AI-based decision-making.” Bridging human intelligence and artificial intelligence. Cham: Springer International Publishing, 2022. 275-285.
12. Akbar, Rehman, and Ali Zafer. “Next-Gen Information Security: AI-Driven Solutions for Real-Time Cyber Threat Detection in Cloud and Network Environments.” (2024).
13. Khayat, Mohamad, et al. “Empowering Security Operation Center with Artificial Intelligence and Machine Learning–A Systematic Literature Review.” IEEE Access (2025).
14. Dinh, Phai Vu and Nguyen, Diep N. and Thai, Hoang Dinh and Nguyen, Quang Uy and Le, Trung Hieu and Bao, Son Pham and Dutkiewicz, Eryk, “A Deep Learning Approach for Outlier Detection in Heterogeneous/Non-IID Data,” GLOBECOM 2024 - 2024 IEEE Global Communications Conference, Cape Town, South Africa, 2024, pp. 1215-1220.
15. Dinh, P. V., Nguyen, D. N., Hoang, D. T., Nguyen, Q. U., & Dutkiewicz, E. (2025). Multiple-Input Variational Auto-Encoder for Anomaly Detection in Heterogeneous Data. arXiv preprint arXiv:2501.08149.
16. P. V. Dinh, Q. U. Nguyen, D. T. Hoang, D. N. Nguyen, S. P. Bao and E. Dutkiewicz, “Constrained Twin Variational Auto-Encoder for Intrusion Detection in IoT Systems,” in IEEE Internet of Things Journal, vol. 11, no. 8, pp. 14789-14803, 15 April 15, 2024.
17. P. V. Dinh, D. N. Nguyen, D. T. Hoang, N. Q. Uy, S. P. Bao and E. Dutkiewicz, “Balanced Twin Auto-Encoder for IoT Intrusion Detection,” GLOBECOM 2022 - 2022 IEEE Global Communications Conference, Rio de Janeiro, Brazil, 2022, pp. 3387-3392
18. Dinh, P. V., Nguyen, Q. U., Dinh, T. H., Nguyen, D. N., Pham, B. S., & Dutkiewicz, E. (2024). Twin Auto-Encoder Model for Learning Separable Representation in Cyberattack Detection. arXiv. https://arxiv.org/abs/2403.15509
19. T. Schlegl, P. Seeböck, S. M. Waldstein, G. Langs, and U. Schmidt Erfurth, “f-anogan: Fast unsupervised anomaly detection with generative adversarial networks,” Medi. Ima. Analys., vol. 54, pp. 30–44, 2019
20. Phai, V. D., Diep Nguyen, D. T. Hoang, N. Q. Uy, and E. Dutkiewicz, “Multiple-Input Variational Auto-Encoder for Anomaly Detection in Heterogeneous Data,” 2025, https://arxiv.org/pdf/2501.08149.

Index

Note: Page references with *Italics* refer to figures, **bold** refer to tables.

E